Praise for *Applied Machine Learning and AI for Engineers*

This book is a fantastic guide to machine learning and AI algorithms. It's succinct while being comprehensive, and the concrete examples with working code show how to take the theory into practice.

—*Mark Russinovich, Azure CTO and Technical Fellow, Microsoft*

When Jeff Prosise is passionate about something (whether it be technology, his pet yellow-nape Amazon "Hawkeye," or his hobby of building and flying radio-controlled jets), you definitely want to listen in. He combines that passion with a clarity of explanation that enables him to communicate and teach complex topics better than anyone I've ever known. He brings you along on a personal journey of learning and understanding. Now Jeff brings these skills to the current and ongoing "technological tsunami" (as he puts it) of machine learning and AI. In this new book, he builds your understanding from the foundations up, always emphasizing an intuitive approach and connecting concepts and solutions to the real world. If you want to understand how AI and ML really work under the hood, and how these technologies have evolved and come to be, READ THIS BOOK.

—*Todd Fine, Chief Strategy Officer, Atmosera*

Jeff distills years of working AI/ML knowledge into a practical and understandable guide for practitioners of all levels.

—*Ken Muse, 4x Azure MVP and Senior DevOps Architect, GitHub*

Applied Machine Learning and AI for Engineers is the book I wish I had when first introduced to machine learning. It's a fantastic introduction for novice ML engineers and a great reference for those more experienced. It is now my go-to source when refreshing and enhancing my understanding of various ML techniques and their applicability. I love that Jeff uses real-world examples and datasets to demonstrate each tool set and approach. Just like a neural net, I have no idea how Jeff came up with this material, but it's tremendously useful all the same.

—Brent Rector, Principal Technical Program Manager, Amazon;
Founder, Wise Owl Consulting, LLC, Wise Owl Aviation Services, LLC,
and Rector Aviation Law PC

I've known Jeff for decades and he has always possessed the ability to plainly explain complicated concepts. He has done it again with *Applied Machine Learning and AI for Engineers*. The examples, analogies, and color figures make the material truly understandable for the beginner to the more advanced reader.

—Jeffrey Richter, Software Architect, Microsoft

This book fills an extremely important gap for engineers and scientists that want to bring the power of AI to bear on their most challenging data analysis problems. It strikes just the right balance between technical depth and practical application, with tools and many hands-on examples, to empower the reader to become an effective AI practitioner in their own application domain (including business, science, and other data-rich fields).

—Shaun S. Gleason, PhD, Director, Cyber Resilience and
Intelligence, Oak Ridge National Laboratory

Applied Machine Learning and AI for Engineers has the potential to become the go-to for ML and AI enthusiasts. What sets this book apart is how relevant the problem statements are in today's fast-paced adoption of machine learning in the tech world. A must-read for newbies and professionals alike!

—Lipi Deepaakshi Patnaik, Software Development
Engineer, Zeta Suite

Jeff has always been able to wrap great stories around deep technical concepts that make learning markedly easier. This might be his best book yet, and perhaps his most relevant subject matter as well. Reading this book will make you want to go build something.

—Doug Turnure, Azure Specialist, Microsoft

Applied Machine Learning and AI for Engineers is a practical handbook for building useful machine learning systems. It provides accessible guidance on how to apply cutting-edge AI algorithms to solve contemporary business problems.

—*Brian Spiering, Data Science Instructor, Metis*

This book is a perfect start for engineers and software developers who want to get into machine learning. It covers good ground in ML and quickly gets you started for a wide variety of problems that are driven by data rather than algorithms.

—*Dr. Manjeet Dahiya, VP and Head,*
AI and Machine Learning, Ecom Express

Whether you're new to applied ML or a seasoned developer looking for a reference, this book is a must-have, up-to-date, comprehensive guide to all major classes of machine learning algorithms (with clean code implementations to really solidify your understanding).

—*Goku Mohandas, Founder, Made With ML*

Applied Machine Learning and AI for Engineers is a brilliant primer for beginner to intermediate readers in AI/ML. The book builds a comfortable flow from conventional ML to deep learning with the desirable bonus of AI implementation on the cloud, making it a complete end-to-end guide for any enthusiastic reader or professional practitioner.

—*Satyarth Praveen, Computing Science Engineer,*
Lawrence Berkeley National Lab

Jeff Prosise is one of the best teachers I have ever encountered, and regardless of the platform (classroom, blog, magazine article, webinar, book, etc.), he has a special talent of taking complex topics and making them accessible for the rest of us. In this book, Jeff goes beyond simply providing a very clear understanding of the basic concepts that undergird machine learning to provide easy-to-follow examples that demonstrate those concepts within the current environment. He delivers a great introduction/overview of a wide variety of topics within machine learning and gives clear guidance on how each can (and should) be used. Jeff is one of the very few that could have taken this complex topic and put it into a form that could be easily absorbed and applied. This book is a MUST READ for engineers and other problem solvers who are looking to use machine learning to augment their skill set.

—*Larry Clement, Assistant Professor, Computing, Software, and*
Data Sciences, and former Department Chair, California Baptist University. Formerly a
senior engineer-scientist with the Boeing Corporation on the C-17 program.

Infused with author Jeff Prosise's iconic teaching style that has helped thousands of developers over the years, *Applied Machine Learning and AI for Engineers* layers context around complex topics and provides easy-to-understand examples and tutorials. Highly recommended for engineers who want to incorporate machine learning concepts and skills in their repertoire.

—Vani Mandava, Head of Engineering, Scientific Software Engineering Center at University of Washington eScience Institute

Applied Machine Learning and AI for Engineers

Solve Business Problems That Can't Be Solved Algorithmically

Jeff Prosise

Foreword by Adam Prosise

Beijing · Boston · Farnham · Sebastopol · Tokyo

Applied Machine Learning and AI for Engineers

by Jeff Prosise

Published by O'Reilly Media, Inc., 1005 Gravenstein Highway North, Sebastopol, CA 95472.

O'Reilly books may be purchased for educational, business, or sales promotional use. Online editions are also available for most titles (*http://oreilly.com*). For more information, contact our corporate/institutional sales department: 800-998-9938 or *corporate@oreilly.com*.

Acquisitions Editor: Nicole Butterfield	**Indexer:** Potomac Indexing, LLC
Development Editor: Jill Leonard	**Interior Designer:** David Futato
Production Editor: Gregory Hyman	**Cover Designer:** Karen Montgomery
Copyeditor: Audrey Doyle	**Illustrator:** Kate Dullea
Proofreader: Piper Editorial Consulting, LLC	

November 2022: First Edition

Revision History for the First Edition

2022-11-10: First Release

See *http://oreilly.com/catalog/errata.csp?isbn=9781492098058* for release details.

978-1-492-09805-8

[LSI]

For my Wintellect family past and present

Table of Contents

Part II.   Deep Learning with Keras and TensorFlow

Foreword

When your dad is insatiably curious about what you are studying, home is not a safe harbor.

I received my master's in analytics in 2018. Through the course of my graduate school program, my cohort learned how to leverage machine learning, AI, and analytics to add value to businesses and develop solutions for real-world challenges. I have a passion for these things—a passion that I share with my dad, Jeff Prosise. In fact, I can't tell you the number of times he asked if he could join me in class (I'm not kidding) or launched a salvo of questions about what we were studying in the kitchen when I escaped the grind at my parents' house.

If you have ever been on an airplane experiencing turbulence and the person next to you coped by striking up a conversation about what a marvel of engineering modern jetliners are because they do not have a single point of failure, you know exactly how I felt.

Our love of data and analytics grew into a shared love of the value it provides. Using the tools and techniques outlined in this book, one can draw certitude in the face of uncertainty. Data science allows you to find underlying truth—to discover what is really happening and how it drives behaviors and outcomes. The ability to peek behind the curtain using analytics, rather than intuition, is an appreciating skill set in our information economy and is vital for people and institutions navigating modern uncertainty.

Equally important is effectively communicating these findings to a nontechnical audience while having a deep technical understanding of just what is going on under the hood. This level of communication can't be faked (I've tried a time or two during those kitchen discussions with my dad).

Analytics, AI, and machine learning do not have insurmountable technical or deployment issues. Rather, the challenge and accessibility of understanding just what is happening and how they work is the impediment, because this understanding is often shrouded in technical jargon and industry shibboleths. These barriers to entry act as a limiting principle and stunt the utilization of data science to address problems and questions.

That is what this book seeks to change: it removes the shroud and jargon, making these tools and resources accessible.

And, to be completely honest, my dad is very good at writing and teaching intimidating technical topics in a manner that makes learning them almost effortless, and he has been for all my life. He has made subjects all the way back to DOS understandable for the movers and shakers in today's business world. He has made subjects that I frankly can't wrap my head around accessible for generations of programmers over the last few decades. To put it simply, he's the best.

The secret is this: his ethos of how he approaches teaching a topic centers around "how would I want this explained to me if I had never heard of it but was interested?" Given the unique challenges data science poses and the myriad perspectives professionals in this space come from, his approach provides a level of accessibility not found in many other places.

Take it from me: you couldn't have a better guide through the complexities of ML and AI. If you're already familiar, then this book will hone your understanding, as it did for me. (Looking at you, Chapter 13.) If you're interested in machine learning, AI, analytics, and the value they add to humanity, you're in the right place.

Regardless of which camp you fall in, you'll come away with deeper knowledge of the subjects he outlines, empowering you to use these tools and then tell the story of what you found.

Usually I would end something like this by saying, "I hope you enjoy the book and learn something," but in this case, that would not be the truth. I don't *hope*—I *know*. I know you will learn from and alongside my dad, as I have.

There's no person I've tried to be more like in my life than Jeff Prosise, and I couldn't be more excited to share this aspect of him with you. And maybe—just maybe—it will ignite a passion for this stuff, as it did with me.

I'll leave you with this. I'll say to you what he apocryphally told me as he held me on the day I was born: "Welcome to the show, kid."

— *Adam Prosise*
Process and Innovation Specialist, Delta Air Lines

Preface

I have witnessed three great technical revolutions in my lifetime: first the personal computer, then the internet, and lastly the smartphone. Machine learning (ML) and AI are just as fundamentally important as all three and will have an equally profound impact on our lives.

I first became interested in machine learning the day my credit card company called to confirm that I was trying to purchase a $700 necklace. I was not, but I was curious: how did they know it wasn't me? I use my card all over the world, and for the record, I *do* buy my wife nice things from time to time. Not once had the credit card company declined a legitimate purchase, but several times they had correctly flagged fraudulent purchases, the one prior to this being an attempt by someone in Brazil to use my card to buy an airline ticket. This time was different: the jewelry store was 2 miles from my house. I tried to imagine an algorithm that could so reliably detect credit card fraud at the point of sale. It didn't take long to realize that something more powerful than a mere algorithm was at work.

It turned out that the credit card company runs every transaction through a sophisticated machine learning model that is incredibly adept at detecting fraud. That moment changed my life. It's a splendid example of how ML and AI are making the world a better place. Moreover, understanding how ML could analyze credit card transactions in real time and pick out the bad ones while allowing legitimate charges to go through became a mountain that I had to climb.

Who Should Read This Book

Recently, I received a call from the head of engineering at a manufacturing company. He started the conversation like this: "Until last week, I didn't know what ML and AI stood for. Now my CEO has tasked me with figuring out how they can improve our business, and to do it before our competitors get ahead of us. I am starting at square one. Can you help?"

The next call came from a government contracting firm interested in using machine learning to detect tax fraud and money laundering. The team there was reasonably well versed in machine learning theory but wondered how best to go about building the models they needed.

Professionals everywhere are realizing that ML and AI represent a technological tsunami, and they're trying to get on top of the wave before it crashes over them. This book is for them: engineers, software developers, IT managers, and others whose goal is to build a practical understanding of ML and AI and put that knowledge to work solving problems that were difficult or even intractable before. It seeks to impart an *intuitive* understanding and resorts to equations only when necessary. Despite what you may have heard, you don't have to be an expert in calculus or linear algebra to build systems that recognize objects in photos, translate English to French, or expose drug traffickers and tax cheats.

Why I Wrote This Book

Inside every author is a tiny gremlin that says they can tell the story in a way that no one else has. I wrote my first computer book more than 30 years ago and my last one more than 20 years ago, and I didn't intend to write another one. But now I have a story to tell. It's an important story—one that every engineer and software developer should hear. I'm not entirely satisfied with the way others have told it, so I wrote the book that I wish I had had when I was learning the craft. It starts with the basics and leads you on a journey to the heights of ML and AI. By the end, you'll understand how credit card companies detect fraud, how aircraft companies use machine learning to perform predictive maintenance on jet engines, how self-driving cars see the world around them, how Google Translate translates text between languages, and how facial recognition systems work. Moreover, you'll be able to build systems like them yourself, or use existing systems to infuse AI into the apps that you write.

Today's most advanced machine learning models are trained on computers equipped with graphics processing units (GPUs) or tensor processing units (TPUs), often at great time and expense. A point of emphasis in this book is presenting examples that can be built on a typical PC or laptop without a GPU. When we tackle computer-vision models that recognize objects in photos, I'll describe how such models work and how they're trained with millions of images on GPU clusters. But then I'll show you how to use a technique called *transfer learning* to repurpose existing models to solve domain-specific problems and train them on an ordinary laptop.

This book draws heavily from the classes and workshops that I teach at companies and research institutions around the world. I love teaching because I love seeing the light bulbs come on. I often kick off classes on ML and AI by saying I'm not here to teach; I'm here to change your life. Here's hoping that your life will be a little bit different, and a little bit better, than it was before you read this book.

Running the Book's Code Samples

Engineers learn best by *doing*, not merely by reading. This book contains numerous code samples that you can run to reinforce the concepts presented in each chapter. Most are written in Python and use popular open source libraries such as Scikit-Learn, Keras, and TensorFlow. All are available in a public GitHub repo (*https:// oreil.ly/applied-machine-learning-code*). It's the single source of truth for the code samples because I can update it at any time.

There are machine learning platforms that allow you to build and train models with no code. But the best way to understand what these platforms do and how they do it is to write code. Python is a simple programming language. It's easy to learn. Engineers today have to be comfortable writing code. You can learn Python as you go by working the examples in this book, and if you're already comfortable with Python (and with programming in general), then you're ahead of the game.

To run my samples on your PC or laptop, you need a 64-bit version of Python 3.7 or higher. You can download a Python runtime from Python.org, or you can install a Python distribution such as Anaconda (*https://oreil.ly/4NCqN*). You also need to make sure the following packages and their dependencies are installed:

- Scikit-Learn and TensorFlow for building machine learning models
- Pandas, Matplotlib, and Seaborn for data wrangling and visualization
- OpenCV and Pillow for handling images
- Flask and Requests for calling REST APIs and building web services
- Sklearn-onnx and Onnxruntime for Open Neural Network Exchange (ONNX) models
- Librosa for generating spectrogram images from audio files
- MTCNN and Keras-vggface for building facial recognition systems
- KerasNLP, Transformers, Datasets, and PyTorch for building natural language processing (NLP) models
- Azure-cognitiveservices-vision-computervision, Azure-ai-textanalytics, and Azure-cognitiveservices-speech for calling Azure Cognitive Services

You can install most of these packages with `pip install` commands. If you installed Anaconda, many of these packages are already there, and you can install the rest using `conda install` commands or an equivalent.

Speaking of environments, it's never a bad idea to use virtual Python environments to prevent package installs from conflicting with other package installs. If you're not familiar with virtual environments, you can read about them at Python.org. If you use Anaconda, virtual environments are baked right in.

Most of my code samples were built for Jupyter notebooks, which provide an interactive platform for writing and executing Python code. Notebooks are incredibly popular in the data science community for exploring data and training machine learning models. You can run Jupyter notebooks locally by installing packages such as Notebook (*https://oreil.ly/ZWQyG*) or JupyterLab (*https://oreil.ly/5A3Ia*), or you can use cloud-hosted environments like Google Colab (*https://oreil.ly/RdRBa*). One of the advantages of Colab is that you don't have to install anything on your computer—not even Python. And in the rare cases in which my samples require a GPU, Colab provides it for you.

Python development environments are notoriously finicky to set up and maintain, especially on Windows. If you'd prefer not to have to create such an environment, or if you tried but failed to get it working, help is only a download away. I packaged a complete development environment suitable for running every sample in this book in a Docker container image (*https://oreil.ly/wzEbA*). Assuming you have the Docker Engine (*https://oreil.ly/XO5GD*) installed on your computer, you can launch the container with the following command:

```
docker run -it -p 8888:8888 jeffpro/applied-machine-learning:latest
```

Navigate in your browser to the URL that appears in the output. You'll land in a full Jupyter environment with all my code samples and everything needed to run them. They're in a folder named *Applied-Machine-Learning* cloned from the GitHub repo of the same name. The downside to using a container is that changes you make aren't persisted by default. One way to remedy that is to use a `-v` switch in the `docker` command to bind to a local directory. For more information, refer to "Use Bind Mounts" (*https://oreil.ly/7wgda*) in the Docker documentation.

Navigating This Book

This book is organized into two parts:

- Part I (Chapters 1 through 7) teaches the ABCs of machine learning and introduce popular learning algorithms such as logistic regression and gradient boosting.
- Part II (Chapters 8 through 14) covers deep learning, which is synonymous with AI today and uses deep neural networks to fit mathematical models to data.

I highly encourage you to work the exercises as you read the book. You'll come away with a deeper understanding of the material, and you'll no doubt think of ways to modify my examples to play "what if?" with the code.

Conventions Used in This Book

The following typographical conventions are used in this book:

Italic
: Indicates new terms, URLs, email addresses, filenames, and file extensions.

`Constant width`
: Used for program listings, as well as within paragraphs to refer to program elements such as variable or function names, databases, data types, environment variables, statements, and keywords.

`Constant width bold`
: Shows commands or other text that should be typed literally by the user.

This element signifies a tip or suggestion.

This element signifies a general note.

This element indicates a warning or caution.

Using Code Examples

As mentioned, supplemental material (code examples, exercises, etc.) is available for download at *https://oreil.ly/applied-machine-learning-code*.

If you have a technical question or a problem using the code examples, please send email to *bookquestions@oreilly.com*.

This book is here to help you get your job done. In general, if example code is offered with this book, you may use it in your programs and documentation. You do not need to contact us for permission unless you're reproducing a significant portion of the code. For example, writing a program that uses several chunks of code from this book does not require permission. Selling or distributing examples from O'Reilly books does require permission. Answering a question by citing this book and quoting example code does not require permission. Incorporating a significant amount of example code from this book into your product's documentation does require permission.

We appreciate, but generally do not require, attribution. An attribution usually includes the title, author, publisher, and ISBN. For example: "*Applied Machine Learning and AI for Engineers* by Jeff Prosise (O'Reilly). Copyright 2023 Jeff Prosise, 978-1-492-09805-8."

If you feel your use of code examples falls outside fair use or the permission given above, feel free to contact us at *permissions@oreilly.com*.

O'Reilly Online Learning

 For more than 40 years, *O'Reilly Media* has provided technology and business training, knowledge, and insight to help companies succeed.

Our unique network of experts and innovators share their knowledge and expertise through books, articles, and our online learning platform. O'Reilly's online learning platform gives you on-demand access to live training courses, in-depth learning paths, interactive coding environments, and a vast collection of text and video from O'Reilly and 200+ other publishers. For more information, visit *https://oreilly.com*.

How to Contact Us

Please address comments and questions concerning this book to the publisher:

O'Reilly Media, Inc.
1005 Gravenstein Highway North
Sebastopol, CA 95472
800-998-9938 (in the United States or Canada)
707-829-0515 (international or local)
707-829-0104 (fax)

We have a web page for this book, where we list errata, examples, and any additional information. You can access this page at *https://oreil.ly/applied-machine-learning*.

Email *bookquestions@oreilly.com* to comment or ask technical questions about this book.

For news and information about our books and courses, visit *https://oreilly.com*.

Find us on LinkedIn: *https://linkedin.com/company/oreilly-media*

Follow us on Twitter: *https://twitter.com/oreillymedia*

Watch us on YouTube: *https://youtube.com/oreillymedia*

Acknowledgments

Writing and publishing a book is a team effort. It starts with the author, but before it lands on shelves, it goes through reviewers, developmental editors, copy editors, artists, and production personnel.

I'd like to thank several friends and family members for reviewing chapters as I wrote them and providing constructive feedback. They are engineers, mathematicians, data analysts, and professors, and they happen to be among the smartest people I know: Larry Clement, Manjeet Dahiya, Tom Marshall, Don Meyer, Goku Mohandas, Ken Muse, Lipi Deepaakshi Patnaik, Charles Petzold, Abby Prosise, Adam Prosise, Jeffrey Richter, Bruce Schecter, Vishwesh Ravi Shrimali, Brian Spiering, and Ron Sumida.

A big thank-you to the team at O'Reilly too, for turning my words into prose and my sketches into art. That includes Jill Leonard, Audrey Doyle, Gregory Hyman, David Futato, Karen Montgomery, and Nicole Butterfield. A special shout-out to Jon Hassell, who heard my vision for this book and said, "Let's do it." I have had the privilege of working with some great publishing teams over the years. None were better than this one.

Finally, to Lori—my wife, travel companion, and partner in crime for the last 40 years. I couldn't have done it without you. And I *promise* this will be the last book I write!

Machine Learning with Scikit-Learn

Machine Learning

Machine learning expands the boundaries of what's possible by allowing computers to solve problems that were intractable just a few short years ago. From fraud detection and medical diagnoses to product recommendations and cars that "see" what's in front of them, machine learning impacts our lives every day. As you read this, scientists are using machine learning to unlock the secrets of the human genome. When we one day cure cancer, we will thank machine learning for making it possible.

Machine learning is revolutionary because it provides an alternative to algorithmic problem-solving. Given a recipe, or *algorithm*, it's not difficult to write an app that hashes a password or computes a monthly mortgage payment. You code up the algorithm, feed it input, and receive output in return. It's another proposition altogether to write code that determines whether a photo contains a cat or a dog. You can try to do it algorithmically, but the minute you get it working, you'll come across a cat or dog picture that breaks the algorithm.

Machine learning takes a different approach to turning input into output. Rather than relying on you to implement an algorithm, it examines a dataset of inputs and outputs and learns how to generate output of its own in a process known as *training*. Under the hood, special algorithms called *learning algorithms* fit mathematical models to the data and codify the relationship between data going in and data coming out. Once trained, a model can accept new inputs and generate outputs consistent with the ones in the training data.

To use machine learning to distinguish between cats and dogs, you don't code a cat-versus-dog algorithm. Instead, you train a machine learning model with cat and dog photos. Success depends on the learning algorithm used and the quality and volume of the training data.

Part of becoming a machine learning engineer is familiarizing yourself with the various learning algorithms and developing an intuition for when to use one versus another. That intuition comes from experience and from an understanding of how machine learning fits mathematical models to data. This chapter represents the first step on that journey. It begins with an overview of machine learning and the most common types of machine learning models, and it concludes by introducing two popular learning algorithms and using them to build simple yet fully functional models.

What Is Machine Learning?

At an existential level, machine learning (ML) is a means for finding patterns in numbers and exploiting those patterns to make predictions. ML makes it possible to train a model with rows or sequences of 1s and 0s, and to *learn* from the data so that, given a new sequence, the model can predict what the result will be. *Learning* is the process by which ML finds patterns that can be used to predict future outputs, and it's where the "learning" in "machine learning" comes from.

As an example, consider the table of 1s and 0s depicted in Figure 1-1. Each number in the fourth column is somehow based on the three numbers preceding it in the same row. What's the missing number?

0	1	0	0
1	1	0	1
1	1	1	1
0	0	0	0
0	1	1	

Figure 1-1. Simple dataset consisting of 0s and 1s

One possible solution is that for a given row, if the first three columns contain more 0s than 1s, then the fourth contains a 0. If the first three columns contain more 1s than 0s, then the answer is 1. By this logic, the empty box should contain a 1. Data scientists refer to the column containing answers (the red column in the figure) as the *label column*. The remaining columns are *feature columns*. The goal of a predictive model is to find patterns in the rows in the feature columns that allow it to predict what the label will be.

If all datasets were this simple, you wouldn't need machine learning. But real-world datasets are larger and more complex. What if the dataset contained millions of rows and thousands of columns, which, as it happens, is common in machine learning? For that matter, what if the dataset resembled the one in Figure 1-2?

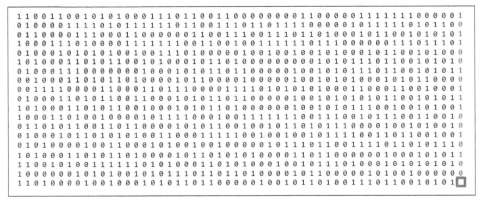

Figure 1-2. A more complex dataset

It's difficult for any human to examine this dataset and come up with a set of rules for predicting whether the red box should contain a 0 or a 1. (And no, it's not as simple as counting 1s and 0s.) Just imagine how much more difficult it would be if the dataset really *did* have millions of rows and thousands of columns.

That's what machine learning is all about: finding patterns in massive datasets of numbers. It doesn't matter whether there are 100 rows or 1,000,000 rows. In many cases, more is better, because 100 rows might not provide enough samples for patterns to be discerned.

It isn't an oversimplification to say that machine learning solves problems by mathematically modeling patterns in sets of numbers. Most any problem can be reduced to a set of numbers. For example, one of the common applications for ML today is *sentiment analysis*: looking at a text sample such as a movie review or a comment left on a website and assigning it a 0 for negative sentiment (for example, "The food was bland and the service was terrible.") or a 1 for positive sentiment ("Excellent food and service. Can't wait to visit again!"). Some reviews might be mixed—for example, "The burger was great but the fries were soggy"—so we use the *probability* that the label is a 1 as a sentiment score. A very negative comment might score a 0.1, while a very positive comment might score a 0.9, as in there's a 90% chance that it expresses positive sentiment.

Sentiment analyzers and other models that work with text are frequently trained on datasets like the one in Figure 1-3, which contains one row for every text sample and one column for every word in the corpus of text (all the words in the dataset). A typical dataset like this one might contain millions of rows and 20,000 or more columns.

Each row contains a 0 for negative sentiment in the label column, or a 1 for positive sentiment. Within each row are word counts—the number of times a given word appears in an individual sample. The dataset is sparse, meaning it is mostly 0s with an occasional nonzero number sprinkled in. But machine learning doesn't care about the makeup of the numbers. If there are patterns that can be exploited to determine whether the next sample expresses positive or negative sentiment, it will find them. Spam filters use datasets such as these with 1s and 0s in the label column denoting spam and nonspam messages. This allows modern spam filters to achieve an astonishing degree of accuracy. Moreover, these models grow smarter over time as they are trained with more and more emails.

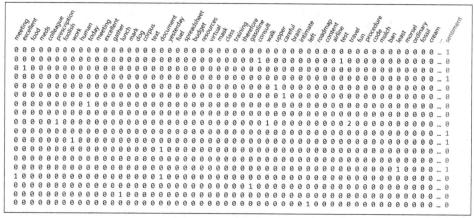

Figure 1-3. Dataset for sentiment analysis

Sentiment analysis is an example of a *text classification* task: analyzing a text sample and classifying it as positive or negative. Machine learning has proven adept at *image classification* as well. A simple example of image classification is looking at photos of cats and dogs and classifying each one as a cat picture (0) or a dog picture (1). Real-world uses for image classification include flagging defective parts coming off an assembly line, identifying objects in view of a self-driving car, and recognizing faces in photos.

Image classification models are trained with datasets like the one in Figure 1-4, in which each row represents an image and each column holds a pixel value. A dataset with 1,000,000 images that are 200 pixels wide and 200 pixels high contains 1,000,000 rows and 40,000 columns. That's 40 billion numbers in all, or 120,000,000,000 if the images are color rather than grayscale. (In color images, pixel values comprise three numbers rather than one.) The label column contains a number representing the class or category to which the corresponding image belongs—in this case, the person whose face appears in the picture: 0 for Gerhard Schroeder, 1 for George W. Bush, and so on.

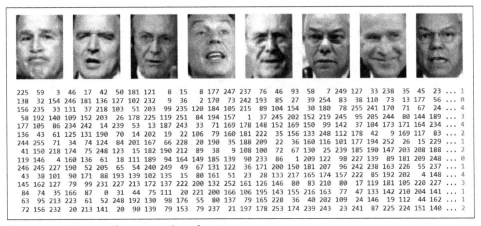

Figure 1-4. Dataset for image classification

These facial images come from a famous public dataset called Labeled Faces in the Wild (*https://oreil.ly/YVIv2*), or LFW for short. It is one of countless labeled datasets that are published in various places for public consumption. Machine learning isn't hard when you have labeled datasets to work with—datasets that others (often grad students) have laboriously spent hours labeling with 1s and 0s. In the real world, engineers sometimes spend the bulk of their time generating these datasets. One of the more popular repositories for public datasets is Kaggle.com, which makes lots of useful datasets available and holds competitions allowing budding ML practitioners to test their skills.

Machine Learning Versus Artificial Intelligence

The terms *machine learning* and *artificial intelligence* (AI) are used almost interchangeably today, but in fact, each term has a specific meaning, as shown in Figure 1-5.

Technically speaking, machine learning is a subset of AI, which encompasses not only machine learning models but also other types of models such as *expert systems* (systems that make decisions based on rules that you define) and *reinforcement learning systems*, which learn behaviors by rewarding positive outcomes while penalizing negative ones. An example of a reinforcement learning system is AlphaGo (*https://oreil.ly/uLwpd*), which was the first computer program to beat a professional human Go player. It trains on games that have already been played and learns strategies for winning on its own.

As a practical matter, what most people refer to as AI today is in fact deep learning, which is a subset of machine learning. *Deep learning* is machine learning performed with neural networks. (There are forms of deep learning that don't involve neural networks—deep Boltzmann machines are one example—but the vast majority of deep learning today involves neural networks.) Thus, ML models can be divided into conventional models that use learning algorithms to model patterns in data, and deep-learning models that use neural networks to do the same.

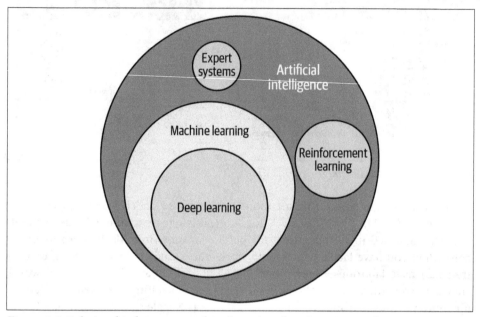

Figure 1-5. Relationship between machine learning, deep learning, and AI

A Brief History of AI

ML and AI have surged in popularity in recent years. AI was a big deal in the 1980s, when it was widely believed that computers would soon be able to mimic the human mind. But excitement waned, and for decades—up until 2010 or so—AI rarely made the news. Then a strange thing happened.

Thanks to the availability of graphics processing units (GPUs) (*https://oreil.ly/tiYd0*) from companies such as NVIDIA, researchers finally had the horsepower they needed to train advanced neural networks. This led to advancements in the state of the art, which led to renewed enthusiasm, which led to additional funding, which precipitated further advancements, and suddenly AI was a thing again. Neural networks have been around (at least in theory) since the 1950s, but researchers lacked the computational power to train them on large datasets. Today anyone can buy a GPU or spin up a GPU cluster in the cloud. AI is advancing more rapidly now than ever before, and with that progress comes the ability to do things in software that engineers could only have dreamed about as recently as a decade ago.

Over time, data scientists have devised special types of neural networks that excel at certain tasks, including tasks involving computer vision—for example, distilling information from images—and tasks that involve human languages such as translating English to French. We'll take a deep dive into neural networks beginning in Chapter 8, and you'll learn specifically how deep learning has elevated machine learning to new heights.

Supervised Versus Unsupervised Learning

Most ML models fall into one of two broad categories: *supervised learning* models and *unsupervised learning* models. The purpose of supervised learning models is to make predictions. You train them with labeled data so that they can take future inputs and predict what the labels will be. Most of the ML models in use today are supervised learning models. A great example is the model that the US Postal Service uses to turn handwritten zip codes into digits that a computer can recognize to sort the mail. Another example is the model that your credit card company uses to authorize purchases.

Unsupervised learning models, by contrast, don't require labeled data. Their purpose is to provide insights into existing data, or to group data into categories and categorize future inputs accordingly. A classic example of unsupervised learning is inspecting records regarding products purchased from your company and the customers who purchased them to determine which customers might be most interested in a new product you are launching and then building a marketing campaign that targets those customers.

A spam filter is a supervised learning model. It requires labeled data. A model that segments customers based on incomes, credit scores, and purchasing history is an unsupervised learning model, and the data that it consumes doesn't have to be labeled. To help drive home the difference, the remainder of this chapter explores supervised and unsupervised learning in greater detail.

Unsupervised Learning with k-Means Clustering

Unsupervised learning frequently employs a technique called *clustering*. The purpose of clustering is to group data by similarity. The most popular clustering algorithm is k-means clustering (*https://oreil.ly/8duYJ*), which takes n data samples and groups them into m clusters, where m is a number you specify.

Grouping is performed using an iterative process that computes a centroid for each cluster and assigns samples to clusters based on their proximity to the cluster centroids. If the distance from a particular sample to the centroid of cluster 1 is 2.0 and the distance from the same sample to the center of cluster 2 is 3.0, then the sample is assigned to cluster 1. In Figure 1-6, 200 samples are loosely arranged in three clusters. The diagram on the left shows the raw, ungrouped samples. The diagram on the right shows the cluster centroids (the red dots) with the samples colored by cluster.

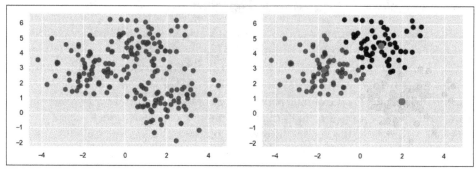

Figure 1-6. Data points grouped using k-means clustering

How do you code up an unsupervised learning model that implements *k*-means clustering? The easiest way to do it is to use the world's most popular machine learning library: Scikit-Learn (*https://oreil.ly/bSQT2*). It's free, it's open source, and it's written in Python. The documentation is great, and if you have a question, chances are you'll find an answer by Googling it. I'll use Scikit for most of the examples in the first half of this book. The book's Preface describes how to install Scikit and configure your computer to run my examples (or use a Docker container to do the same), so if you haven't done so already, now's a great time to set up your environment.

To get your feet wet with *k*-means clustering, start by creating a new Jupyter notebook and pasting the following statements into the first cell:

```
%matplotlib inline
import matplotlib.pyplot as plt
import seaborn as sns
sns.set()
```

Run that cell, and then run the following code in the next cell to generate a semirandom assortment of *x* and *y* coordinate pairs. This code uses Scikit's `make_blobs` function (*https://oreil.ly/h5sIB*) to generate the coordinate pairs, and Matplotlib's `scatter` function (*https://oreil.ly/bnmNw*) to plot them:

```
from sklearn.datasets import make_blobs

points, cluster_indexes = make_blobs(n_samples=300, centers=4,
                                     cluster_std=0.8, random_state=0)

x = points[:, 0]
y = points[:, 1]

plt.scatter(x, y, s=50, alpha=0.7)
```

Your output should be identical to mine, thanks to the `random_state` parameter that seeds the random-number generator used internally by `make_blobs`:

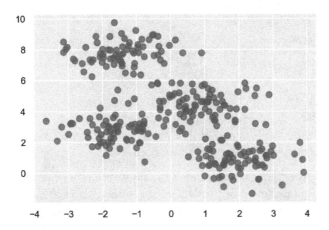

Next, use *k*-means clustering to divide the coordinate pairs into four groups. Then render the cluster centroids in red and color-code the data points by cluster. Scikit's `KMeans` class (*https://oreil.ly/wgm9z*) does the heavy lifting, and once it's fit to the coordinate pairs, you can get the locations of the centroids from `KMeans`' `cluster_centers_` attribute:

```
from sklearn.cluster import KMeans

kmeans = KMeans(n_clusters=4, random_state=0)
kmeans.fit(points)
predicted_cluster_indexes = kmeans.predict(points)

plt.scatter(x, y, c=predicted_cluster_indexes, s=50, alpha=0.7, cmap='viridis')

centers = kmeans.cluster_centers_
plt.scatter(centers[:, 0], centers[:, 1], c='red', s=100)
```

Here is the result:

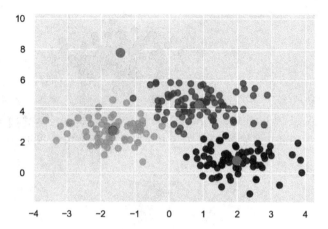

Try setting n_clusters to other values, such as 3 and 5, to see how the points are grouped with different cluster counts. Which begs the question: how do you know what the *right* number of clusters is? The answer isn't always obvious from looking at a plot, and if the data has more than three dimensions, you can't plot it anyway.

One way to pick the right number is with the elbow method, which plots *inertias* (the sum of the squared distances of the data points to the closest cluster center) obtained from KMeans.inertia_ as a function of cluster counts. Plot inertias this way and look for the sharpest elbow in the curve:

```
inertias = []

for i in range(1, 10):
    kmeans = KMeans(n_clusters=i, random_state=0)
    kmeans.fit(points)
    inertias.append(kmeans.inertia_)

plt.plot(range(1, 10), inertias)
plt.xlabel('Number of clusters')
plt.ylabel('Inertia')
```

In this example, it appears that 4 is the right number of clusters:

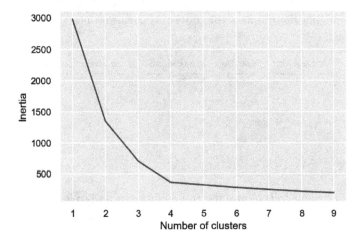

In real life, the elbow might not be so distinct. That's OK, because by clustering the data in different ways, you sometimes obtain insights that you wouldn't obtain otherwise.

Applying k-Means Clustering to Customer Data

Let's use *k*-means clustering to tackle a real problem: segmenting customers to identify ones to target with a promotion to increase their purchasing activity. The dataset that you'll use is a sample customer segmentation dataset named *customers.csv*. Start by creating a subdirectory named *Data* in the folder where your notebooks reside, downloading *customers.csv* (*https://oreil.ly/Md86Y*), and copying it into the *Data* subdirectory. Then use the following code to load the dataset into a Pandas `DataFrame` and display the first five rows:

```
import pandas as pd

customers = pd.read_csv('Data/customers.csv')
customers.head()
```

From the output, you learn that the dataset contains five columns, two of which describe the customer's annual income and spending score. The latter is a value from 0 to 100. The higher the number, the more this customer has spent with your company in the past:

	CustomerID	Gender	Age	Annual Income (k$)	Spending Score (1-100)
0	1	Male	19	15	39
1	2	Male	21	15	81
2	3	Female	20	16	6
3	4	Female	23	16	77
4	5	Female	31	17	40

Now use the following code to plot the annual incomes and spending scores:

```
%matplotlib inline
import matplotlib.pyplot as plt
import seaborn as sns
sns.set()

points = customers.iloc[:, 3:5].values
x = points[:, 0]
y = points[:, 1]

plt.scatter(x, y, s=50, alpha=0.7)
plt.xlabel('Annual Income (k$)')
plt.ylabel('Spending Score')
```

From the results, it appears that the data points fall into roughly five clusters:

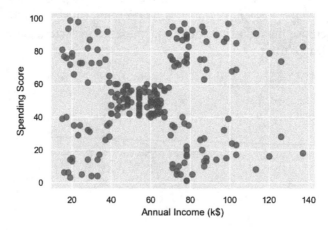

Use the following code to segment the customers into five clusters and highlight the clusters:

```
from sklearn.cluster import KMeans

kmeans = KMeans(n_clusters=5, random_state=0)
kmeans.fit(points)
predicted_cluster_indexes = kmeans.predict(points)

plt.scatter(x, y, c=predicted_cluster_indexes, s=50, alpha=0.7, cmap='viridis')
plt.xlabel('Annual Income (k$)')
plt.ylabel('Spending Score')

centers = kmeans.cluster_centers_
plt.scatter(centers[:, 0], centers[:, 1], c='red', s=100)
```

Here is the result:

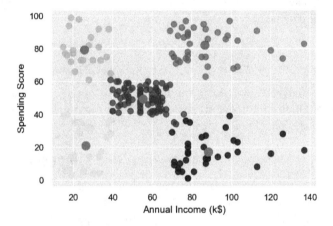

The customers in the lower-right quadrant of the chart might be good ones to target with a promotion to increase their spending. Why? Because they have high incomes but low spending scores. Use the following statements to create a copy of the Data Frame and add a column named Cluster containing cluster indexes:

```
df = customers.copy()
df['Cluster'] = kmeans.predict(points)
df.head()
```

Here is the output:

	CustomerID	Gender	Age	Annual Income (k$)	Spending Score (1-100)	Cluster
0	1	Male	19	15	39	4
1	2	Male	21	15	81	3
2	3	Female	20	16	6	4
3	4	Female	23	16	77	3
4	5	Female	31	17	40	4

Now use the following code to output the IDs of customers who have high incomes but low spending scores:

```
import numpy as np

# Get the cluster index for a customer with a high income and low spending score
cluster = kmeans.predict(np.array([[120, 20]]))[0]

# Filter the DataFrame to include only customers in that cluster
clustered_df = df[df['Cluster'] == cluster]

# Show the customer IDs
clustered_df['CustomerID'].values
```

You could easily use the resulting customer IDs to extract names and email addresses from a customer database:

```
array([125, 129, 131, 135, 137, 139, 141, 145, 147, 149, 151, 153, 155,
       157, 159, 161, 163, 165, 167, 169, 171, 173, 175, 177, 179, 181,
       183, 185, 187, 189, 191, 193, 195, 197, 199], dtype=int64)
```

The key here is that you used clustering to group customers by annual income and spending score. Once customers are grouped in this manner, it's a simple matter to enumerate the customers in each cluster.

Segmenting Customers Using More Than Two Dimensions

The previous example was an easy one because you used just two variables: annual incomes and spending scores. You could have done the same without help from machine learning. But now let's segment the customers again, this time using everything except the customer IDs. Start by replacing the strings "Male" and "Female" in the Gender column with 1s and 0s, a process known as *label encoding*. This is necessary because machine learning can only deal with numerical data:

```
from sklearn.preprocessing import LabelEncoder

df = customers.copy()
encoder = LabelEncoder()
df['Gender'] = encoder.fit_transform(df['Gender'])
df.head()
```

The Gender column now contains 1s and 0s:

	CustomerID	Gender	Age	Annual Income (k$)	Spending Score (1-100)
0	1	1	19	15	39
1	2	1	21	15	81
2	3	0	20	16	6
3	4	0	23	16	77
4	5	0	31	17	40

Extract the gender, age, annual income, and spending score columns. Then use the elbow method to determine the optimum number of clusters based on these features:

```
points = df.iloc[:, 1:5].values
inertias = []

for i in range(1, 10):
    kmeans = KMeans(n_clusters=i, random_state=0)
    kmeans.fit(points)
    inertias.append(kmeans.inertia_)

plt.plot(range(1, 10), inertias)
plt.xlabel('Number of Clusters')
plt.ylabel('Inertia')
```

The elbow is less distinct this time, but 5 appears to be a reasonable number:

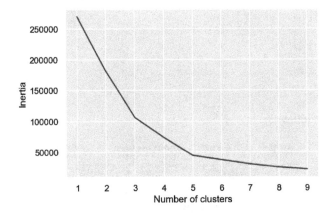

Segment the customers into five clusters and add a column named Cluster containing the index of the cluster (0-4) to which the customer was assigned:

```
kmeans = KMeans(n_clusters=5, random_state=0)
kmeans.fit(points)

df['Cluster'] = kmeans.predict(points)
df.head()
```

Here is the output:

	CustomerID	Gender	Age	Annual Income (k$)	Spending Score (1-100)	Cluster
0	1	1	19	15	39	0
1	2	1	21	15	81	4
2	3	0	20	16	6	0
3	4	0	23	16	77	4
4	5	0	31	17	40	0

You have a cluster number for each customer, but what does it mean? You can't plot gender, age, annual income, and spending score in a two-dimensional chart the way you plotted annual income and spending score in the previous example. But you *can* get the mean (average) of these values for each cluster from the cluster centroids. Create a new `DataFrame` with columns for average age, average income, and so on, and then show the results in a table:

```
results = pd.DataFrame(columns = ['Cluster', 'Average Age', 'Average Income',
                                  'Average Spending Index', 'Number of Females',
                                  'Number of Males'])

for i, center in enumerate(kmeans.cluster_centers_):
    age = center[1]    # Average age for current cluster
    income = center[2] # Average income for current cluster
    spend = center[3]  # Average spending score for current cluster

    gdf = df[df['Cluster'] == i]
    females = gdf[gdf['Gender'] == 0].shape[0]
    males = gdf[gdf['Gender'] == 1].shape[0]

    results.loc[i] = ([i, age, income, spend, females, males])

results.head()
```

The output is as follows:

	Cluster	Average Age	Average Income	Average Spending Index	Number of Females	Number of Males
0	0.0	45.217391	26.304348	20.913043	14.0	9.0
1	1.0	32.692308	86.538462	82.128205	21.0	18.0
2	2.0	43.088608	55.291139	49.569620	46.0	33.0
3	3.0	40.666667	87.750000	17.583333	17.0	19.0
4	4.0	25.521739	26.304348	78.565217	14.0	9.0

Based on this, if you were going to target customers with high incomes but low spending scores for a promotion, which group of customers (which cluster) would you choose? Would it matter whether you targeted males or females? For that matter, what if your goal was to create a loyalty program rewarding customers with high spending scores, but you wanted to give preference to younger customers who might be loyal customers for a long time? Which cluster would you target then?

Among the more interesting insights that clustering reveals is that some of the biggest spenders are young people (average age = 25.5) with modest incomes. Those customers are more likely to be female than male. All of this is useful information to have if you're growing a company and want to better understand the demographics that you serve.

k-means might be the most commonly used clustering algorithm, but it's not the only one. Others include agglomerative clustering (*https://oreil.ly/zQpxZ*), which clusters data points in a hierarchical manner, and DBSCAN (*https://oreil.ly/TanDh*), which stands for *density-based spatial clustering of applications with noise*. DBSCAN doesn't require the cluster count to be specified ahead of time. It can also identify points that fall outside the clusters it identifies, which is useful for detecting *outliers*—anomalous data points that don't fit in with the rest. Scikit-Learn provides implementations of both algorithms in its `AgglomerativeClustering` (*https://oreil.ly/ 7CztS*) and `DBSCAN` (*https://oreil.ly/D13gs*) classes.

Do real companies use clustering to extract insights from customer data? Indeed they do. During grad school, my son, now a data analyst for Delta Air Lines, interned at a pet supplies company. He used *k*-means clustering to determine that the number one reason that leads coming in through the company's website weren't converted to sales was the length of time between when the lead came in and Sales first contacted the customer. As a result, his employer introduced additional automation to the sales workflow to ensure that leads were acted on quickly. That's unsupervised learning at work. And it's a splendid example of a company using machine learning to improve its business processes.

Supervised Learning

Unsupervised learning is an important branch of machine learning, but when most people hear the term *machine learning* they think about supervised learning. Recall that supervised learning models make predictions. For example, they predict whether a credit card transaction is fraudulent or a flight will arrive on time. They're also trained with labeled data.

Supervised learning models come in two varieties: *regression models* and *classification models*. The purpose of a regression model is to predict a numeric outcome such as the price that a home will sell for or the age of a person in a photo. Classification models, by contrast, predict a *class* or category from a finite set of classes defined in the training data. Examples include whether a credit card transaction is legitimate or fraudulent and what number a handwritten digit represents. The former is a *binary classification* model because there are just two possible outcomes: the transaction is legitimate or it's not. The latter is an example of *multiclass classification*. Because there are 10 digits (0–9) in the Western Arabic numeral system, there are 10 possible classes that a handwritten digit could represent.

The two types of supervised learning models are pictured in Figure 1-7. On the left, the goal is to input an *x* and predict what *y* will be. On the right, the goal is to input an *x* and a *y* and predict what class the point corresponds to: a triangle or an ellipse.

In both cases, the purpose of applying machine learning to the problem is to build a model for making predictions. Rather than build that model yourself, you train a machine learning model with labeled data and allow it to devise a mathematical model for you.

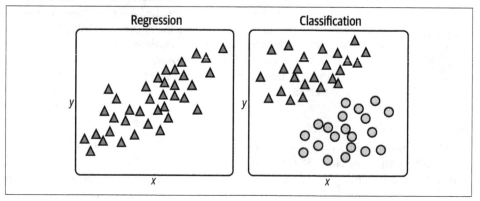

Figure 1-7. Regression versus classification

For these datasets, you could easily build mathematical models without resorting to machine learning. For a regression model, you could draw a line through the data points and use the equation of that line to predict a y given an x (Figure 1-8). For a classification model, you could draw a line that cleanly separates triangles from ellipses—what data scientists call a *classification boundary*—and predict which class a new point represents by determining whether the point falls above or below the line. A point just above the line would be a triangle, while a point just below it would classify as an ellipse.

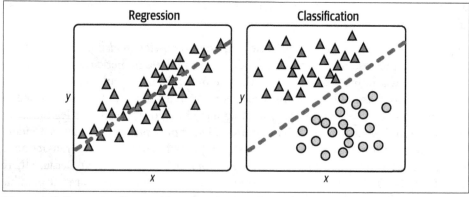

Figure 1-8. Regression line and linear separation boundary

In the real world, datasets are rarely this orderly. They typically look more like the ones in Figure 1-9, in which there is no single line you can draw to correlate the x and y values on the left or cleanly separate the classes on the right. The goal, therefore, is to build the best model you can. That means picking the learning algorithm that produces the most accurate model.

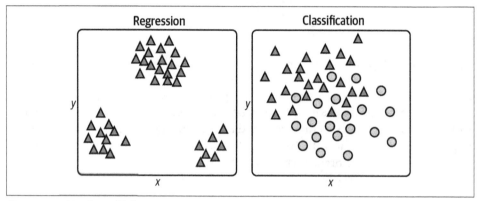

Figure 1-9. Real-world datasets

There are many supervised learning algorithms. They go by names such as linear regression, random forests, gradient-boosting machines (GBMs), and support vector machines (SVMs). Many, but not all, can be used for regression *and* classification. Even seasoned data scientists frequently experiment to determine which learning algorithm produces the most accurate model. These and other learning algorithms will be covered in subsequent chapters.

k-Nearest Neighbors

One of the simplest supervised learning algorithms is *k*-nearest neighbors (*https:// oreil.ly/dPhKi*). The premise behind it is that given a set of data points, you can predict a label for a new point by examining the points nearest it. For a simple regression problem in which each data point is characterized by x and y coordinates, this means that given an x, you can predict a y by finding the n points with the nearest xs and averaging their ys. For a classification problem, you find the n points closest to the point whose class you want to predict and choose the class with the highest occurrence count. If $n = 5$ and the five nearest neighbors include three triangles and two ellipses, then the answer is a triangle, as pictured in Figure 1-10.

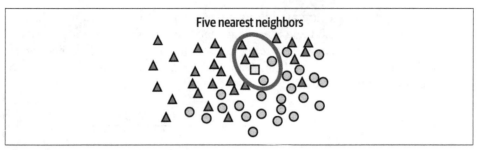

Figure 1-10. Classification with k-nearest neighbors

Here's an example involving regression. Suppose you have 20 data points describing how much programmers earn per year based on years of experience. Figure 1-11 plots years of experience on the x-axis and annual income on the y-axis. Your goal is to predict what someone with 10 years of experience should earn. In this example, $x = 10$, and you want to predict what y should be.

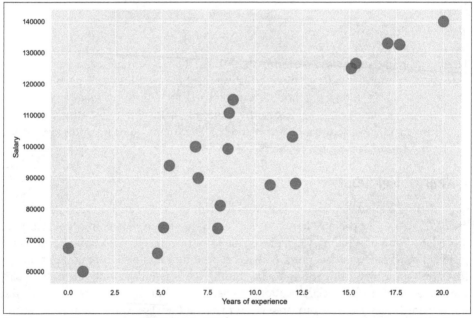

Figure 1-11. Programmers' salaries in dollars versus years of experience

Applying *k*-nearest neighbors with $n = 10$ identifies the points highlighted in orange in Figure 1-12 as the nearest neighbors—the 10 whose *x* coordinates are closest to $x = 10$. The average of these points' *y* coordinates is 94,838. Therefore, *k*-nearest neighbors with $n = 10$ predicts that a programmer with 10 years of experience will earn $94,838, as indicated by the red dot.

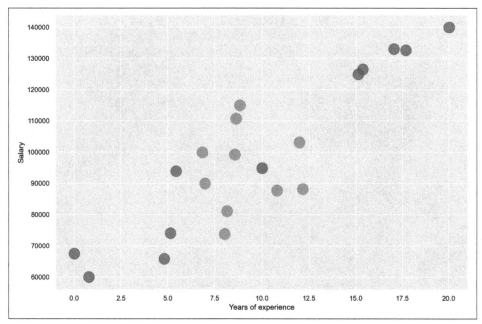

Figure 1-12. Regression with k-nearest neighbors and n = 10

The value of *n* that you use with *k*-nearest neighbors frequently influences the outcome. Figure 1-13 shows the same solution with *n* = 5. The answer is slightly different this time because the average *y* for the five nearest neighbors is 98,713.

In real life, it's a little more nuanced because while the dataset has just one label column, it probably has several *feature columns*—not just *x*, but x_1, x_2, x_3, and so on. You can compute distances in *n*-dimensional space easily enough, but there are several ways to measure distances to identify a point's nearest neighbors, including Euclidean distance, Manhattan distance, and Minkowski distance (*https://oreil.ly/36K7A*). You can even use weights so that nearby points contribute more to the outcome than faraway points. And rather than find the *n* nearest neighbors, you can select all the neighbors within a given radius, a technique known as *radius neighbors*. Still, the principle is the same regardless of the number of dimensions in the dataset, the method used to measure distance, or whether you choose *n* nearest neighbors or all the neighbors within a specified radius: find data points that are similar to the target point and use them to regress or classify the target.

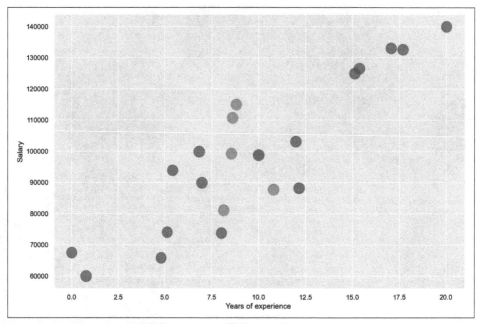

Figure 1-13. Regression with k-nearest neighbors and n = 5

Using k-Nearest Neighbors to Classify Flowers

Scikit-Learn includes classes named KNeighborsRegressor (*https://oreil.ly/feXD0*) and KNeighborsClassifier (*https://oreil.ly/8o7Uv*) to help you train regression and classification models using the *k*-nearest neighbors learning algorithm. It also includes classes named RadiusNeighborsRegressor (*https://oreil.ly/FP3wH*) and RadiusNeighborsClassifier (*https://oreil.ly/g1TfE*) that accept a radius rather than a number of neighbors. Let's look at an example that uses KNeighborsClassifier to classify flowers using the famous Iris dataset (*https://oreil.ly/lpXFR*). That dataset includes 150 samples, each representing one of three species of iris. Each row contains four measurements—sepal length, sepal width, petal length, and petal width, all in centimeters—plus a label: 0 for a setosa iris, 1 for versicolor, and 2 for virginica. Figure 1-14 shows an example of each species and illustrates the difference between petals and sepals.

Figure 1-14. Iris dataset (Middle panel: "Blue Flag Flower Close-Up [Iris Versicolor]" by Danielle Langlois is licensed under CC BY-SA 2.5, https://creativecommons.org/licenses/by-sa/2.5/deed.en; rightmost panel: "Image of Iris Virginica Shrevei BLUE FLAG" by Frank Mayfield is licensed under CC BY-SA 2.0, https://creativecommons.org/licenses/by-sa/2.0/deed.en)

To train a machine learning model to differentiate between species of iris based on sepal and petal measurements, begin by running the following code in a Jupyter notebook to load the dataset, add a column containing the class name, and show the first five rows:

```
import pandas as pd
from sklearn.datasets import load_iris

iris = load_iris()
df = pd.DataFrame(iris.data, columns=iris.feature_names)
df['class'] = iris.target
df['class name'] = iris.target_names[iris['target']]
df.head()
```

The Iris dataset is one of several sample datasets (*https://oreil.ly/BMQYL*) included with Scikit. That's why you can load it by calling Scikit's `load_iris` function (*https://oreil.ly/NmpjR*) rather than reading it from an external file. Here's the output from the code:

	sepal length (cm)	sepal width (cm)	petal length (cm)	petal width (cm)	class	class name
0	5.1	3.5	1.4	0.2	0	setosa
1	4.9	3.0	1.4	0.2	0	setosa
2	4.7	3.2	1.3	0.2	0	setosa
3	4.6	3.1	1.5	0.2	0	setosa
4	5.0	3.6	1.4	0.2	0	setosa

Before you train a machine learning model from the data, you need to split the dataset into two datasets: one for training and one for testing. That's important, because if you don't test a model with data it hasn't seen before—that is, data it wasn't trained with—you have no idea how accurate it is at making predictions.

Fortunately, Scikit's `train_test_split` function (*https://oreil.ly/6TuKC*) makes it easy to split a dataset using a fractional split that you specify. Use the following statements to perform an 80/20 split with 80% of the rows set aside for training and 20% reserved for testing:

```
from sklearn.model_selection import train_test_split

x_train, x_test, y_train, y_test = train_test_split(
    iris.data, iris.target, test_size=0.2, random_state=0)
```

Now, `x_train` and `y_train` hold 120 rows of randomly selected measurements and labels, while `x_test` and `y_test` hold the remaining 30. Although 80/20 splits are customary for small datasets like this one, there's no rule saying you *have* to split 80/20. The more data you train with, the more accurate the model is. (That's not strictly true, but generally speaking, you always want as much training data as you can get.) The more data you test with, the more confidence you have in measurements of the model's accuracy. For a small dataset, 80/20 is a reasonable place to start.

The next step is to train a machine learning model. Thanks to Scikit, that requires just a few lines of code:

```
from sklearn.neighbors import KNeighborsClassifier

model = KNeighborsClassifier()
model.fit(x_train, y_train)
```

In Scikit, you create a machine learning model by instantiating the class encapsulating the learning algorithm you selected—in this case, `KNeighborsClassifier`. Then you call `fit` on the model to train it by fitting it to the training data. With just 120 rows of training data, training happens very quickly.

The final step is to use the 30 rows of test data split off from the original dataset to measure the model's accuracy. In Scikit, that's accomplished by calling the model's `score` method:

```
model.score(x_test, y_test)
```

In this example, `score` returns 0.966667, which means the model got it right about 97% of the time when making predictions with the features in `x_test` and comparing the predicted labels to the actual labels in `y_test`.

Of course, the whole purpose of training a predictive model is to make predictions with it. In Scikit, you make a prediction by calling the model's `predict` method. Use the following statements to predict the class—0 for setosa, 1 for versicolor, and 2 for virginica—identifying the species of an iris whose sepal length is 5.6 cm, sepal width is 4.4 cm, petal length is 1.2 cm, and petal width is 0.4 cm:

```
model.predict([[5.6, 4.4, 1.2, 0.4]])
```

The `predict` method can make multiple predictions in a single call. That's why you pass it a list of lists rather than just a list. It returns a list whose length equals the number of lists you passed in. Since you passed just one list to `predict`, the return value is a list with one value. In this example, the predicted class is 0, meaning the model predicted that an iris whose sepal length is 5.6 cm, sepal width is 4.4 cm, petal length is 1.2 cm, and petal width is 0.4 cm is mostly likely a setosa iris.

When you create a `KNeighborsClassifier` without specifying the number of neighbors, it defaults to 5. You can specify the number of neighbors this way:

```
model = KNeighborsClassifier(n_neighbors=10)
```

Try fitting (training) and scoring the model again using `n_neighbors=10`. Does the model score the same? Does `predict` still predict class 0? Feel free to experiment with other `n_neighbors` values to get a feel for their effect on the outcome.

KNeighborsClassifier Internals

k-nearest neighbors is sometimes referred to as a *lazy* learning algorithm because most of the work is done when you call `predict` rather than when you call `fit`. In fact, training technically doesn't have to do anything except make a copy of the training data for when `predict` is called. So what happens inside `KNeighborsClassifier`'s `fit` method?

In most cases, `fit` constructs a binary tree in memory that makes `predict` faster by preventing it from having to perform a brute-force search for neighboring samples. If it determines that a binary tree won't help, `KNeighborsClassifier` resorts to brute force when making predictions. This typically happens when the training data is sparse—that is, mostly zeros with a few nonzero values sprinkled in.

One of the wonderful things about Scikit-Learn is that it is open source. If you care to know more about how a particular class or method works, you can go straight to the source code on GitHub. You'll find the source code for `KNeighborsClassifier` and `RadiusNeighborsClassifier` on GitHub (*https://oreil.ly/mC4Rt*).

The process employed here—load the data, split the data, create a classifier or regressor, call `fit` to fit it to the training data, call `score` to assess the model's accuracy using test data, and finally, call `predict` to make predictions—is one that you will use over and over with Scikit. In the real world, data frequently requires cleaning before it's used for training and testing. For example, you might have to remove rows with missing values or dedupe the data to eliminate redundant rows. You'll see plenty of examples of this later, but in this example, the data was complete and well structured right out of the box, and therefore required no further preparation.

Summary

Machine learning offers engineers and software developers an alternative approach to problem-solving. Rather than use traditional computer algorithms to transform input into output, machine learning relies on learning algorithms to build mathematical models from training data. Then it uses those models to turn future inputs into outputs.

Most machine learning models fall into either of two categories. Unsupervised learning models are widely used to analyze datasets by highlighting similarities and differences. They don't require labeled data. Supervised learning models learn from labeled data in order to make predictions—for example, to predict whether a credit card transaction is legitimate. Supervised learning can be used to solve regression problems or classification problems. Regression models predict numeric outcomes, while classification models predict classes (categories).

k-means clustering is a popular unsupervised learning algorithm, while k-nearest neighbors is a simple yet effective supervised learning algorithm. Many, but not all, supervised learning algorithms can be used for regression *and* for classification. Scikit-Learn's `KNeighborsRegressor` class, for example, applies k-nearest neighbors to regression problems, while `KNeighborsClassifier` applies the same algorithm to classification problems.

Educators often use k-nearest neighbors to introduce supervised learning because it's easily understood and it performs reasonably well in a variety of problem domains. With k-nearest neighbors under your belt, the next step on the road to machine learning proficiency is getting to know other supervised learning algorithms. That's the focus of Chapter 2, which introduces several popular learning algorithms in the context of regression modeling.

CHAPTER 2
Regression Models

You learned in Chapter 1 that supervised learning models come in two varieties: regression models and classification models. You also learned that regression models predict numeric outcomes, such as the price that a home will sell for or the number of visitors a website will attract. Regression modeling is a vital and sometimes underappreciated aspect of machine learning. Retailers use it to forecast demand (*https://oreil.ly/pqs2a*). Banks use it to screen loan applications, factoring in variables such as credit scores, debt-to-income ratios, and loan-to-value ratios. Insurance companies use it to set premiums. Whenever you need numerical predictions, regression modeling is the right tool for the job.

When building a regression model, the first and most important decision you make is what learning algorithm to use. Chapter 1 presented a simple three-class classification model that used the *k*-nearest neighbors learning algorithm to identify a species of iris given the flower's sepal and petal measurements. *k*-nearest neighbors can be used for regression too, but it's one of many you can choose from for making numerical predictions. Other learning algorithms frequently produce more accurate models.

This chapter introduces common regression algorithms, many of which can be used for classification also, and guides you through the process of building a regression model that predicts taxi fares using data published by the New York City Taxi and Limousine Commission. It also describes various means for assessing a regression model's accuracy and introduces an important technique for measuring accuracy called cross-validation.

Linear Regression

Next to *k*-nearest neighbors, linear regression is perhaps the simplest learning algorithm of all. It works best with data that is relatively linear—that is, data points that fall roughly along a line. Thinking back to high school math class, you'll recall that the equation for a line in two dimensions is:

$$y = mx + b$$

where *m* is the slope of the line and *b* is where the line intersects the y-axis. The income-versus-years-of-experience dataset in Figure 1-11 lends itself well to linear regression. Figure 2-1 shows a regression line fit to the data points. Predicting the income for a programmer with 10 years of experience is as simple as finding the point on the line where *x* = 10. The equation of the line is $y = 3,984x + 60,040$. Plugging 10 into that equation for *x*, the predicted income is $99,880.

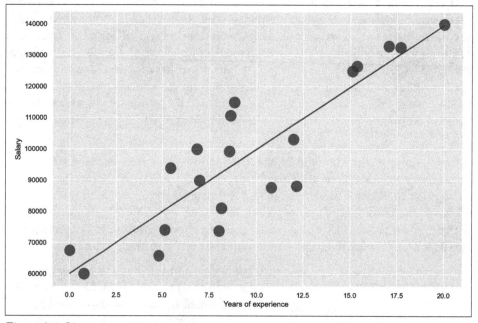

Figure 2-1. Linear regression

The goal when training a linear regression model is to find values for *m* and *b* that produce the most accurate predictions. This is typically done using an iterative process that starts with assumed values for *m* and *b* and repeats until it converges on suitable values.

The most common technique for fitting a line to a set of points is *ordinary least squares* regression (*https://oreil.ly/n5fx9*), or OLS for short. It works by squaring the distance in the *y* direction between each point and the regression line, summing the squares, and dividing by the number of points to compute the *mean squared error*, or MSE. (Squaring each distance prevents negative distances from offsetting positive distances.) Then it adjusts *m* and *b* to reduce the MSE the next time around and repeats until the MSE is sufficiently low. I won't go into the details of how it determines in which direction to adjust *m* and *b* (it's not hard, but it involves a smidgeon of calculus—specifically, using partial derivatives (*https://oreil.ly/dnOhg*) of the MSE function to determine whether to increase or decrease *m* and *b* in the next iteration), but OLS can often fit a line to a set of points with a dozen or fewer iterations.

Scikit-Learn has a number of classes to help you build linear regression models, including the `LinearRegression` class (*https://oreil.ly/e8r8p*), which embodies OLS, and the `PolynomialFeatures` class (*https://oreil.ly/s3rJN*), which fits a polynomial curve rather than a straight line to the training data. Training a linear regression model can be as simple as this:

```
model = LinearRegression()
model.fit(x, y)
```

Scikit has other linear regression classes with names such as `Ridge` (*https://oreil.ly/7I3ON*) and `Lasso` (*https://oreil.ly/Fnj2v*). One scenario in which they're useful is when the training data contains outliers. Recall from Chapter 1 that outliers are data points that don't conform with the rest. Outliers can bias a model or make it less accurate. `Ridge` and `Lasso` add *regularization* (*https://oreil.ly/x4Dt2*), which mitigates the effect of outliers by lessening their influence on the outcome as coefficients are adjusted during training. An alternate approach to dealing with outliers is to remove them altogether, which is what you'll do in the taxi-fare example at the end of this chapter.

Lasso regression has a secondary benefit too. If the training data suffers from multicollinearity (*https://oreil.ly/qNDA0*), a condition in which two or more input variables are linearly correlated so that one can be predicted from another with a reasonable degree of accuracy, `Lasso` effectively ignores the redundant data.

A classic example of multicollinearity occurs when a dataset includes one column specifying the number of rooms in a house and another column specifying the square footage. More rooms generally means more area, so the two variables are correlated to some degree.

Linear regression isn't limited to two dimensions (x and y values); it works with any number of dimensions. Linear regression with one independent variable (x) is known as *simple linear regression*, while linear regression with two or more independent variables—for example, x_1, x_2, x_3, and so on—is called *multiple linear regression*. If a dataset is two dimensional, it's simple enough to plot the data to determine its shape. You can plot three-dimensional data too, but plotting datasets with four or five dimensions is more challenging, and datasets with hundreds or thousands of dimensions are impossible to visualize.

How do you determine whether a high-dimensional dataset might lend itself to linear regression? One way to do it is to reduce n dimensions to two or three using techniques such as principal component analysis (PCA) (*https://oreil.ly/jzc37*) and t-distributed stochastic neighbor embedding (t-SNE) (*https://oreil.ly/6mbe6*) so that you can plot them. These techniques are covered in Chapter 6. Both reduce the dimensionality of a dataset without incurring a commensurate loss of information. With PCA, for example, it isn't uncommon to reduce the number of dimensions by 90% while retaining 90% of the information in the original dataset. It might sound like magic, but it's not. It's math.

If the number of dimensions is relatively small, a simpler technique for visualizing high-dimensional datasets is *pair plots* (*https://oreil.ly/kHiBd*), which plot pairs of dimensions in conventional 2D charts. Figure 2-2 shows a pair plot charting sepal length versus petal length, sepal width versus petal width, and other parameter pairs for the Iris dataset (*https://oreil.ly/TjQb3*) introduced in Chapter 1.

Seaborn's `pairplot` function (*https://oreil.ly/2T9OS*) makes it easy to create pair plots. The plot in Figure 2-2 was generated with one line of code:

```
sns.pairplot(df)
```

The pair plot not only helps you visualize relationships in the dataset, but in this example, the histogram in the lower-right corner reveals that the dataset is balanced too. There is an equal number of samples of all three classes, and for reasons you'll learn in Chapter 3, you always prefer to train classification models with balanced datasets.

Linear regression is a *parametric* learning algorithm, which means that its purpose is to examine a dataset and find the optimum values for parameters in an equation—for example, m and b. k-nearest neighbors, by contrast, is a *nonparametric* learning algorithm because it doesn't fit data to an equation. Why does it matter whether a learning algorithm is parametric or nonparametric? Because datasets used to train parametric models frequently need to be *normalized*. At its simplest, normalizing data means making sure all the values in all the columns have consistent ranges. I'll cover normalization in Chapter 5, but for now, realize that training parametric models with unnormalized data—for example, a dataset that contains values from 0 to 1 in one column and 0 to 1,000,000 in another—can make those models less accurate or prevent them from converging on a solution altogether. This is particularly true with support vector machines and neural networks, but it applies to other parametric models as well. Even k-nearest neighbors models work best with normalized data because while the learning algorithm isn't parametric, it uses distance-based calculations internally.

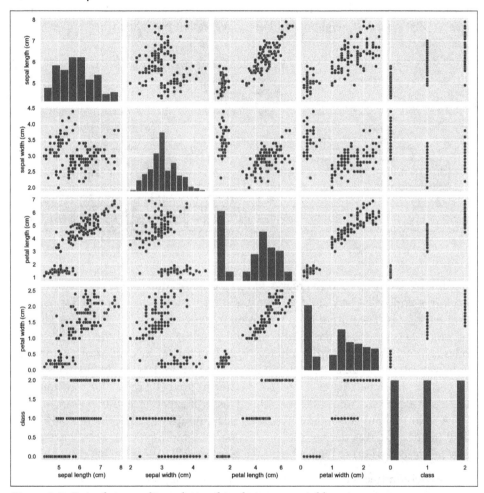

Figure 2-2. Pair plot revealing relationships between variable pairs

Decision Trees

Even if you've never taken a computer science course, you probably know what a binary tree (*https://oreil.ly/pE6EI*) is. In machine learning, a *decision tree* is a tree structure that predicts an outcome by answering a series of questions. Most decision trees are binary trees, in which case the questions require simple yes-or-no answers.

Figure 2-3 shows a decision tree built by Scikit from the income-versus-experience dataset introduced in Chapter 1. The tree is simple because the dataset contains just one feature column (years of experience) and I limited the tree's depth to 3, but the technique extends to trees of unlimited size and complexity. In this example, predicting a salary for a programmer with 10 years of experience requires just three yes/no decisions, as indicated by the red arrows. The answer is about $100K, which is pretty close to what *k*-nearest neighbors and linear regression predicted when applied to the same dataset.

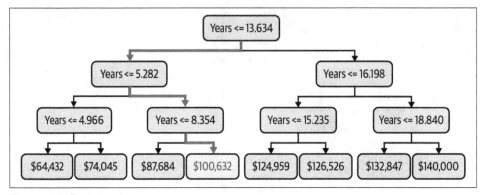

Figure 2-3. Decision tree

Decision trees can be used for regression and classification. For a regressor, the leaf nodes (the nodes that lack children) represent regression values. For a classifier, they represent classes. The output from a decision tree regressor isn't continuous. The output will always be one of the values assigned to a leaf node, and the number of leaf nodes is finite. The output from a linear regression model, by contrast, is continuous. It can assume any value along the line fit to the training data. In the previous example, you get the same answer if you ask the tree to predict a salary for someone with 10 years of experience and someone with 13 years of experience. Bump years of experience up to 14, however, and the predicted salary jumps to $125K (Figure 2-4). If you allow the tree to grow deeper, the answers become more refined. But allowing it to grow *too* deep can lead to big problems for reasons we'll cover momentarily.

Once a decision tree model is trained—that is, once the tree is built—predictions are made quickly. But how do you decide what decisions to make at each node? For example, why is the number of years represented by the root node in Figure 2-3 equal to 13.634? Why not 10.000 or 8.742 or some other number? For that matter, if the dataset has multiple feature columns, how do you decide which column to break on at each decision node?

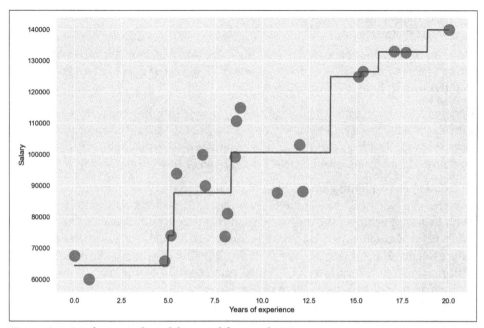

Figure 2-4. Mathematical model created from a decision tree

Decision trees are built by recursively splitting the training data. The fundamental decisions that the splitting algorithm makes when it adds a node to the tree are 1) which column will this node split, and 2) what is the value that the split is based upon. In each iteration, the goal is to select a column and split value that does the most to reduce the "impurity" of the remaining data for classification problems or the variance of the remaining data for regression problems. A common impurity measure for classifiers is Gini (*https://oreil.ly/bCuJx*), which roughly quantifies the percentage of samples that a split value would misclassify. For regressors, the sum of the squared error or absolute error, where "error" is the difference between the split value and the values on either side of the split, is typically used instead. The tree-building process starts at the root node and works its way recursively downward until the tree is fully leafed out or external constraints (such as a limit on maximum depth) prevent further growth.

Scikit's `DecisionTreeRegressor` class (*https://oreil.ly/jLyVQ*) and `DecisionTree Classifier` class (*https://oreil.ly/Ylrq4*) make building decision trees easy. Each implements the well-known CART algorithm (*https://oreil.ly/wQGjx*) for building binary trees, and each lets you choose from a handful of criteria for measuring impurity or variance. Each also supports parameters such as `max_depth`, `min _samples_split`, and `min_samples_leaf` that let you constrain a decision tree's growth. If you accept the default values, building a decision tree can be as simple as this:

```
model = DecisionTreeRegressor()
model.fit(x, y)
```

Decision trees are nonparametric. Training a decision tree model involves building a binary tree, not fitting an equation to a dataset. This means data used to build a decision tree doesn't have to be normalized.

Decision trees have a big upside: they work as well with nonlinear data as they do with linear data. In fact, they largely don't care how the data is shaped. But there's a downside too. It's a big one, and it's one of the reasons standalone decision trees are rarely used in machine learning. That reason is *overfitting* (*https://oreil.ly/ycffk*).

Decision trees are highly prone to overfitting. If allowed to grow large enough, a decision tree can essentially memorize the training data. It might *appear* to be accurate, but if it's fit too tightly to the training data, it might not *generalize* well. That means it won't be as accurate when it's asked to make predictions with data it hasn't seen before. Figure 2-5 shows a decision tree fit to the income-versus-experience dataset with no constraints on depth. The jagged path followed by the red line as it passes through all the points is a clear sign of overfitting. Overfitting is the bane of data scientists. The only thing worse than a model that's inaccurate is one that appears to be accurate but in reality is not.

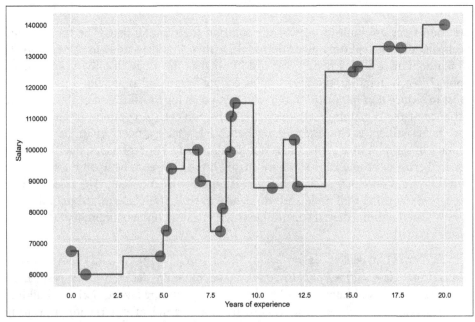

Figure 2-5. Decision tree overfit to the training data

One way to prevent overfitting when using decision trees is to constrain their growth so that they *can't* memorize the training data. Another way is to use groups of decision trees called *random forests*.

Random Forests

A *random forest* is a collection of decision trees (often hundreds of them), each trained differently on the same data, as depicted in Figure 2-6. Typically, each tree is trained on randomly selected rows in the dataset, and branching is based on columns that are randomly selected at every split. The model can't fit too tightly to the training data because every tree trains on a different subset of the data. The trees are built independently, and when the model makes a prediction, it runs the input through all the decision trees and averages the result. Because the trees are constructed independently, training can be parallelized on hardware that supports it.

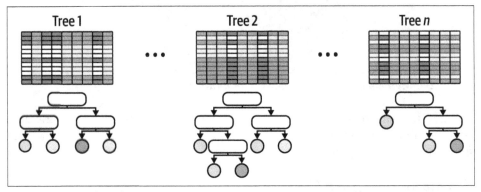

Figure 2-6. Random forests

It's a simple concept, and one that works well in practice. Random forests can be used for both regression and classification, and Scikit provides classes such as Random Forest Regressor (*https://oreil.ly/QwaN1*) and RandomForestClassifier (*https://oreil.ly/xIXEp*) to help out. They feature a number of tunable parameters, including n_estimators, which specifies the number of trees in the random forest (default = 100); max_depth, which limits the depth of each tree; and max_samples, which specifies the fraction of the rows in the training data used to build individual trees. Figure 2-7 shows how RandomForestRegressor fits to the income-versus-experience dataset with max_depth=3 and max_samples=0.5, meaning no tree sees more than 50% of the rows in the dataset.

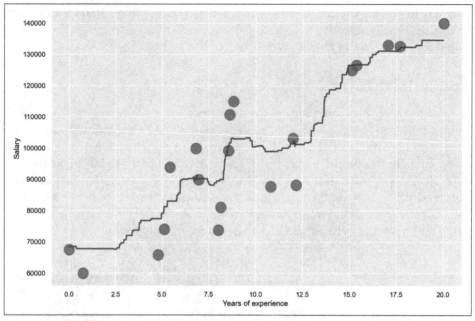

Figure 2-7. Mathematical model created from a random forest

Because decision trees are nonparametric, random forests are nonparametric also. And even though Figure 2-7 shows how a random forest fits a linear dataset, random forests are perfectly capable of modeling nonlinear datasets too.

Gradient-Boosting Machines

Random forests are proof of the supposition that you can take many *weak learners*— models that by themselves are not strong predictors—and combine them to form accurate models. No individual tree in a random forest can predict an outcome with a great deal of accuracy. But put all the trees together and average the results and they often outperform other models. Data scientists refer to this as *ensemble modeling* or *ensemble learning* (*https://oreil.ly/V9fqa*).

Another way to exploit ensemble modeling is gradient boosting (*https://oreil.ly/dzOiH*). Models that use it are called *gradient-boosting machines*, or GBMs. Most GBMs use decision trees and are sometimes referred to as *gradient-boosted decision trees* (GBDTs). Like random forests, GBDTs comprise collections of decision trees. But rather than build independent decision trees from random subsets of the data, GBDTs build *dependent* decision trees, one after another, training each using output from the last. The first decision tree models the dataset. The second decision tree models the error in the output from the first, the third models the error in the output from the second, and so on. To make a prediction, a GBDT runs the input through each decision tree and sums all the outputs to arrive at a result. With each addition, the result becomes slightly more accurate, giving rise to the term *additive modeling* (*https://oreil.ly/vyJyq*). It's like driving a golf ball down the fairway and hitting successively shorter shots until you finally reach the hole.

Each decision tree in a GBDT model is a weak learner. In fact, GBDTs typically use *decision tree stumps*, which are decision trees with depth 1 (a root node and two child nodes), as shown in Figure 2-8. During training, you start by taking the mean of all the target values in the training data to create a baseline for predictions. Then you subtract the mean from the target values to generate a new set of target values or *residuals* for the first tree to predict. After training the first tree, you run the input through it to generate a set of predictions. Then you add the predictions to the previous set of predictions, generate a new set of residuals by subtracting the sum from the original (actual) target values, and train a second tree to predict those residuals. Repeating this process for *n* trees, where *n* is typically 100 or more, produces an ensemble model. To help ensure that each decision tree is a weak learner, GBDT models multiply the output from each decision tree by a *learning rate* to reduce their influence on the outcome. The learning rate is usually a small number such as 0.1 and is a parameter that you can specify when using classes that implement GBMs.

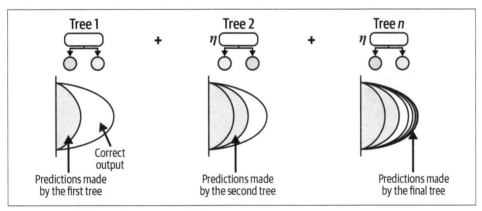

Figure 2-8. Gradient-boosting machines

Scikit includes classes named GradientBoostingRegressor (*https://oreil.ly/ZhSop*) and GradientBoostingClassifier (*https://oreil.ly/9RlW5*) to help you build GBDTs. But if you really want to understand how GBDTs work, you can build one yourself with Scikit's DecisionTreeRegressor class. The code in Example 2-1 implements a GBDT with 100 decision tree stumps and predicts the annual income of a programmer with 10 years of experience.

Example 2-1. Gradient-boosted decision tree implementation

```
learning_rate = 0.1 # Learning rate
n_trees = 100 # Number of decision trees
trees = [] # Trees that comprise the model

# Compute the mean of all the target values
y_pred = np.array([y.mean()] * len(y))
baseline = y_pred

# Create n_trees and train each with the error
# in the output from the previous tree
for i in range(n_trees):
    error = y - y_pred
    tree = DecisionTreeRegressor(max_depth=1, random_state=0)
    tree.fit(x, error)
    predictions = tree.predict(x)
    y_pred = y_pred + (learning_rate * predictions)
    trees.append(tree)

# Predict a y for x=10
y_pred = np.array([baseline[0]] * len(x))

for tree in trees:
    y_pred = y_pred + (learning_rate * tree.predict([[10.0]]))

y_pred[0]
```

The diagram on the left in Figure 2-9 shows the output from a single decision tree stump applied to the income-versus-experience dataset. That model is such a weak learner that it can predict only two different income levels. The diagram on the right shows the output from the model in Example 2-1. The additive effect of the weak learners produces a strong learner that predicts a programmer with 10 years of experience should earn $99,082 per year, which is consistent with the predictions made by other models.

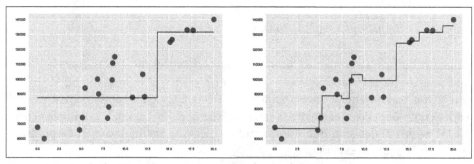

Figure 2-9. Single decision tree versus gradient-boosted decision trees

GBDTs can be used for regression and classification, and they are nonparametric. Aside from neural networks and support vector machines, GBDTs are frequently the ones that data scientists find most capable of modeling complex datasets.

Unlike linear regression models and random forests, GBDTs are susceptible to overfitting. One way to mitigate overfitting when using `GradientBoostingRegressor` and `GradientBoostingClassifier` is to use the `subsample` parameter to prevent individual trees from seeing the entire dataset, analogous to what `max_samples` does for random forests. Another way is to use the `learning_rate` parameter to lower the learning rate, which defaults to 0.1.

Support Vector Machines

I will save a full treatment of support vector machines (SVMs) for Chapter 5, but along with GBMs, they represent the cutting edge of statistical machine learning. They can often fit models to highly nonlinear datasets that other learning algorithms cannot. They're so important that they merit separate treatment from all other algorithms. They work by employing a mathematical device called *kernel tricks* to simulate the effect of adding dimensions to data. The idea is that data that isn't separable in *m* dimensions might be separable in *n* dimensions. Here's a quick example.

The classes in the two-dimensional dataset on the left in Figure 2-10 can't be separated with a line. But if you add a third dimension so that points closer to the center have higher *z* values and points farther from the center have lower *z* values, as shown on the right, you can slide a plane between the red points and the purple points and achieve 100% separation of the classes. That is the principle by which SVMs work. It is mathematically complex when generalized to work with arbitrary datasets, but it is an extremely powerful technique that is vastly simplified by Scikit.

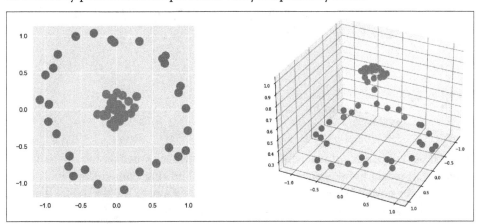

Figure 2-10. Support vector machines

SVMs are primarily used for classification, but they can be used for regression as well. Scikit includes classes for doing both, including SVC (*https://oreil.ly/MWsSX*) for classification problems and SVR (*https://oreil.ly/YPvZa*) for regression problems. You will learn all about these classes in Chapter 5. For now, drop the term *support vector machine* at the next machine learning gathering you attend and you will instantly become the life of the party.

Accuracy Measures for Regression Models

As you learned in Chapter 1, you need one set of data for training a model and another set for testing it, and you can score a model for accuracy by passing test data to the model's `score` method. Testing quantifies how accurate the model is at making predictions. It is incredibly important to test a model with a dataset other than the one it was trained with because it will probably learn the training data reasonably well, but that doesn't mean it will generalize well—that is, make accurate predictions. And if you don't test a model, you don't know how accurate it is.

Engineers frequently use Scikit's `train_test_split` function (*https://oreil.ly/AkvO3*) to split a dataset into a training dataset and a test dataset. But when you split a small dataset this way, you can't necessarily trust the score returned by the model's `score` method. And what does the score mean, anyway? The answer is different for regression models and classification models, and while there are a handful of ways to score regression models, there are many ways to score classification models. Let's take a moment to understand the number that `score` returns for a regression model, and what you can do to have more faith in that number.

To demonstrate why you need to be somewhat skeptical of the value returned by `score` when dealing with small datasets, try a simple experiment. Use the following code to load Scikit's California housing dataset (*https://oreil.ly/mt7ch*), shuffle the rows, and extract the first 1,000 rows:

```
from sklearn.utils import shuffle
from sklearn.datasets import fetch_california_housing

df = fetch_california_housing(as_frame=True).frame
df = shuffle(df, random_state=0)
df = df.head(1000)
df.head()
```

The dataset contains columns with names such as `MedInc` (median income) and `Med HouseVal` (median home value), as shown in the following figure. The details aren't important for now. What is important is that you are going to build a model that uses the values in all the other columns to predict `MedHouseVal` values:

	MedInc	HouseAge	AveRooms	AveBedrms	Population	AveOccup	Latitude	Longitude	MedHouseVal
14740	4.1518	22.0	5.663073	1.075472	1551.0	4.180593	32.58	-117.05	1.369
10101	5.7796	32.0	6.107226	0.927739	1296.0	3.020979	33.92	-117.97	2.413
20566	4.3487	29.0	5.930712	1.026217	1554.0	2.910112	38.65	-121.84	2.007
2670	2.4511	37.0	4.992958	1.316901	390.0	2.746479	33.20	-115.60	0.725
15709	5.0049	25.0	4.319261	1.039578	649.0	1.712401	37.79	-122.43	4.600

Use the following code to split the data 80/20, train a linear regression model with 80% of the data to predict the price of a house, and score the model with 20% of the data split off for testing:

```
from sklearn.linear_model import LinearRegression
from sklearn.model_selection import train_test_split

x = df.drop(['MedHouseVal'], axis=1)
y = df['MedHouseVal']
x_train, x_test, y_train, y_test = train_test_split(x, y, test_size=0.2,
                                                    random_state=0)

model = LinearRegression()
model.fit(x_train, y_train)
model.score(x_test, y_test)
```

In this example, `score` returns 0.5863, which ostensibly indicates that the model was about 59% accurate using the features in the test data to make predictions. So far, so good.

Now change the `random_state` value passed to `train_test_split` from 0 to 1 and run the code again:

```
x_train, x_test, y_train, y_test = train_test_split(x, y, test_size=0.2,
                                                    random_state=1)

model = LinearRegression()
model.fit(x_train, y_train)
model.score(x_test, y_test)
```

This time, `score` returns 0.6255. So which is it? Is the model 59% accurate at making predictions or 63% accurate? Why did `score` return two different values?

When `train_test_split` splits a dataset, it randomly selects rows for the training dataset and the test dataset. The `random_state` parameter seeds the random-number generator used to make selections. By specifying two different seed values, you train the model with two different datasets and test it with two different datasets too. Sure, there is overlap between them. But the fact remains that the number you seed `train_test_split`'s random-number generator with affects the outcome. The smaller the dataset, the greater that effect is likely to be.

The solution is *cross-validation*. To cross-validate a model, you partition the dataset into *folds*, as pictured in Figure 2-11. Using five folds is common, but you can use any number of folds you like. Then you train the model five times—once for each fold—using a different 80% of the dataset for training and a different 20% for testing each time, and average the scores to generate a cross-validated score. That score is more reliable than the score returned by score because it is less sensitive to how the data is split. This process is known as *k-fold cross-validation*.

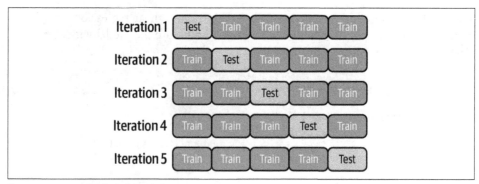

Figure 2-11. 5-fold cross-validation

The downside to cross-validation is that it takes longer. With 5-fold cross-validation, you're training the model five times. The good news is that cross-validation is generally applied only to small datasets, and if the dataset is small, the model probably trains quickly.

You could write the code to do cross-validation yourself, but you don't have to because Scikit does it for you. Cross-validating a model requires just one line of code:

```
cross_val_score(model, x, y, cv=5).mean()
```

The cross-validated score in this example should come out to about 0.61, which lies between the two values generated by the model's score method. It is a more accurate measure of the model's accuracy than either of the other scores.

You don't have to train a model before cross-validating it since cross_val_score trains it for you. However, cross_val_score trains a *copy* of the model, not the model itself, so once you've used cross-validation to gauge the model's accuracy, you still need to call fit before making predictions:

```
model.fit(x, y)
```

Note that you pass the entire dataset to `fit` rather than a subset split off for training. That's another benefit of cross-validation: you train your model with *all* of the data, which is a big deal if the dataset is small to begin with. There's no longer a hard requirement to hold some of it out for testing. However, even with cross-validation, it's still useful to score the model with data reserved exclusively for testing if such data is available. Remember: *you don't truly know how accurate a model is until you know how it responds to data it wasn't trained with.*

> As a rule of thumb, you should reserve cross-validation for small datasets and use train/test splits for large datasets. The larger the dataset, the less sensitive it is to how the data is split.

Which begs the question: precisely what does the value returned when you score or cross-validate a model represent? For a regression model, it's the *coefficient of determination* (*https://oreil.ly/W1UHd*), also known as the *R-squared* score or simply R^2. The coefficient of determination is usually a value from 0 to 1 ("usually" because it can, in certain circumstances, go negative) which quantifies the variance in the output that can be explained by the variables in the input. A simple way to think of it is that an R^2 score of 0.8 means that the model should, on average, be about 80% accurate in making predictions, or get the answer right within 20%. The higher the R^2 score, the more accurate the model. There are other ways to measure the accuracy of a regression model, including mean squared error (MSE) and mean absolute error (MAE). Those numbers are meaningful only in the context of the range of output values, whereas R^2 gives you one simple number that is independent of range. You can read more about regression metrics and methods for retrieving them in the Scikit documentation (*https://oreil.ly/G3LUI*).

The value returned by a model's `score` method is completely different for a classification model. I will address the various ways to quantify the accuracy of classification models in Chapter 3.

Using Regression to Predict Taxi Fares

Imagine that you work for a taxi company, and one of your customers' biggest complaints is that they don't know how much a ride will cost until it's over. That's because distance isn't the only variable that determines a fare amount. You decide to do something about it by building a mobile app that customers can use when they climb into a taxi to estimate what the fare will be. To provide the intelligence for the app, you intend to use the massive amounts of fare data the company has collected over the years to build a machine learning model.

Let's train a regression model to predict a fare amount given the time of day, the day of the week, and the pickup and drop-off locations. Start by downloading the CSV file containing the dataset (*https://oreil.ly/qgx9X*) and copying it into the *Data* subdirectory where your Jupyter notebooks are hosted. Then use the following code to load the dataset into a notebook. The dataset contains about 55,000 rows and is a subset of a much larger dataset used in Kaggle's New York City Taxi Fare Prediction competition (*https://oreil.ly/Dy9Ye*). The data requires a fair amount of prep work before it's of any use—something that's not uncommon in machine learning. Data scientists often find that collecting and preparing data accounts for 90% or more of their time:

```
import pandas as pd

df = pd.read_csv('Data/taxi-fares.csv', parse_dates=['pickup_datetime'])
df.head()
```

Note the use of the `read_csv` function's `parse_dates` parameter to parse the strings in the `pickup_datetime` column into Python `datetime` objects (*https://oreil.ly/msqMQ*). Here's the output from the code:

	key	fare_amount	pickup_datetime	pickup_longitude	pickup_latitude	dropoff_longitude	dropoff_latitude	passenger_count
0	2014-06-15 17:11:00.000000107	7.0	2014-06-15 17:11:00+00:00	-73.995420	40.759662	-73.987607	40.751247	1
1	2011-03-14 22:43:00.00000095	4.9	2011-03-14 22:43:00+00:00	-73.993552	40.731110	-73.998497	40.737200	5
2	2011-02-14 15:14:00.00000067	6.1	2011-02-14 15:14:00+00:00	-73.972380	40.749527	-73.990638	40.745328	1
3	2009-10-29 11:29:00.00000040	6.9	2009-10-29 11:29:00+00:00	-73.973703	40.763542	-73.984253	40.758603	5
4	2011-07-02 10:38:00.00000028	10.5	2011-07-02 10:38:00+00:00	-73.921262	40.743615	-73.967383	40.765162	1

Each row represents a taxi ride and contains information such as the fare amount, the pickup and drop-off locations (expressed as latitudes and longitudes), and the passenger count. It's the fare amount that we want to predict. Use the following code to draw a histogram showing how many rows contain a passenger count of 1, how many contain a passenger count of 2, and so on:

```
%matplotlib inline
import matplotlib.pyplot as plt
import seaborn as sns
sns.set()

sns.countplot(x=df['passenger_count'])
```

Here is the output:

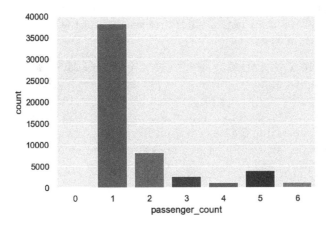

Most of the rows in the dataset have a passenger count of 1. Since we're interested in predicting the fare amount only for single passengers, use the following code to remove all rows with multiple passengers and remove the key column from the dataset since that column isn't needed—in other words, it's not one of the features that we will try to predict on:

```
df = df[df['passenger_count'] == 1]
df = df.drop(['key', 'passenger_count'], axis=1)
df.head()
```

That leaves 38,233 rows in the dataset, which you can see for yourself with the following statement:

```
df.shape
```

Now use Pandas' corr method (*https://oreil.ly/BNX2X*) to find out how much influence input variables such as latitude and longitude have on the values in the fare_amount column:

```
corr_matrix = df.corr()
corr_matrix['fare_amount'].sort_values(ascending=False)
```

The output looks like this:

```
fare_amount        1.000000
dropoff_longitude  0.020438
pickup_longitude   0.015742
pickup_latitude    -0.015915
dropoff_latitude   -0.021711
Name: fare_amount, dtype: float64
```

The numbers don't look very encouraging. Latitudes and longitudes have little to do with fare amounts, at least in their present form. And yet, intuitively, they should have *a lot* to do with fare amounts since they specify starting and ending points, and longer rides incur higher fares.

Now comes the fun part: creating whole new columns of data that have more impact on the outcome—columns whose values are computed from values in other columns. Add columns specifying the day of the week (0=Monday, 1=Sunday, and so on), the hour of the day that the passenger was picked up (0–23), and the distance (by air, not on the street) in miles that the ride covered. To compute distances, this code assumes that most rides are short and that it is therefore safe to ignore the curvature of the Earth:

```python
from math import sqrt

for i, row in df.iterrows():
    dt = row['pickup_datetime']
    df.at[i, 'day_of_week'] = dt.weekday()
    df.at[i, 'pickup_time'] = dt.hour
    x = (row['dropoff_longitude'] - row['pickup_longitude']) * 54.6
    y = (row['dropoff_latitude'] - row['pickup_latitude']) * 69.0
    distance = sqrt(x**2 + y**2)
    df.at[i, 'distance'] = distance

df.head()
```

You no longer need all the columns, so use these statements to remove the ones that won't be used:

```python
df.drop(columns=['pickup_datetime', 'pickup_longitude', 'pickup_latitude',
                 'dropoff_longitude', 'dropoff_latitude'], inplace=True)
df.head()
```

Let's check the correlation again:

```python
corr_matrix = df.corr()
corr_matrix['fare_amount'].sort_values(ascending=False)
```

There still isn't a strong correlation between distance traveled and fare amount. Perhaps this will explain why:

```python
df.describe()
```

Here is the output:

	fare_amount	day_of_week	pickup_time	distance
count	38233.000000	38233.000000	38233.000000	38233.000000
mean	11.214115	2.951534	13.387989	12.018397
std	9.703149	1.932809	6.446519	217.357022
min	-22.100000	0.000000	0.000000	0.000000
25%	6.000000	1.000000	9.000000	0.762116
50%	8.500000	3.000000	14.000000	1.331326
75%	12.500000	5.000000	19.000000	2.402226
max	256.000000	6.000000	23.000000	4923.837280

The dataset contains outliers, and outliers frequently skew the results of machine learning models. Filter the dataset by eliminating negative fare amounts and placing reasonable limits on fares and distance, and then run a correlation again:

```
df = df[(df['distance'] > 1.0) & (df['distance'] < 10.0)]
df = df[(df['fare_amount'] > 0.0) & (df['fare_amount'] < 50.0)]

corr_matrix = df.corr()
corr_matrix['fare_amount'].sort_values(ascending=False)
```

Once more, here is the output:

```
fare_amount    1.000000
distance       0.851913
day_of_week   -0.003570
pickup_time   -0.023085
Name: fare_amount, dtype: float64
```

That looks better! Most (85%) of the variance in fare amounts is explained by the distance traveled. The correlation between the day of the week, the hour of the day, and the fare amount is still weak, but that's not surprising since distance traveled is the main factor that drives taxi fares. Let's leave those columns in since it makes sense that it might take longer to get from point A to point B during rush hour, or that traffic at 5:00 p.m. on Friday might be different than traffic at 5:00 p.m. on Saturday.

Now it's time to train a regression model. Let's try three different learning algorithms to determine which one yields the most accurate fit, and use cross-validation to gauge accuracy. Start with a linear regression model:

```
from sklearn.linear_model import LinearRegression
from sklearn.model_selection import cross_val_score

x = df.drop(['fare_amount'], axis=1)
y = df['fare_amount']
```

```
model = LinearRegression()
cross_val_score(model, x, y, cv=5).mean()
```

Try a RandomForestRegressor with the same dataset and see how its accuracy compares. Recall that random-forest models train multiple decision trees on the data and average the results of all the trees to make a prediction:

```
from sklearn.ensemble import RandomForestRegressor

model = RandomForestRegressor(random_state=0)
cross_val_score(model, x, y, cv=5).mean()
```

Finally, try GradientBoostingRegressor. Gradient-boosting machines use multiple decision trees, each of which is trained to compensate for the error in the output from the previous one:

```
from sklearn.ensemble import GradientBoostingRegressor

model = GradientBoostingRegressor(random_state=0)
cross_val_score(model, x, y, cv=5).mean()
```

Assuming the GradientBoostingRegressor produced the highest cross-validated coefficient of determination, train it using the entire dataset:

```
model.fit(x, y)
```

Finish up by using the trained model to make a pair of predictions. First, estimate what it will cost to hire a taxi for a 2-mile trip at 5:00 p.m. on Friday afternoon. Because you passed DataFrames containing column names to the fit method, recent versions of Scikit will display a warning if you pass lists or NumPy arrays to predict. Therefore, input a DataFrame instead:

```
model.predict(pd.DataFrame({ 'day_of_week': [4], 'pickup_time': [17],
                             'distance': [2.0] }))
```

Now predict the fare amount for a 2-mile trip taken at 5:00 p.m. one day later (on Saturday):

```
model.predict(pd.DataFrame({ 'day_of_week': [5], 'pickup_time': [17],
                             'distance': [2.0] }))
```

Does the model predict a higher or lower fare amount for the same trip on Saturday afternoon? Do the answers make sense given that the data comes from New York City cabs? Consider that rush-hour traffic is likely to be heavier on Friday afternoon than on Saturday afternoon.

Summary

Regression models are supervised learning models that predict numeric outcomes such as the cost of a taxi ride. Prominent learning algorithms used for regression include the following:

Linear regression

    Models training data by fitting it to the equation of a line

Decision trees

    Use binary trees to predict an outcome by answering a series of yes-and-no questions

Random forests

    Use multiple independent decision trees to model the data and are resistant to overfitting

Gradient-boosting machines

    Use multiple *dependent* decision trees, each modeling the error in the output from the last

Support vector machines

    Take an entirely different approach to modeling data by adding dimensionality under the supposition that data that isn't linearly separable in the original problem space might be linearly separable in higher-dimensional space

Scikit provides convenient implementations of these and other learning algorithms in classes such as `LinearRegression`, `RandomForestRegressor`, and `GradientBoosting Regressor`.

A common metric for quantifying the accuracy of regression models is the R^2 score, also known as the coefficient of determination. It's typically a value from 0 to 1, with higher numbers indicating higher accuracy. Technically, it's a measure of the variance in the output that can be explained by the values in the input. For small datasets, k-fold cross-validation gives you more confidence in the R^2 score than simply splitting the data once for training and testing. k-fold trains the model k times, each time with the dataset split differently.

Real-world datasets tend to be messy and often require further preparation to be useful for machine learning. As the taxi-fare example demonstrated, outliers in training data can affect a model's accuracy or prevent the model from being useful at all. One solution is to identify the outliers and remove them before training the model. Another is to employ a learning algorithm such as ridge regression or lasso regression that supports regularization.

Regression models are common in machine learning, but classification models are more common still. Chapter 3 tackles classification models head-on, introduces another leading learning algorithm called logistic regression, and builds on what you learned in this chapter.

## CHAPTER 3

Classification Models

The machine learning model featured in the previous chapter used various forms of regression to predict taxi fares based on distance to travel, the day of the week, and the time of day. Regression models predict numerical outcomes and are widely used in industry to forecast sales, prices, demand, and other numbers that drive business decisions. Equally important are classification models, which predict categorical outcomes such as whether a credit card transaction is fraudulent or which letter of the alphabet a handwritten character represents.

Most classification models fall into two categories: *binary classification models*, in which there are just two possible outcomes, and *multiclass classification models*, in which there are more than two possible outcomes. In both instances, the model assigns a single class, or *class label*, to an input. Less common are *multilabel classification models*, which can classify a single input as belonging to several classes—for example, predicting that a document is both a paper on machine learning and a paper on genomics. Some can predict that an input belongs to none of the possible classes too.

Much of what you know about regression models also applies to classification models. For example, many of the learning algorithms that power regression models work equally well with classification models. One substantive difference between regression and classification is how you measure a model's accuracy. There's no such thing as an R^2 score for a classification model. In its place are an abundance of measures, such as precision, recall, specificity, sensitivity, and F1 score, to name but a few. One of the keys to becoming proficient with classification models is getting comfortable with the various accuracy metrics and, more importantly, understanding which one (or ones) to use based on the model's intended application.

You've seen one example of multiclass classification in the iris tutorial in Chapter 1. It's time to delve deeper into machine learning classifiers, starting with one of the most tried-and-true learning algorithms of all, one that works *only* for classification models: logistic regression.

Logistic Regression

Many learning algorithms exist for classification problems. In Chapter 2, you learned how decision trees, random forests, and gradient-boosting machines (GBMs) fit regression models to training data. These algorithms can be used for classification as well, and Scikit helps out by offering classes such as `DecisionTreeClassifier` (*https://oreil.ly/8N6MG*), `RandomForestClassifier` (*https://oreil.ly/3mNhv*), and `GradientBoostingClassifier` (*https://oreil.ly/ilXW4*). In Chapter 1, you used Scikit's `KNeighborsClassifier` class (*https://oreil.ly/zdDCH*) to build a three-class classification model with *k*-nearest neighbors as the learning algorithm.

These are important learning algorithms, and they see use in many contemporary machine learning models. But one of the most popular classification algorithms is *logistic regression* (*https://oreil.ly/o5rBl*), which analyzes a distribution of data and fits an equation to it that defines the probability that a given sample belongs to each of two possible classes. It might determine, for example, that there's a 10% chance the values in a sample correspond to class 0 and a 90% chance they correspond to class 1. In this case, logistic regression will predict that the sample corresponds to class 1. Despite the name, logistic regression is a *classification* algorithm, not a regression algorithm. Its purpose is not to create regression models but to quantify probabilities for the purpose of classifying input samples.

As an example, consider the data points in Figure 3-1, which belong to two classes: 0 (blue) and 1 (red). Let's assume that the x-axis specifies the number of hours a person studied for an exam and the y-axis indicates whether they passed (1) or failed (0). The blues fall in the range $x = 0$ to $x = 10$, while the reds fall in the range $x = 5$ to $x = 15$. You can't pick a value for x that separates the classes since both have values between $x = 5$ and $x = 10$. (Try drawing a vertical line that has only reds on one side and only blues on the other.) But you *can* draw a curve that, given an x, shows the probability that a point with that x belongs to class 1. As x increases, so too does the likelihood that the point represents class 1 (pass) rather than class 0 (fail). From the curve, you can see that if $x = 2$, there is less than a 5% chance that the point corresponds to class 1. But if $x = 10$, there is about a 76% chance that it's class 1. If asked to classify that point as a red or a blue, you would conclude that it's a red because it's much more likely to be red than blue.

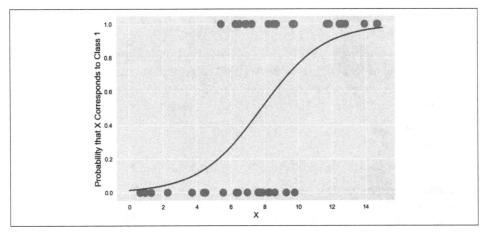

Figure 3-1. Logistic regression

The curve in Figure 3-1 is a *sigmoid curve*. It charts a function known as the *logistic function* (also known as the *logit function*) (*https://oreil.ly/tZTvE*) that has been used in statistics for decades. For logistic regression, the function is defined this way, where *x* is the input value and *m* and *b* are parameters learned during training:

$$y = \frac{1}{1 + e^{-(mx + b)}}$$

This equation reveals why logistic regression is called logistic *regression*, despite the fact that it's a classification algorithm. The exponent of *e* happens to be the equation for linear regression.

The logistic regression learning algorithm fits the logistic function to a distribution of data and uses the resulting *y* values as probabilities to classify data points. It works with any number of features (not just *x*, but x_1, x_2, x_3, and so on), and it is a parametric learning algorithm since it uses training data to find optimum values for *m* and *b*. *How* it finds the optimum values is an implementational detail that libraries such as Scikit-Learn handle for you. Scikit defaults to a numerical optimization algorithm known as *Limited-memory Broyden–Fletcher–Goldfarb–Shanno* (L-BFGS) (*https://oreil.ly/hhJw0*) but supports other optimization methods as well. This, incidentally, is one reason why Scikit is so popular in the machine learning community. It's not difficult to calculate *m* and *b* from the training data for a linear regression model, but it's harder to do it for a logistic regression model, not to mention more sophisticated parametric models such as support vector machines.

Scikit's `LogisticRegression` class (*https://oreil.ly/wpwGs*) is logistic regression in a box. With it, training a logistic regression model can be as simple as this:

```
model = LogisticRegression()
model.fit(x, y)
```

Once the model is trained, you can call its `predict` method to predict which class the input belongs to, or its `predict_proba` method to get the computed probabilities for each class. If you fit a `LogisticRegression` model to the dataset in Figure 3-1, the following statement predicts whether $x = 10$ corresponds to class 0 or class 1:

```
predicted_class = model.predict([[10.0]])[0]
print(predicted_class) # Outputs 1
```

And these statements show the probabilities computed for each class:

```
predicted_probabilities = model.predict_proba([[10.0]])[0]
print(f'Class 0: {predicted_probabilities[0]}') # 0.23508543966167028
print(f'Class 1: {predicted_probabilities[1]}') # 0.7649145603383297
```

Scikit also includes the `LogisticRegressionCV` class (*https://oreil.ly/EPjhI*) for training logistic regression models with built-in cross-validation. (If you need a refresher, cross-validation was introduced in Chapter 2.) At the expense of additional training time, the following statements train a logistic regression model using five folds:

```
model = LogisticRegressionCV(cv=5)
model.fit(x, y)
```

Logistic regression is technically a binary classification algorithm, but it can be used for multiclass classification too. I'll say more about this toward the end of the chapter. For now, think of logistic regression as a machine learning algorithm that uses the well-known logistic function to quantify the probability that an input corresponds to each of two classes, and you'll have an accurate conceptual understanding of what logistic regression is.

Accuracy Measures for Classification Models

You can quantify the accuracy of a classification model the same way you do for a regression model: by calling the model's `score` method. For a classifier, `score` returns the sum of the true positives and the true negatives divided by the total number of samples. If the test data includes 10 positives (samples of class 1) and 10 negatives (samples of class 0) and the model correctly identifies 8 of the positives and 7 of the negatives, then the score is (8 + 7) / 20, or 0.75. This is sometimes referred to as the model's *accuracy score*.

There are many other ways to score a classification model, and which one is "right" often depends on how the model will be used. Rather than compute an accuracy score, data scientists sometimes measure a classification model's *precision* and *recall* (*https://oreil.ly/YJwCM*) instead:

Precision

Computed by dividing the number of true positives by the sum of the true positives and false positives

Recall

Computed by dividing the number of true positives by the sum of the true positives and false negatives

In effect, precision imposes a penalty on false positives (instances in which the model incorrectly predicts a 1), while recall penalizes false negatives by lowering the score when a model incorrectly predicts a 0.

Figure 3-2 illustrates the difference. Suppose you train a model to differentiate between polar bear images and walrus images, and to test it you submit three polar bear images and three walrus images. Furthermore, assume that the model correctly classifies two of the polar bear images, but incorrectly classifies two walrus images as polar bear images, as indicated by the red boxes. In this case, the model's precision in identifying polar bears is 50% because only two of the four images the model classified as polar bears were in fact polar bears. But recall is 67% since the model correctly identified two of the three polar bear images. That's precision and recall in a nutshell. The former quantifies how confident you can be that a positive prediction is accurate, while the latter quantifies the model's ability to accurately identify positive samples. The two can be combined into one score called the *F1 score* (also known as the F-score) (*https://oreil.ly/qGnR0*) using a simple formula.

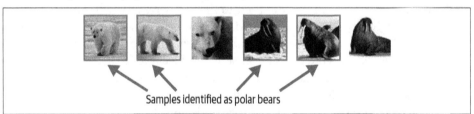

Samples identified as polar bears

Figure 3-2. Using precision and recall to measure the accuracy of a classifier

Scikit provides helpful functions such as precision_score (*https://oreil.ly/2x7yM*), recall_score (*https://oreil.ly/xp7tO*), and f1_score (*https://oreil.ly/fUu9A*) for retrieving classification metrics. Whether you prefer precision or recall depends on which is higher: the cost of false positives or the cost of false negatives. Use precision when the cost of false positives is high—for example, when "positive" means an email is spam. You would rather a spam filter send a few spam mails to your inbox than route a legitimate (and potentially important) email to your junk folder. By contrast, use recall if the cost of false negatives is high. A great example is when using machine learning to spot tumors in X-rays and MRI scans. You would much rather mistakenly send a patient to a doctor due to a false positive than tell that patient there are no tumors when there really are.

Spotting Polar Bears in the Wild

The polar bear versus walrus example was taken from a tutorial I wrote for Microsoft that began with the following introduction:

> You're the leader of a group of climate scientists who are concerned about the dwindling polar bear population in the Arctic. To address the problem, your team has placed hundreds of motion-activated cameras at strategic locations throughout the region. Instead of manually examining each photo that's taken to determine whether it contains a polar bear, your challenge is to devise an automated system that processes data from these cameras in real time and displays an alert on a map when one of your cameras photographs a polar bear. You need a solution that uses artificial intelligence (AI) to determine with a high degree of accuracy whether a photo contains a polar bear. And you need it fast, because climate change won't wait.

The tutorial combines several Azure services to form an end-to-end solution, and it uses Microsoft's Power BI for visualizations. If you're interested, you can check it out online (*https://oreil.ly/Q8Bwb*).

Accuracy score, precision, recall, and F1 score apply to binary classification *and* multiclass classification models. An additional metric—one that applies to binary classification only—is the *receiver operating characteristic* (ROC) curve (*https://oreil.ly/BDvv9*), which plots the true-positive rate (TPR) against the false-positive rate (FPR) at various probability thresholds. A sample ROC curve is shown in Figure 3-3. A straight line stretching from the lower left to the upper right would indicate that the model gets it right just 50% of the time, which is no better than guessing for a binary classifier. The more the curve arches toward the upper-left corner, the more accurate the model. Data scientists often use the area under the curve (AUC, or ROC AUC) as an overall measure of accuracy. Scikit provides a class named `RocCurveDisplay` (*https://oreil.ly/mJePr*) for plotting ROC curves, and a function named `roc_auc_score` (*https://oreil.ly/ryuxL*) for retrieving ROC AUC scores. Scores returned by this function are values from 0.0 to 1.0. The higher the score, the more accurate the model.

Yet another way to assess the accuracy of a classification model is to plot a *confusion matrix* (*https://oreil.ly/YxADc*) like the one in Figure 3-4. It works for binary and multiclass classification, and it shows for each class how the model performed during testing. In this example, the model was asked to differentiate between images containing masked faces and images containing unmasked faces. It got it right 78 out of 85 times when presented with pictures of people wearing masks, and 41 out of 58 times when presented with pictures of people not wearing masks. Scikit offers a `confusion_matrix` function (*https://oreil.ly/U4JBb*) for computing a confusion matrix, and a `ConfusionMatrixDisplay` class (*https://oreil.ly/CLnP2*) with methods named `from_estimator` (*https://oreil.ly/2XdMZ*) and `from_predictions` (*https://oreil.ly/6sXhx*) for plotting confusion matrices.

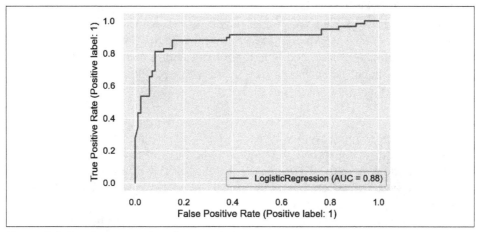

Figure 3-3. Receiver operating characteristic curve

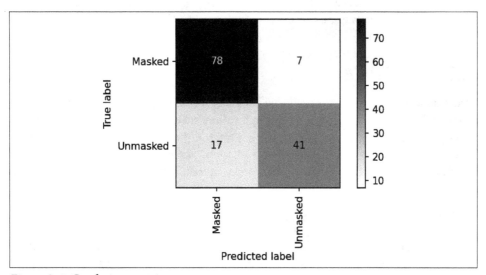

Figure 3-4. Confusion matrix

 Countless code samples online use Scikit's `plot_confusion_matrix` function to display confusion matrices. `ConfusionMatrixDisplay` was introduced in Scikit 1.0 and is the proper way to generate confusion matrices. `plot_confusion_matrix` is slated to be removed from the library in version 1.2.

Other terms you might come across when discussing the accuracy of classification models include *sensitivity* and *specificity* (*https://oreil.ly/EhfoW*). Sensitivity is identical to recall, so Scikit doesn't include a separate method for computing it. Specificity is recall for the negative class rather than the positive class and is calculated by dividing the number of true negatives by the sum of the true negatives and false positives. Scikit doesn't provide a dedicated function for calculating specificity, either, but you can do it easily enough by calling `recall_score` with a `pos_label` parameter indicating that 0 (rather than 1) is the positive label:

```
recall_score(y_test, y_predicted, pos_label=0)
```

Sensitivity and specificity are frequently used in drug testing and cancer screening. Suppose you're traveling abroad and require a negative COVID test before returning home. If you don't have COVID, what are the chances that a test will incorrectly say you do? (I have asked myself that question many times recently while traveling overseas.) The answer is the test's specificity—a measure of how accurate the test is at identifying negative samples. Sensitivity, on the other hand, reveals how likely the test is to be correct if it says you *do* have COVID. It's a subtle distinction, but an important one if you're concerned that a faulty test might keep you from going home (specificity) or if you want to be certain you don't have COVID before visiting an elderly relative (sensitivity).

Categorical Data

Machine learning finds patterns in numbers. It works *only* with numbers. Yet many datasets have columns containing string values such as `"male"` and `"female"` or `"red"`, `"green"`, and `"blue"`. Data scientists refer to these as *categorical values* and the columns that contain them as *categorical columns*. Machine learning can't handle categorical values directly. To use them in a model, you must convert them into numbers.

Two popular techniques exist for converting categorical values into numerical values. One is *label encoding*, which you briefly saw in the *k*-means clustering example in Chapter 1. Label encoding replaces categorical values with integers. If there are three unique values in a column, label encoding replaces them with 0s, 1s, and 2s. To demonstrate, run the following code in a Jupyter notebook:

```
import pandas as pd
from sklearn.preprocessing import LabelEncoder

data = [[10, 'red'], [20, 'blue'], [12, 'red'], [16, 'green'], [22, 'blue']]
df = pd.DataFrame(data, columns=['Length', 'Color'])

encoder = LabelEncoder()
df['Color'] = encoder.fit_transform(df['Color'])
df.head()
```

This code creates a DataFrame containing a numeric column named Length and a categorical column named Color, which contains three different categorical values. Here's what the dataset looks like before encoding:

	Length	Color
0	10	red
1	20	blue
2	12	red
3	16	green
4	22	blue

And here's how it looks after the values in the Color column are label-encoded using Scikit's LabelEncoder class:

	Length	Color
0	10	2
1	20	0
2	12	2
3	16	1
4	22	0

The encoded dataset can be used to train a machine learning model. The unencoded dataset cannot. You can get an ordered list of the classes that were encoded from the encoder's classes_ attribute.

The other, more common means for converting categorical values into numeric values is *one-hot encoding*, which adds one column to the dataset for each unique value in a categorical column and fills the encoded columns with 1s and 0s. One-hot encoding can be performed with Scikit's OneHotEncoder class (*https://oreil.ly/fkDgF*) or by calling get_dummies (*https://oreil.ly/hqMPs*) on a Pandas DataFrame. Here is how the latter is used to encode the dataset:

```
data = [[10, 'red'], [20, 'blue'], [12, 'red'], [16, 'green'], [22, 'blue']]
df = pd.DataFrame(data, columns=['Length', 'Color'])

df = pd.get_dummies(df, columns=['Color'])
df.head()
```

And here are the results:

	Length	Color_blue	Color_green	Color_red
0	10	0	0	1
1	20	1	0	0
2	12	0	0	1
3	16	0	1	0
4	22	1	0	0

Label encoding and one-hot encoding are used with regression problems and classification problems. The obvious question is, which one should you use? Generally speaking, data scientists prefer one-hot encoding to label encoding. The former gives every unique value an equal weight, whereas the latter implies that some values may be more important than others—for example, that `"red"` (2) is more important than `"blue"` (0). Label encoding, on the other hand, is more memory efficient. The number of columns is the same before and after encoding, whereas one-hot encoding adds one column per unique value. For very large datasets with thousands of unique values in a categorical column, label encoding requires substantially less memory.

The accuracy of a machine learning model is rarely impacted by the encoding method you choose. If you're in doubt, you'll rarely go wrong with one-hot encoding. If you want to be certain, you can encode the data both ways and compare the results after training a machine learning model.

Binary Classification

Binary classifiers are supervised learning models trained with labeled data: 0s for the negative class and 1s for the positive. The predictions they make are 0s and 1s too. They also divulge a probability for each class that you can factor into your conclusions. For example, a credit card company might decide that a transaction will be declined only if the model predicts with at least 99% certainty that the transaction is fraudulent. In that case, the probability that the model computed is more important than the raw prediction that it made.

To help bring home everything presented thus far regarding binary classification, let's use Scikit to build a couple of models: first a simple one that demonstrates core principles, followed by a second one that solves a genuine business problem.

Classifying Passengers Who Sailed on the Titanic

One of the more famous public datasets in machine learning is the Titanic dataset (*https://oreil.ly/JtxlV*), which contains information regarding hundreds of passengers who sailed on the ill-fated voyage of the *RMS Titanic*, including which ones survived and which ones did not. Let's use logistic regression to build a binary classification model from the dataset and see if we can predict the odds that a passenger will survive given that person's gender, age, and fare class (whether they traveled in first, second, or third class).

The first step is to download the dataset (*https://oreil.ly/vMnPI*) and copy it to the *Data* subdirectory of the directory that hosts your Jupyter notebooks. Then run the following code in a notebook to load the dataset and get a feel for its contents:

```
import pandas as pd

df = pd.read_csv('Data/titanic.csv')
df.head()
```

Here is the output:

	PassengerId	Survived	Pclass	Name	Sex	Age	SibSp	Parch	Ticket	Fare	Cabin	Embarked
0	1	0	3	Braund, Mr. Owen Harris	male	22.0	1	0	A/5 21171	7.2500	NaN	S
1	2	1	1	Cumings, Mrs. John Bradley (Florence Briggs Th...	female	38.0	1	0	PC 17599	71.2833	C85	C
2	3	1	3	Heikkinen, Miss. Laina	female	26.0	0	0	STON/O2. 3101282	7.9250	NaN	S
3	4	1	1	Futrelle, Mrs. Jacques Heath (Lily May Peel)	female	35.0	1	0	113803	53.1000	C123	S
4	5	0	3	Allen, Mr. William Henry	male	35.0	0	0	373450	8.0500	NaN	S

The dataset contains 891 rows and 12 columns. Some of the columns, such as PassengerId and Name, aren't relevant to a machine learning model. Others are very relevant. The ones we'll focus on are:

Survived
    Indicates whether the passenger survived the voyage (1) or did not (0)

Pclass
    Indicates whether the passenger was traveling in first class (1), second class (2), or third class (3)

Sex
    Indicates the passenger's gender

Age
    Indicates the passenger's age

The Survived column is the label column—the one we'll try to predict. The other columns are relevant because first-class passengers were more likely to survive the sinking (*https://oreil.ly/i0N6s*) because their cabins were closer to the top deck of the ship and nearer the lifeboats. Plus, women and children were more likely to be given space in lifeboats.

Now use the following statement to see if the dataset is missing any values:

```
df.info()
```

Here's the output:

```
<class 'pandas.core.frame.DataFrame'>
RangeIndex: 891 entries, 0 to 890
Data columns (total 12 columns):
 #   Column       Non-Null Count  Dtype
---  ------       --------------  -----
 0   PassengerId  891 non-null    int64
 1   Survived     891 non-null    int64
 2   Pclass       891 non-null    int64
 3   Name         891 non-null    object
 4   Sex          891 non-null    object
 5   Age          714 non-null    float64
 6   SibSp        891 non-null    int64
 7   Parch        891 non-null    int64
 8   Ticket       891 non-null    object
 9   Fare         891 non-null    float64
 10  Cabin        204 non-null    object
 11  Embarked     889 non-null    object
dtypes: float64(2), int64(5), object(5)
memory usage: 83.7+ KB
```

The Cabin column is missing a lot of values, but we don't care since we're not using that column. We *will* use the Age column, and that column is missing some values as well. We could replace the missing values with the mean of all the other ages—an approach that data scientists refer to as *imputing* missing values—but we'll take the simpler approach of removing rows with missing values. Use the following statements to remove the columns that aren't needed, drop rows with missing values, and one-hot-encode the values in the Sex and Pclass columns:

```
df = df[['Survived', 'Age', 'Sex', 'Pclass']]
df = pd.get_dummies(df, columns=['Sex', 'Pclass'])
df.dropna(inplace=True)
df.head()
```

Here is the resulting dataset:

	Survived	Age	Sex_female	Sex_male	Pclass_1	Pclass_2	Pclass_3
0	0	22.0	0	1	0	0	1
1	1	38.0	1	0	1	0	0
2	1	26.0	1	0	0	0	1
3	1	35.0	1	0	1	0	0
4	0	35.0	0	1	0	0	1

The next task is to split the dataset for training and testing:

```
from sklearn.model_selection import train_test_split

x = df.drop('Survived', axis=1)
y = df['Survived']

x_train, x_test, y_train, y_test = train_test_split(x, y, test_size=0.2,
                                            stratify=y, random_state=0)
```

Note the `stratify=y` parameter passed to `train_test_split`. That's important, because of the 714 samples remaining after rows with missing values are removed, 290 represent passengers who survived and 424 represent passengers who did not. We want the training dataset and the test dataset to contain similar proportions of both classes, and `stratify=y` accomplishes that. Without stratification, the model might appear to be more or less accurate than it really is.

Now create a logistic regression model, train it with the data split off for training, and score it with the test data:

```
from sklearn.linear_model import LogisticRegression

model = LogisticRegression(random_state=0)
model.fit(x_train, y_train)
model.score(x_test, y_test)
```

Score the model again using cross-validation in order to have more confidence in the score. Remember that this is the *accuracy score* computed by summing the true positives and true negatives and dividing by the total number of samples:

```
from sklearn.model_selection import cross_val_score

cross_val_score(model, x, y, cv=5).mean()
```

Use the following statements to display a confusion matrix showing precisely how the model performed during testing:

```
%matplotlib inline
from sklearn.metrics import ConfusionMatrixDisplay as cmd

cmd.from_estimator(model, x_test, y_test,
                display_labels=['Perished', 'Survived'],
                cmap='Blues', xticks_rotation='vertical')
```

Observe that the model is more accurate when predicting that passengers *won't* survive than when predicting that they will. That's because the dataset used to train the model contained more examples of passengers who perished than of passengers who survived. You always prefer to train a binary classification model with a perfectly balanced dataset containing an equal number of positive and negative samples, but it's acceptable to train with an imbalanced dataset if you take the imbalance into account when assessing the model's accuracy.

Now use Scikit's `precision_score` and `recall_score` functions to compute the model's precision, recall, sensitivity, and specificity:

```
from sklearn.metrics import precision_score, recall_score

y_pred = model.predict(x_test)
precision = precision_score(y_test, y_pred)
recall = recall_score(y_test, y_pred)
sensitivity = recall
specificity = recall_score(y_test, y_pred, pos_label=0)

print(f'Precision: {precision}')
print(f'Recall: {recall}')
print(f'Sensitivity: {sensitivity}')
print(f'Specificity: {specificity}')
```

Is the high specificity score consistent with the observation that the model is more adept at identifying passengers who won't survive than those who will? How would you explain the relatively low recall and sensitivity scores?

Now let's use the trained model to make some predictions. First, find out whether a 30-year-old female traveling in first class is likely to survive the voyage. Since the model was trained with a `DataFrame` containing column names, we'll use the same column names to formulate an input:

```
female = pd.DataFrame({ 'Age': [30], 'Sex_female': [1], 'Sex_male': [0],
                        'Pclass_1': [1], 'Pclass_2': [0], 'Pclass_3': [0] })

model.predict(female)[0]
```

The model predicts she will survive, but what are the *odds* that she we will survive?

```
probability = model.predict_proba(female)[0][1]
print(f'Probability of survival: {probability:.1%}')
```

A 30-year-old female traveling in first class is more than 90% likely to survive the voyage, but what about a 60-year-old male traveling in third class?

```
male = pd.DataFrame({ 'Age': [60], 'Sex_female': [0], 'Sex_male': [1],
                      'Pclass_1': [0], 'Pclass_2': [0], 'Pclass_3': [1] })

probability = model.predict_proba(male)[0][1]
print(f'Probability of survival: {probability:.1%}')
```

Feel free to experiment with other inputs to see what the model says. How likely, for example, is a 12-year-old boy traveling in second class to survive the sinking of the *Titanic*?

Detecting Credit Card Fraud

One of the most compelling uses for machine learning today is spotting fraudulent financial transactions. Credit card companies apply machine learning at the point of sale to decide whether to accept or decline individual charges. While these companies are understandably reluctant to publish the details of how they do it or the data they use to train their models, at least one such dataset has been published for public consumption. The data in it was anonymized using a technique called *principal component analysis* (*PCA*), which I'll introduce in Chapter 6.

The dataset is pictured in Figure 3-5. The data comes from real transactions made by European credit card holders in September 2013. Most of the columns have uninformative names, such as V1 and V2, and contain similarly opaque values. Three columns—Time, Amount, and Class—have real names and unaltered values revealing when the transaction took place, the amount of the transaction, and whether the transaction was legitimate (Class=0) or fraudulent (Class=1).

	Time	V1	V2	V3	...	V27	V28	Amount	Class
0	0.0	-1.359807	-0.072781	2.536347	...	0.133558	-0.021053	149.62	0
1	0.0	1.191857	0.266151	0.166480	...	-0.008983	0.014724	2.69	0
2	1.0	-1.358354	-1.340163	1.773209	...	-0.055353	-0.059752	378.66	0
3	1.0	-0.966272	-0.185226	1.792993	...	0.062723	0.061458	123.50	0
4	2.0	-1.158233	0.877737	1.548718	...	0.219422	0.215153	69.99	0
5	2.0	-0.425966	0.960523	1.141109	...	0.253844	0.081080	3.67	0
6	4.0	1.229658	0.141004	0.045371	...	0.034507	0.005168	4.99	0
7	7.0	-0.644269	1.417964	1.074380	...	-1.206921	-1.085339	40.80	0
8	7.0	-0.894286	0.286157	-0.113192	...	0.011747	0.142404	93.20	0
9	9.0	-0.338262	1.119593	1.044367	...	0.246219	0.083076	3.68	0

Figure 3-5. The fraud-detection dataset

Each row represents one transaction. Of the 284,807 transactions in the dataset, only 492 are fraudulent. The dataset is highly imbalanced, so you would expect a machine learning model trained on it to be much better at classifying legitimate transactions than fraudulent ones. That's not necessarily a problem, because credit card companies would rather misclassify fraudulent transactions and allow 100 of them to slide through than misclassify one legitimate transaction and anger a customer.

Begin by downloading a ZIP file (*https://oreil.ly/EYbNK*) containing the dataset. Copy *creditcard.csv* from the ZIP file into your notebooks' *Data* subdirectory, and then run the following code in a Jupyter notebook to load the dataset and show the first several rows:

```
import pandas as pd

df = pd.read_csv('Data/creditcard.csv')
df.head()
```

Find out how many rows the dataset contains and whether any of those rows have missing values:

```
df.info()
```

The dataset contains 284,807 rows, and none of them are missing values. Split the data for training and testing, and use `train_test_split`'s `stratify` parameter to ensure that the ratio of legitimate and fraudulent transactions is consistent in the training dataset and the testing dataset:

```
from sklearn.model_selection import train_test_split

x = df.drop(['Time', 'Class'], axis=1)
y = df['Class']

x_train, x_test, y_train, y_test = train_test_split(x, y, test_size=0.2,
                                        stratify=y, random_state=0)
```

Train a logistic regression model to separate the classes:

```
from sklearn.linear_model import LogisticRegression

lr_model = LogisticRegression(random_state=0, max_iter=5000)
lr_model.fit(x_train, y_train)
```

Note the `max_iter=5000` parameter passed to the `LogisticRegression` function. `max_iter` specifies the maximum number of iterations allowed to converge on a solution when fitting the logistic function to a dataset. The default is 100, which isn't enough in this example. Raising the limit to 5,000 gives the internal solver the headroom it needs to find a solution.

A typical accuracy score computed by dividing the sum of the true positives and true negatives by the number of test samples isn't very helpful because the dataset is so imbalanced. Fraudulent transactions represent less than 0.2% of all the samples, which means that the model could simply guess that every transaction is legitimate and get it right about 99.8% of the time. Use a confusion matrix to visualize how the model performs during testing:

```
%matplotlib inline
from sklearn.metrics import ConfusionMatrixDisplay as cmd

labels = ['Legitimate', 'Fraudulent']
cmd.from_estimator(lr_model, x_test, y_test, display_labels=labels,
                cmap='Blues', xticks_rotation='vertical')
```

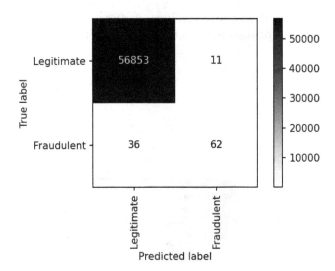

A logistic regression model correctly identified 56,853 transactions as legitimate while misclassifying legitimate transactions as fraudulent just 11 times. We want to minimize the latter number because we don't want to annoy customers by declining legitimate transactions. Let's see if a random-forest classifier can do better:

```
from sklearn.ensemble import RandomForestClassifier

rf_model = RandomForestClassifier(random_state=0)
rf_model.fit(x_train, y_train)

cmd.from_estimator(rf_model, x_test, y_test, display_labels=labels,
                   cmap='Blues', xticks_rotation='vertical')
```

A random forest mistook just four legitimate transactions as fraudulent. That's an improvement over logistic regression. Let's see if a gradient-boosting classifier can do better still:

```
from sklearn.ensemble import GradientBoostingClassifier

gbm_model = GradientBoostingClassifier(random_state=0)
gbm_model.fit(x_train, y_train)

cmd.from_estimator(gbm_model, x_test, y_test, display_labels=labels,
                   cmap='Blues', xticks_rotation='vertical')
```

The GBM misclassified more legitimate transactions than the random forest, so we'll stick with the random forest. Out of 56,864 legitimate transactions, the random forest correctly classified 56,860 of them. This means that legitimate transactions are classified correctly more than 99.99% of the time. Meanwhile, the model caught about 75% of the fraudulent transactions.

Use the following statements to measure the random-forest classifier's precision, recall, sensitivity, and specificity:

```
from sklearn.metrics import precision_score, recall_score

y_pred = rf_model.predict(x_test)
precision = precision_score(y_test, y_pred)
recall = recall_score(y_test, y_pred)
sensitivity = recall
specificity = recall_score(y_test, y_pred, pos_label=0)

print(f'Precision: {precision}')
print(f'Recall: {recall}')
print(f'Sensitivity: {sensitivity}')
print(f'Specificity: {specificity}')
```

Here is the output:

```
Precision: 0.9466666666666667
Recall: 0.7244897959183674
Sensitivity: 0.7244897959183674
Specificity: 0.9999296567248172
```

Given credit card companies' desire to keep customers happy and spending money, which of these metrics do you think they're most interested in? If you answered specificity, you answered correctly. Specificity is a measure of how reliable the test is at *not* falsely classifying a negative sample as positive—in this case, at not classifying a legitimate transaction as fraudulent.

Unfortunately, you can't make predictions with this model because you don't know the meaning of the numbers in the V1 through V28 columns, and you can't generate columnar values from a new transaction because you don't have the transform applied to the original dataset. You don't even know what the original dataset looks like. Most likely each row contains information about the card holder—for example, annual income, credit score, age, country of residence, and amount of money spent on the card last year—plus information about the product that was purchased and where it was purchased. The feature-engineering aspect of machine learning—figuring out what data is relevant to the model you're attempting to build—is just as challenging, if not more so, than data preparation and choosing a learning algorithm.

In real life, the models that credit card companies use to detect fraud are more sophisticated than this, and they often incorporate several models since no single model is 100% accurate. A company might build three models, for example, and allow them to vote on whether a given transaction is legitimate. Regardless, you have proven the principle that, given the right features, you can build a classification model that is reasonably accurate at detecting credit card fraud. And you have seen firsthand how easy Scikit makes it to experiment with different learning algorithms to determine which produces the most useful model.

Multiclass Classification

Now it's time to tackle multiclass classification, in which there are *n* possible outcomes rather than just two. A great example of multiclass classification is performing optical character recognition: examining a handwritten digit and predicting which digit 0 through 9 it corresponds to. Another example is looking at a facial photo and identifying the person in the photo by running it through a model trained to recognize hundreds of people.

Virtually everything you learned about binary classification applies to multiclass classification too. In Scikit, any classifier that works with binary classification also works with multiclass classification models. The importance of this can't be overstated. Some learning algorithms, such as logistic regression, work only in binary classification scenarios. Many machine learning libraries make you write explicit code to extend logistic regression to perform multiclass classification—or use a different form of logistic regression altogether. Scikit doesn't. Instead, it makes sure classifiers such as `LogisticRegression` work in either scenario, and when necessary, it does extra work behind the scenes to make it happen.

For logistic regression, Scikit uses one of two strategies to extend the algorithm to work in multiclass scenarios. (You can specify which strategy to use with the `LogisticRegression` class's `multi_class` parameter, or accept the default of `'auto'` and allow Scikit to choose.) One is *multinomial logistic regression*, which replaces the logistic function with a softmax function (*https://oreil.ly/BXtMC*) that yields multiple probabilities—one per class. The other is *one-vs-rest*, also known as *one-vs-all*, which trains n binary classification models, where n is the number of classes that the model can predict. Each of the n models pairs one class against all the other classes, and when the model is asked to make a prediction, it runs the input through all n models and uses the output from the one that yields the highest probability. This strategy is depicted in Figure 3-6.

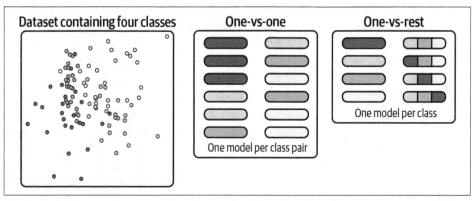

Figure 3-6. Strategies for extending binary classification algorithms to support multiclass classification

The one-vs-rest approach works well for logistic regression, but for some binary-only classification algorithms, Scikit uses a *one-vs-one* approach instead. When you use Scikit's `SVC` class (a support vector machine classifier that you'll learn about in Chapter 5) to perform multiclass classification, for example, Scikit builds one model for each pair of classes. If the model includes four possible classes, Scikit builds no fewer than seven models under the hood.

You don't have to know any of this to build a multiclass classification model. But it does explain why some multiclass classification models require more memory and train more slowly than others. Some classification algorithms, such as random forests and GBMs, support multiclass classification natively. For algorithms that don't, Scikit has your back. It fills the gap and does so as transparently as possible.

To reiterate: all Scikit classifiers are capable of performing binary classification and multiclass classification. This simplifies the code you write and lets you focus on building and training models rather than understanding the underlying mechanics of the algorithms.

Building a Digit Recognition Model

Want to experience multiclass classification firsthand? How about a model that examines scanned, handwritten digits and predicts what digits 0–9 they correspond to? The US Postal Service built a similar model many years ago to recognize handwritten zip codes as part of an effort to automate mail sorting. We'll use a sample dataset that's built into Scikit: the University of California Irvine's Optical Recognition of Handwritten Digits dataset (*https://oreil.ly/i55Ob*), which contains almost 1,800 handwritten digits. Each digit is represented by an 8 × 8 array of numbers from 0 to 16, with higher numbers indicating darker pixels. We will use logistic regression to make predictions from the data. Figure 3-7 shows the first 50 digits in the dataset.

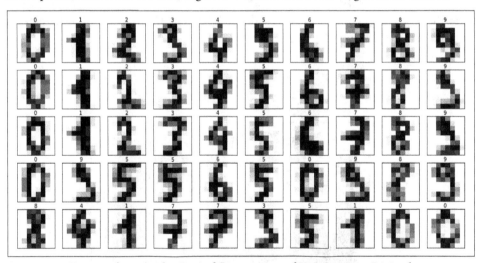

Figure 3-7. First 50 digits in the Optical Recognition of Handwritten Digits dataset

Start by creating a Jupyter notebook and executing the following statements in the first cell:

```
from sklearn import datasets

digits = datasets.load_digits()
print('digits.images: ' + str(digits.images.shape))
print('digits.target: ' + str(digits.target.shape))
```

Here's what the first digit looks like in numerical form:

```
digits.images[0]
```

And here's how it looks to the eye:

```
%matplotlib inline
import matplotlib.pyplot as plt

plt.tick_params(axis='both', which='both', bottom=False, top=False, left=False,
                right=False, labelbottom=False, labelleft=False)
plt.imshow(digits.images[0], cmap=plt.cm.gray_r)
```

It's obviously a 0, but you can confirm that from its label:

```
digits.target[0]
```

Plot the first 50 images and show the corresponding labels:

```
fig, axes = plt.subplots(5, 10, figsize=(12, 7),
                         subplot_kw={'xticks': [], 'yticks': []})

for i, ax in enumerate(axes.flat):
    ax.imshow(digits.images[i], cmap=plt.cm.gray_r)
    ax.text(0.45, 1.05, str(digits.target[i]), transform=ax.transAxes)
```

Classification models work best with balanced datasets. Use the following statements to plot the distribution of the samples:

```
plt.xticks([])
plt.hist(digits.target, rwidth=0.9)
```

Here is the output:

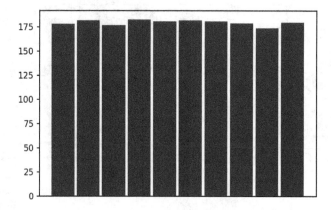

The dataset is almost perfectly balanced, so let's split it and train a logistic regression model:

```
from sklearn.linear_model import LogisticRegression
from sklearn.model_selection import train_test_split
```

```
x_train, x_test, y_train, y_test = train_test_split(
    digits.data, digits.target, test_size=0.2, random_state=0)

model = LogisticRegression(max_iter=5000)
model.fit(x_train, y_train)
```

Use the **score** method to quantify the model's accuracy:

```
model.score(x_test, y_test)
```

Use a confusion matrix to see how the model performs on the test dataset:

```
%matplotlib inline
import matplotlib.pyplot as plt
from sklearn.metrics import ConfusionMatrixDisplay as cmd
import seaborn as sns
sns.set()

fig, ax = plt.subplots(figsize=(8, 8))
ax.grid(False)
cmd.from_estimator(model, x_test, y_test, cmap='Blues', colorbar=False, ax=ax)
```

The resulting output paints an encouraging picture: large numbers and dark colors along the diagonal, and small numbers and light colors outside the diagonal. A perfect model would have all zeros outside the diagonal, but of course, perfect models don't exist:

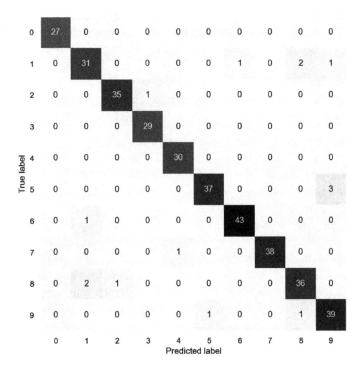

Pick one of the digits from the dataset and plot it to see what it looks like:

```
sns.reset_orig() # Undo sns.set()
plt.tick_params(axis='both', which='both', bottom=False, top=False, left=False,
                right=False, labelbottom=False, labelleft=False)
plt.imshow(digits.images[100], cmap=plt.cm.gray_r)
```

Pass it to the model and see what digit the model predicts it is:

```
model.predict([digits.data[100]])[0]
```

What probabilities does the model predict for each possible digit?

```
model.predict_proba([digits.data[100]])
```

What is the probability that the digit is a 4?

```
model.predict_proba([digits.data[100]])[0][4]
```

When used for binary classification, `predict_proba` returns two probabilities: one for the negative class and one for the positive class. For multiclass classification, `predict_proba` returns probabilities for each possible class. This permits you to assess the model's confidence in the prediction returned by `predict`. Not surprisingly, `predict` returns the class assigned the highest probability.

Summary

Classification models are widely used in industry to predict categorical outcomes such as whether a credit card transaction should be accepted or declined. Binary classification models predict either of two outcomes, while multiclass classification models predict more than two outcomes. One of the most widely used learning algorithms for classification models is logistic regression, which fits an equation to training data and uses it to predict outcomes by computing the possibility that each is the correct one and selecting the outcome with the highest probability.

There are many ways to score classification models. Which method is correct depends on how you intend to use the model. For example, when the cost of false positives is high, use precision to assess the model's accuracy. Precision is computed by dividing the number of true positives by the sum of the true positives and false positives. On the other hand, if the cost of false negatives is high, use recall instead. Recall is computed by dividing the number of true positives by the sum of the true positives and false negatives. Closely related to precision and recall are sensitivity and specificity. Sensitivity is identical to recall, while specificity is recall for the negative class rather than the positive class. Confusion matrices offer a convenient way to visualize how a model performs on test data without reducing the accuracy to a single number.

Some learning algorithms work only with binary classification, but Scikit works some magic under the hood to make sure *any* learning algorithm can be used for binary or multiclass classification. This isn't true of all machine learning libraries. Some restrict their learning algorithms to specific scenarios or require you to write extra code to use a binary classification algorithm in a multiclass model.

All the models in this chapter were trained with numerical data, even though some of the datasets contained categorical values—values that are strings rather than numbers—that had to be converted to numbers using one-hot encoding. You may wonder how to build a classification model that works *solely* on text—for example, a model that scores restaurant reviews for sentiment or classifies emails as spam or not spam. It's a perfectly legitimate question to ask. And it happens to be the subject of Chapter 4.

Text Classification

One of the more novel uses for binary classification is sentiment analysis, which examines a sample of text such as a product review, a tweet, or a comment left on a website and scores it on a scale of 0.0 to 1.0, where 0.0 represents negative sentiment and 1.0 represents positive sentiment. A review such as "great product at a great price" might score 0.9, while "overpriced product that barely works" might score 0.1. The score is the probability that the text expresses positive sentiment. Sentiment analysis models are difficult to build algorithmically but are relatively easy to craft with machine learning. For examples of how sentiment analysis is used in business today, see the article "8 Sentiment Analysis Real-World Use Cases" (*https://oreil.ly/nWq4a*) by Nicholas Bianchi.

Sentiment analysis is one example of a task that involves classifying textual data rather than numerical data. Because machine learning works with numbers, you must convert text to numbers before training a sentiment analysis model, a model that identifies spam emails, or any other model that classifies text. A common approach is to build a table of word frequencies called a *bag of words* (*https://oreil.ly/W4M6O*). Scikit-Learn provides classes to help. It also includes support for normalizing text so that, for example, "awesome" and "Awesome" don't count as two different words.

This chapter begins by describing how to prepare text for use in classification models. After building a sentiment analysis model, you'll learn about another popular learning algorithm called Naive Bayes that works particularly well with text and use it to build a model that distinguishes between legitimate emails and spam emails. Finally, you'll learn about a mathematical technique for measuring the similarity of two text samples and use it to build an app that recommends movies based on other movies you enjoy.

Preparing Text for Classification

Before you train a model to classify text, you must convert the text into numbers, a process known as *vectorization*. Chapter 1 presented the illustration reproduced in Figure 4-1, which demonstrates a common technique for vectorizing text. Each row represents a text sample such as a tweet or a movie review, and each column represents a word in the training text. The numbers in the rows are word counts, and the final number in each row is a label: 0 for negative and 1 for positive.

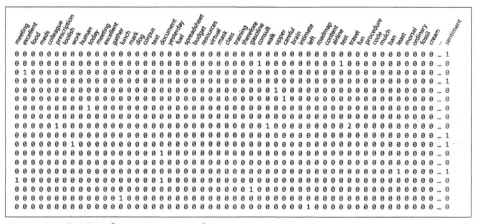

Figure 4-1. Dataset for sentiment analysis

Text is typically *cleaned* before it's vectorized. Examples of cleaning include converting characters to lowercase (so, for example, "Excellent" is equivalent to "excellent"), removing punctuation symbols, and optionally removing *stop words*—common words such as *the* and *and* that are likely to have little impact on the outcome. Once cleaned, sentences are divided into individual words (*tokenized*) and the words are used to produce datasets like the one in Figure 4-1.

Scikit-Learn has three classes that handle the bulk of the work of cleaning and vectorizing text:

CountVectorizer *(https://oreil.ly/00918)*
> Creates a dictionary (*vocabulary*) from the corpus of words in the training text and generates a matrix of word counts like the one in Figure 4-1

HashingVectorizer *(https://oreil.ly/vTJ4b)*
> Uses word hashes rather than an in-memory vocabulary to produce word counts and is therefore more memory efficient

TfidfVectorizer *(https://oreil.ly/xzaYZ)*
> Creates a dictionary from words provided to it and generates a matrix similar to the one in Figure 4-1, but rather than containing integer word counts, the matrix

contains term frequency-inverse document frequency (TFIDF) (*https://oreil.ly/NQrDJ*) values between 0.0 and 1.0 reflecting the relative importance of individual words

All three classes are capable of converting text to lowercase, removing punctuation symbols, removing stop words, splitting sentences into individual words, and more. They also support *n-grams*, which are combinations of two or more consecutive words (you specify the number *n*) that should be treated as a single word. The idea is that words such as *credit* and *score* might be more meaningful if they appear next to each other in a sentence than if they appear far apart. Without *n*-grams, the relative proximity of words is ignored. The downside to using *n*-grams is that it increases memory consumption and training time. Used judiciously, however, it can make text classification models more accurate.

Neural networks have other, more powerful ways of taking word order into account that don't require related words to occur next to each other. A conventional machine learning model can't connect the words *blue* and *sky* in the sentence "I like blue, for it is the color of the sky," but a neural network can. I will shed more light on this in Chapter 13.

Here's an **example** demonstrating what CountVectorizer does and how it's used:

```
import pandas as pd
from sklearn.feature_extraction.text import CountVectorizer

lines = [
    'Four score and 7 years ago our fathers brought forth,',
    '... a new NATION, conceived in liberty $$$,',
    'and dedicated to the PrOpOsItIoN that all men are created equal',
    'One nation\'s freedom equals #freedom for another $nation!'
]

# Vectorize the lines
vectorizer = CountVectorizer(stop_words='english')
word_matrix = vectorizer.fit_transform(lines)

# Show the resulting word matrix
feature_names = vectorizer.get_feature_names_out()
line_names = [f'Line {(i + 1):d}' for i, _ in enumerate(word_matrix)]

df = pd.DataFrame(data=word_matrix.toarray(), index=line_names,
                  columns=feature_names)

df.head()
```

Here's the output:

	ago	brought	conceived	created	dedicated	equal	equals	fathers	forth	freedom	liberty	men	nation	new	proposition	score	years
Line 1	1	1	0	0	0	0	0	1	1	0	0	0	0	0	0	1	1
Line 2	0	0	1	0	0	0	0	0	0	0	1	0	1	1	0	0	0
Line 3	0	0	0	1	1	1	0	0	0	0	0	1	0	0	1	0	0
Line 4	0	0	0	0	0	0	1	0	0	2	0	0	2	0	0	0	0

The corpus of text in this case is four strings in a Python list. CountVectorizer broke the strings into words, removed stop words and symbols, and converted all remaining words to lowercase. Those words comprise the columns in the dataset, and the numbers in the rows show how many times a given word appears in each string. The stop_words='english' parameter tells CountVectorizer to remove stop words using a built-in dictionary of more than 300 English-language stop words. If you prefer, you can provide your own list of stop words in a Python list. (Or you can leave the stop words in there; it often doesn't matter.) And if you're training with text written in another language, you can get lists of multilanguage stop words from other Python libraries such as the Natural Language Toolkit (NLTK) (*https://oreil.ly/ 2WzKr*) and Stop-words (*https://oreil.ly/Z4mRJ*).

Observe from the output that equal and equals count as separate words, even though they have similar meaning. Data scientists sometimes go a step further when preparing text for machine learning by *stemming* (*https://oreil.ly/q5ZhR*) or *lemmatizing* (*https://oreil.ly/BbiUx*) words. If the preceding text were stemmed, all occurrences of equals would be converted to equal. Scikit lacks support for stemming and lemmatization, but you can get it from other libraries such as NLTK.

CountVectorizer removes punctuation symbols, but it doesn't remove numbers. It ignored the 7 in line 1 because it ignores single characters. But if you changed 7 to 777, the term *777* would appear in the vocabulary. One way to fix that is to define a function that removes numbers and pass it to CountVectorizer via the preprocessor parameter:

```
import re

def preprocess_text(text):
    return re.sub(r'\d+', '', text).lower()

vectorizer = CountVectorizer(stop_words='english', preprocessor=preprocess_text)
word_matrix = vectorizer.fit_transform(lines)
```

Note the call to lower to convert the text to lowercase. CountVectorizer doesn't convert text to lowercase if you provide a preprocessing function, so the preprocessing function must convert it itself. It still removes punctuation characters, however.

Another useful parameter to `CountVectorizer` is `min_df`, which ignores words that appear fewer than the specified number of times. It can be an integer specifying a minimum count (for example, ignore words that appear fewer than five times in the training text, or `min_df=5`), or it can be a floating-point value from 0.0 to 1.0 specifying the minimum percentage of samples in which a word must appear—for example, ignore words that appear in less than 10% of the samples (`min_df=0.1`). It's great for filtering out words that probably aren't meaningful anyway, and it reduces memory consumption and training time by decreasing the size of the vocabulary. `Count Vectorizer` also supports a `max_df` parameter for eliminating words that appear too frequently.

The preceding examples use `CountVectorizer`, which probably leaves you wondering when (and why) you would use `HashingVectorizer` or `TfidfVectorizer` instead. `HashingVectorizer` is useful when dealing with large datasets. Rather than store words in memory, it hashes each word and uses the hash as an index into an array of word counts. It can therefore do more with less memory and is very useful for reducing the size of vectorizers when serializing them so that you can restore them later—a topic I'll say more about in Chapter 7. The downside to `HashingVectorizer` is that it doesn't let you work backward from vectorized text to the original text. `Count Vectorizer` does, and it provides an `inverse_transform` method (*https://oreil.ly/jlJxb*) for that purpose.

`TfidfVectorizer` is frequently used to perform *keyword extraction*: examining a document or set of documents and extracting keywords that characterize their content. It assigns words numerical weights reflecting their importance, and it uses two factors to determine the weights: how often a word appears in individual documents, and how often it appears in the overall document set. Words that appear more frequently in individual documents but occur in fewer documents receive higher weights. I won't go further into it here, but if you're curious to learn more, this book's GitHub repo contains a notebook (*https://oreil.ly/WNOiU*) that uses `Tfidf Vectorizer` to extract keywords from the manuscript of Chapter 1.

Sentiment Analysis

To train a sentiment analysis model, you need a labeled dataset. Several such datasets (*https://oreil.ly/qYYGM*) are available in the public domain. One of those is the IMDB movie review dataset (*https://oreil.ly/INB8o*), which contains 25,000 samples of negative reviews and 25,000 samples of positive reviews posted on the Internet Movie Database website (*https://imdb.com*). Each review is meticulously labeled with a 0 for negative sentiment or a 1 for positive sentiment. To demonstrate how sentiment analysis works, let's build a binary classification model and train it with this dataset. We'll use logistic regression as the learning algorithm. A sentiment analysis score yielded by this model is simply the probability that the input expresses positive sentiment, which is easily retrieved by calling `LogisticRegression`'s `predict_proba` method.

Start by downloading the dataset (*https://oreil.ly/uex7A*) and copying it to the *Data* subdirectory of the directory that hosts your Jupyter notebooks. Then run the following code in a notebook to load the dataset and show the first five rows:

```python
import pandas as pd

df = pd.read_csv('Data/reviews.csv', encoding='ISO-8859-1')
df.head()
```

The `encoding` attribute is necessary because the CSV file uses ISO-8859-1 character encoding rather than UTF-8. The output is as follows:

	Text	Sentiment
0	Once again Mr. Costner has dragged out a movie...	0
1	This is an example of why the majority of acti...	0
2	First of all I hate those moronic rappers, who...	0
3	Not even the Beatles could write songs everyon...	0
4	Brass pictures (movies is not a fitting word f...	0

Find out how many rows the dataset contains and confirm that there are no missing values:

```python
df.info()
```

Use the following statement to see how many instances there are of each class (0 for negative and 1 for positive):

```python
df.groupby('Sentiment').describe()
```

Here is the output:

	Text			
	count	unique	top	freq
Sentiment				
0	25000	24697	Nickelodeon has gone down the toilet. They hav...	3
1	25000	24884	Loved today's show!!! It was a variety and not...	5

There is an even number of positive and negative samples, but in each case, the number of unique samples is less than the number of samples for that class. That means the dataset has duplicate rows, and duplicate rows could bias a machine learning model. Use the following statements to delete the duplicate rows and check for balance again:

```
df = df.drop_duplicates()
df.groupby('Sentiment').describe()
```

Now there are no duplicate rows, and the number of positive and negative samples is roughly equal.

Next, use CountVectorizer to prepare and vectorize the text in the Text column. Set min_df to 20 to ignore words that appear infrequently in the training text. This reduces the likelihood of out-of-memory errors and will probably make the model more accurate as well. Also use the ngram_range parameter to allow Count Vectorizer to include word pairs as well as individual words:

```
from sklearn.feature_extraction.text import CountVectorizer

vectorizer = CountVectorizer(ngram_range=(1, 2), stop_words='english',
                             min_df=20)

x = vectorizer.fit_transform(df['Text'])
y = df['Sentiment']
```

Now split the dataset for training and testing. We'll use a 50/50 split since there are almost 50,000 samples in total:

```
from sklearn.model_selection import train_test_split

x_train, x_test, y_train, y_test = train_test_split(x, y, test_size=0.5,
                                                    random_state=0)
```

The next step is to train a classifier. We'll use Scikit's LogisticRegression class, which uses logistic regression to fit a model to the data:

```
from sklearn.linear_model import LogisticRegression

model = LogisticRegression(max_iter=1000, random_state=0)
model.fit(x_train, y_train)
```

Validate the trained model with the 50% of the dataset set aside for testing and show the results in a confusion matrix:

```
%matplotlib inline
from sklearn.metrics import ConfusionMatrixDisplay as cmd

cmd.from_estimator(model, x_test, y_test,
                   display_labels=['Negative', 'Positive'],
                   cmap='Blues', xticks_rotation='vertical')
```

The confusion matrix reveals that the model correctly identified 10,795 negative reviews while misclassifying 1,574 of them. It correctly identified 10,966 positive reviews and got it wrong 1,456 times:

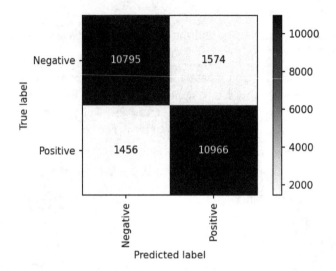

Now comes the fun part: analyzing text for sentiment. Use the following statements to produce a sentiment score for the sentence "The long lines and poor customer service really turned me off":

```
text = 'The long lines and poor customer service really turned me off'
model.predict_proba(vectorizer.transform([text]))[0][1]
```

Here's the output:

```
0.09183447847778639
```

Now do the same for "The food was great and the service was excellent!":

```
text = 'The food was great and the service was excellent!'
model.predict_proba(vectorizer.transform([text]))[0][1]
```

If you expected a higher score for this one, you won't be disappointed:

```
0.8536277207125618
```

Feel free to try sentences of your own and see if you agree with the sentiment scores the model predicts. It's not perfect, but it's good enough that if you run hundreds of reviews or comments through it, you should get a reliable indication of the sentiment expressed in the text.

Sometimes `CountVectorizer`'s built-in list of stop words *lowers* the accuracy of a model because the list is so broad. As an experiment, remove `stop_words='english'` from `CountVectorizer` and run the code again. Check the confusion matrix. Does the accuracy increase or decrease? Feel free to vary other parameters such as `min_df` and `ngram_range` too. In the real world, data scientists often try many different parameter combinations to determine which one produces the best results.

Naive Bayes

Logistic regression is a go-to algorithm for classification models and is often very effective at classifying text. But in scenarios involving text classification, data scientists often turn to another learning algorithm called *Naive Bayes* (*https://oreil.ly/ MqGwH*). It's a classification algorithm based on Bayes' theorem (*https://oreil.ly/ dZxN3*), which provides a means for calculating conditional probabilities. Mathematically, Bayes' theorem is stated this way:

$$P(A|B) = \frac{P(B|A) \cdot P(A)}{P(B)}$$

This says the probability that A is true given that B is true is equal to the probability that B is true given that A is true multiplied by the probability that A is true divided by the probability that B is true. That's a mouthful, and while accurate, it doesn't explain why Naive Bayes is so useful for classifying text—or how you apply it, for example, to a collection of emails to determine which ones are spam.

Let's start with a simple example. Suppose 10% of all the emails you receive are spam. That's $P(A)$. Analysis reveals that 5% of the spam emails you receive contain the word *congratulations*, but just 1% of *all* your emails contain the same word. Therefore, $P(B|A)$ is 0.05 and $P(B)$ is 0.01. The probability of an email being spam if it contains the word *congratulations* is $P(A|B)$, which is (0.05 x 0.10) / 0.01, or 0.50.

Of course, a spam filter must consider all the words in an email, not just one. It turns out that if you make some simple (naive) assumptions—that the order of the words in an email doesn't matter, and that every word has equal weight—you can write Bayes' equation this way for a spam classifier:

$$P(S|message) = P(S) \cdot P(word_1|S) \cdot P(word_2|S)...P(word_n|S)$$

In plain English, the probability that a message is spam is proportional to the product of:

- The probability that any message in the dataset is spam, or $P(S)$
- The probability that each word in the message appears in a spam message, or $P(word|S)$

$P(S)$ can be calculated easily enough: it's simply the fraction of the messages in the dataset that are spam messages. If you train a machine learning model with 1,000 messages and 500 of them are spam, then $P(S) = 0.5$. For a given word, $P(word|S)$ is simply the number of times the word appears in spam messages divided by the number of words in all the spam messages. The entire problem reduces to word counts. You can do a similar calculation to compute the probability that the message is *not* spam, and then use the higher of the two probabilities to make a prediction.

Here's an example involving four sample emails. The emails are:

Text	Spam
Raise your credit score in minutes	1
Here are the minutes from yesterday's meeting	0
Meeting tomorrow to review yesterday's scores	0
Score tomorrow's meds at yesterday's prices	1

If you remove stop words, convert characters to lowercase, and stem the words such that *tomorrow's* becomes *tomorrow*, you're left with this:

Text	Spam
raise credit score minute	1
minute yesterday meeting	0
meeting tomorrow review yesterday score	0
score tomorrow med yesterday price	1

Because two of the four messages are spam and two are not, the probability that any message is spam ($P(S)$) is 0.5. The same goes for the probability that any message is not spam ($P(N) = 0.5$). In addition, the spam messages contain nine unique words, while the nonspam messages contain a total of eight.

The next step is to build the following table of word frequencies. Take the word *yesterday* as an example. It appears once in a message that's labeled as spam, so P(*yesterday*|S) is 1/9, or 0.111. It appears twice in nonspam messages, so P(*yesterday*|N) is 2/8, or 0.250:

Word	P(*word*|S)	P(*word*|N)
raise	1/9 = 0.111	0/8 = 0.000
credit	1/9 = 0.111	0/8 = 0.000
score	2/9 = 0.222	1/8 = 0.125
minute	1/9 = 0.111	1/8 = 0.125
yesterday	1/9 = 0.111	2/8 = 0.250
meeting	0/9 = 0.000	2/8 = 0.250
tomorrow	1/9 = 0.111	1/8 = 0.125
review	0/9 = 0.000	1/8 = 0.125
med	1/9 = 0.111	0/8 = 0.000
price	1/9 = 0.111	0/8 = 0.000

This works up to a point, but the zeros in the table are a problem. Let's say you want to determine whether "Scores must be reviewed by tomorrow" is spam. Removing stop words leaves you with "score review tomorrow." You can compute the probability that the message is spam this way:

$P(S|score\ review\ tomorrow) = P(S) \cdot P(score|S) \cdot P(review|S) \cdot P(tomorrow|S)$
$P(S|score\ review\ tomorrow) = 0.5 \cdot 0.222 \cdot 0.0 \cdot 0.111 = 0$
$P(S|score\ review\ tomorrow) = 0$

The result is 0 because *review* doesn't appear in a spam message, and 0 times anything is 0. The algorithm simply can't assign a spam probability to "Scores must be reviewed by tomorrow."

A common way to resolve this is to apply *Laplace smoothing* (*https://oreil.ly/iRt2y*), also known as *additive smoothing*. Typically, this involves adding 1 to each numerator and the number of unique words in the dataset (in this case, 10) to each denominator. Now, P(*review*|S) evaluates to (0 + 1) / (9 + 10), which equals 0.053. It's not much, but it's better than nothing (literally). Here are the word frequencies again, this time revised with Laplace smoothing:

Word	P(word|S)	P(word|N)
raise	(1 + 1) / (9 + 10) = 0.105	(0 + 1) / (8 + 10) = 0.056
credit	(1 + 1) / (9 + 10) = 0.105	(0 + 1) / (8 + 10) = 0.056
score	(2 + 1) / (9 + 10) = 0.158	(1 + 1) / (8 + 10) = 0.111
minute	(1 + 1) / (9 + 10) = 0.105	(1 + 1) / (8 + 10) = 0.111
yesterday	(1 + 1) / (9 + 10) = 0.105	(2 + 1) / (8 + 10) = 0.167
meeting	(0 + 1) / (9 + 10) = 0.053	(2 + 1) / (8 + 10) = 0.167
tomorrow	(1 + 1) / (9 + 10) = 0.105	(1 + 1) / (8 + 10) = 0.111
review	(0 + 1) / (9 + 10) = 0.053	(1 + 1) / (8 + 10) = 0.111
med	(1 + 1) / (9 + 10) = 0.105	(0 + 1) / (8 + 10) = 0.056
price	(1 + 1) / (9 + 10) = 0.105	(0 + 1) / (8 + 10) = 0.056

Now you can determine whether "Scores must be reviewed by tomorrow" is spam by performing two simple calculations:

$$P(S \mid score\ review\ tomorrow) = 0.5 \cdot 0.158 \cdot 0.053 \cdot 0.105 = 0.000440$$
$$P(N \mid score\ review\ tomorrow) = 0.5 \cdot 0.111 \cdot 0.111 \cdot 0.111 = 0.000684$$

By this measure, "Scores must be reviewed by tomorrow" is likely not to be spam. The probabilities are relative, but you could normalize them and conclude there's about a 40% chance the message is spam and a 60% chance it's not based on the emails the model was trained with.

Fortunately, you don't have to do these computations by hand. Scikit-Learn provides several classes to help out, including the `MultinomialNB` class (*https://oreil.ly/twFtY*), which works great with tables of word counts produced by `CountVectorizer`.

Spam Filtering

It's no coincidence that modern spam filters are remarkably adept at identifying spam. Virtually all of them rely on machine learning. Such models are difficult to implement algorithmically because an algorithm that uses keywords such as *credit* and *score* to determine whether an email is spam is easily fooled. Machine learning, by contrast, looks at a body of emails and uses what it learns to classify the next email. Such models often achieve more than 99% accuracy. And they get smarter over time as they're trained with more and more emails.

The previous example used logistic regression to predict whether text input to it expresses positive or negative sentiment. It used the probability that the text expresses positive sentiment as a sentiment score, and you saw that expressions such as "The long lines and poor customer service really turned me off" score close to 0.0, while expressions such as "The food was great and the service was excellent" score close to 1.0. Now let's build a binary classification model that classifies emails as spam or not spam and use Naive Bayes to fit the model to the training data.

There are several spam classification datasets available in the public domain. Each contains a collection of emails with samples labeled with 1s for spam and 0s for not spam. We'll use a relatively small dataset containing 1,000 samples. Begin by downloading the dataset (*https://oreil.ly/hljvo*) and copying it into your notebooks' *Data* subdirectory. Then load the data and display the first five rows:

```python
import pandas as pd

df = pd.read_csv('Data/ham-spam.csv')
df.head()
```

Now check for duplicate rows in the dataset:

```python
df.groupby('IsSpam').describe()
```

The dataset contains one duplicate row. Let's remove it and check for balance:

```python
df = df.drop_duplicates()
df.groupby('IsSpam').describe()
```

The dataset now contains 499 samples that are not spam, and 500 that are. The next step is to use `CountVectorizer` to vectorize the emails. Once more, we'll allow `Count Vectorizer` to consider word pairs as well as individual words and remove stop words using Scikit's built-in dictionary of English stop words:

```python
from sklearn.feature_extraction.text import CountVectorizer

vectorizer = CountVectorizer(ngram_range=(1, 2), stop_words='english')
x = vectorizer.fit_transform(df['Text'])
y = df['IsSpam']
```

Split the dataset so that 80% can be used for training and 20% for testing:

```python
from sklearn.model_selection import train_test_split

x_train, x_test, y_train, y_test = train_test_split(x, y, test_size=0.2,
                                                    random_state=0)
```

The next step is to train a Naive Bayes classifier using Scikit's `MultinomialNB` class (*https://oreil.ly/0CtOh*):

```python
from sklearn.naive_bayes import MultinomialNB

model = MultinomialNB()
model.fit(x_train, y_train)
```

Validate the trained model with the 20% of the dataset set aside for testing using a confusion matrix:

```
%matplotlib inline
from sklearn.metrics import ConfusionMatrixDisplay as cmd

cmd.from_estimator(model, x_test, y_test,
                   display_labels=['Not Spam', 'Spam'],
                   cmap='Blues', xticks_rotation='vertical')
```

The model correctly identified 101 of 102 legitimate emails as not spam, and 95 of 98 spam emails as spam:

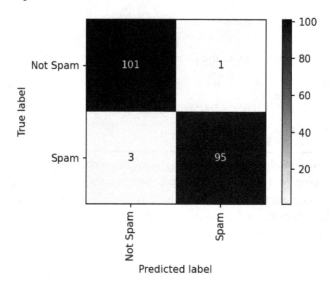

Use the `score` method to get a rough measure of the model's accuracy:

```
model.score(x_test, y_test)
```

Now use Scikit's `RocCurveDisplay` class to visualize the ROC curve:

```
from sklearn.metrics import RocCurveDisplay as rcd
import seaborn as sns
sns.set()

rcd.from_estimator(model, x_test, y_test)
```

The results are encouraging. Trained with just 999 samples, the area under the ROC curve (AUC) indicates the model is more than 99.9% accurate at classifying emails as spam or not spam:

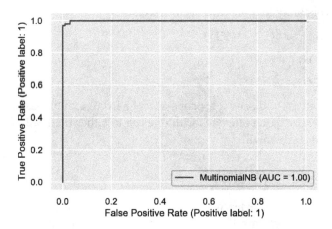

Let's see how the model classifies a few emails that it hasn't seen before, starting with one that isn't spam. The model's `predict` method predicts a class—0 for not spam, or 1 for spam:

```
msg = 'Can you attend a code review on Tuesday to make sure the logic is solid?'
input = vectorizer.transform([msg])
model.predict(input)[0]
```

The model says this message is not spam, but what's the probability that it's not spam? You can get that from `predict_proba`, which returns an array containing two values: the probability that the predicted class is 0, and the probability that the predicted class is 1, in that order:

```
model.predict_proba(input)[0][0]
```

The model seems very sure that this email is legitimate:

```
0.9999497111473539
```

Now test the model with a spam message:

```
msg = 'Why pay more for expensive meds when you can order them online ' \
      'and save $$$?'

input = vectorizer.transform([msg])
model.predict(input)[0]
```

What is the probability that the message is not spam?

```
model.predict_proba(input)[0][0]
```

The answer is:

```
0.00021423891260677753
```

What is the probability that the message *is* spam?

```
model.predict_proba(input)[0][1]
```

And the answer is:

```
0.9997857610873945
```

Observe that `predict` and `predict_proba` accept a list of inputs. Based on that, could you classify an entire batch of emails with one call to either method? How would you get the results for each email?

Recommender Systems

Another branch of machine learning that has proven its mettle in recent years is *recommender systems*—systems that recommend products or services to customers. Amazon's recommender system reportedly drives 35% of its sales (*https://oreil.ly/ue81Q*). The good news is that you don't have to be Amazon to benefit from a recommender system, nor do you have to have Amazon's resources to build one. They're relatively simple to create once you learn a few basic principles.

Recommender systems come in many forms. *Popularity-based systems* present options to customers based on what products and services are popular at the time—for example, "Here are this week's bestsellers." *Collaborative systems* make recommendations based on what others have selected, as in "People who bought this book also bought these books." Neither of these systems requires machine learning.

Content-based systems, by contrast, benefit greatly from machine learning. An example of a content-based system is one that says "if you bought this book, you might like these books also." These systems require a means for quantifying similarity between items. If you liked the movie *Die Hard*, you might or might not like *Monty Python and the Holy Grail*. If you liked *Toy Story*, you'll probably like *A Bug's Life* too. But how do you make that determination algorithmically?

Content-based recommenders require two ingredients: a way to *vectorize*—convert to numbers—the attributes that characterize a service or product, and a means for calculating similarity between the resulting vectors. The first one is easy. `Count Vectorizer` converts text into tables of word counts. All you need is a way to measure similarity between rows of word counts and you can build a recommender system. And one of the simplest and most effective ways to do that is a technique called *cosine similarity*.

Cosine Similarity

Cosine similarity (*https://oreil.ly/948eP*) is a mathematical means for computing the similarity between pairs of vectors (or rows of numbers treated as vectors). The basic idea is to take each value in a sample—for example, word counts in a row of vectorized text—and use them as endpoint coordinates for a vector, with the other endpoint at the origin of the coordinate system. Do that for two samples, and then compute the cosine between vectors in m-dimensional space, where m is the number of values in each sample. Because the cosine of 0 is 1, two identical vectors have a similarity of 1. The more dissimilar the vectors, the closer the cosine will be to 0.

Here's an example in two-dimensional space to illustrate. Suppose you have three rows containing two values each:

1	2
2	3
3	1

You want to determine whether row 2 is more similar to row 1 or row 3. It's hard to tell just by looking at the numbers, and in real life, there are *many* more numbers. If you simply added the numbers in each row and compared the sums, you would conclude that row 2 is more similar to row 3. But what if you treated each row as a vector, as shown in Figure 4-2?

- Row 1: $(0, 0) \rightarrow (1, 2)$
- Row 2: $(0, 0) \rightarrow (2, 3)$
- Row 3: $(0, 0) \rightarrow (3, 1)$

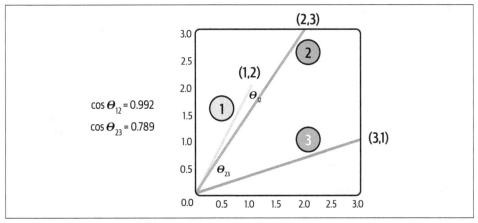

Figure 4-2. Cosine similarity

Now you can plot each row as a vector, compute the cosines of the angles formed by 1 and 2 and 2 and 3, and determine that row 2 is more like row 1 than row 3. That's cosine similarity in a nutshell.

Cosine similarity isn't limited to two dimensions; it works in higher-dimensional space as well. To help compute cosine similarities regardless of the number of dimensions, Scikit offers the `cosine_similarity` function (*https://oreil.ly/vc1Uv*). The following code computes the cosine similarities of the three samples in the preceding example:

```
data = [[1, 2], [2, 3], [3, 1]]
cosine_similarity(data)
```

The return value is a *similarity matrix* containing the cosines of every vector pair. The width and height of the matrix equals the number of samples:

```
array([[1.        , 0.99227788, 0.70710678],
       [0.99227788, 1.        , 0.78935222],
       [0.70710678, 0.78935222, 1.        ]])
```

From this, you can see that the similarity of rows 1 and 2 is 0.992, while the similarity of rows 2 and 3 is 0.789. In other words, row 2 is more similar to row 1 than it is to row 3. There is also more similarity between rows 2 and 3 (0.789) than there is between rows 1 and 3 (0.707).

Building a Movie Recommendation System

Let's put cosine similarity to work building a content-based recommender system for movies. Start by downloading the dataset (*https://oreil.ly/ydE3v*), which is one of several movie datasets available from Kaggle.com. This one has information for about 4,800 movies, including title, budget, genres, keywords, cast, and more. Place the CSV file in your Jupyter notebooks' *Data* subdirectory. Then load the dataset and peruse its contents:

```
import pandas as pd

df = pd.read_csv('Data/movies.csv')
df.head()
```

The dataset contains 24 columns, only a few of which are needed to describe a movie. Use the following statements to extract key columns such as `title` and `genres` and fill missing values with empty strings:

```
df = df[['title', 'genres', 'keywords', 'cast', 'director']]
df = df.fillna('') # Fill missing values with empty strings
df.head()
```

Next, add a column named `features` that combines all the words in the other columns:

```
df['features'] = df['title'] + ' ' + df['genres'] + ' ' + \
            df['keywords'] + ' ' + df['cast'] + ' ' + \
            df['director']
```

Use CountVectorizer to vectorize the text in the features column:

```
from sklearn.feature_extraction.text import CountVectorizer

vectorizer = CountVectorizer(stop_words='english', min_df=20)
word_matrix = vectorizer.fit_transform(df['features'])
word_matrix.shape
```

The table of word counts contains 4,803 rows—one for each movie—and 918 columns. The next task is to compute cosine similarities for each row pair:

```
from sklearn.metrics.pairwise import cosine_similarity

sim = cosine_similarity(word_matrix)
```

Ultimately, the goal of this system is to input a movie title and identify the n movies that are most similar to that movie. To that end, define a function named get _recommendations that accepts a movie title, a DataFrame containing information about all the movies, a similarity matrix, and the number of movie titles to return:

```
def get_recommendations(title, df, sim, count=10):
    # Get the row index of the specified title in the DataFrame
    index = df.index[df['title'].str.lower() == title.lower()]

    # Return an empty list if there is no entry for the specified title
    if (len(index) == 0):
        return []

    # Get the corresponding row in the similarity matrix
    similarities = list(enumerate(sim[index[0]]))

    # Sort the similarity scores in that row in descending order
    recommendations = sorted(similarities, key=lambda x: x[1], reverse=True)

    # Get the top n recommendations, ignoring the first entry in the list since
    # it corresponds to the title itself (and thus has a similarity of 1.0)
    top_recs = recommendations[1:count + 1]

    # Generate a list of titles from the indexes in top_recs
    titles = []

    for i in range(len(top_recs)):
        title = df.iloc[top_recs[i][0]]['title']
        titles.append(title)

    return titles
```

This function sorts the cosine similarities in descending order to identify the count movies most like the one identified by the title parameter. Then it returns the titles of those movies.

Now use `get_recommendations` to search the database for similar movies. First ask for the 10 movies that are most similar to the James Bond thriller *Skyfall*:

```
get_recommendations('Skyfall', df, sim)
```

Here is the output:

```
['Spectre',
 'Quantum of Solace',
 'Johnny English Reborn',
 'Clash of the Titans',
 'Die Another Day',
 'Diamonds Are Forever',
 'Wrath of the Titans',
 'I Spy',
 'Sanctum',
 'Blackthorn']
```

Call `get_recommendations` again to list movies that are like *Mulan*:

```
get_recommendations('Mulan', df, sim)
```

Feel free to try other movies as well. Note that you can only input movie titles that are in the dataset. Use the following statements to print a complete list of titles:

```
pd.set_option('display.max_rows', None)
print(df['title'])
```

I think you'll agree that the system does a pretty credible job of picking similar movies. Not bad for about 20 lines of code!

Summary

Machine learning models that classify text are common and see a variety of uses in industry and in everyday life. What rational human being doesn't wish for a magic wand that eradicates all spam mails, for example?

Text used to train a text classification model must be prepared and vectorized prior to training. Preparation includes converting characters to lowercase and removing punctuation characters, and may include removing stop words, removing numbers, and stemming or lemmatizing. Once prepared, text is vectorized by converting it into a table of word frequencies. Scikit's `CountVectorizer` class makes short work of the vectorization process and handles some of the preparation duties too.

Logistic regression and other popular classification algorithms can be used to classify text once it's converted to numerical form. For text classification tasks, however, the Naive Bayes learning algorithm frequently outperforms other algorithms. By making a few "naive" assumptions such as that the order in which words appear in a text sample doesn't matter, Naive Bayes reduces to a process of word counting. Scikit's `MultinomialNB` class provides a handy Naive Bayes implementation.

Cosine similarity is a mathematical means for computing the similarity between two rows of numbers. One use for it is building systems that recommend products or services based on other products or services that a customer has purchased. Word frequency tables produced from textual descriptions by `CountVectorizer` can be combined with cosine similarity to create intelligent recommender systems intended to supplement a company's bottom line.

Feel free to use this chapter's examples as a starting point for experiments of your own. For instance, see if you can tweak the parameters passed to `CountVectorizer` in any of the examples and increase the accuracy of the resulting model. Data scientists call the search for the optimum parameter combination *hyperparameter tuning*, and it's a subject you'll learn about in the next chapter.

## CHAPTER 5

Support Vector Machines

Support vector machines (SVMs) represent the cutting edge of machine learning. They are most often used to solve classification problems, but they can also be used for regression. Due to the unique way in which they fit mathematical models to data, SVMs often succeed at finding separation between classes when other models do not. They technically perform binary classification only, but Scikit-Learn enables them to do multiclass classification as well using techniques discussed in Chapter 3.

Scikit-Learn makes building SVMs easy with classes such as SVC (short for support vector classifier) (*https://oreil.ly/IAiWs*) for classification models and SVR (support vector regressor) (*https://oreil.ly/f8B8K*) for regression models. You can use these classes without understanding how SVMs work, but you'll get more out of them if you *do* understand how they work. It's also important to know how to tune SVMs for individual datasets and how to prepare data before you train a model. Toward the end of this chapter, we'll build an SVM that performs facial recognition. But first, let's look behind the scenes and discover why SVMs are often the go-to mechanism for modeling real-world datasets.

How Support Vector Machines Work

First, why are they called support vector machines? The purpose of an SVM classifier is the same as any other classifier: to find a decision boundary that cleanly separates the classes. SVMs do this by finding a line in 2D space, a plane in 3D space, or a hyperplane in higher-dimensional space that allows them to distinguish between different classes with the greatest certainty possible. In the example in Figure 5-1, there are an infinite number of lines you can draw to separate the two classes, but the best line is the one that produces the widest margin (the one shown on the right). The width of the margin is the distance between the points closest to the boundary in each class along a line perpendicular to the boundary. These points are called *support vectors* and are circled in red.

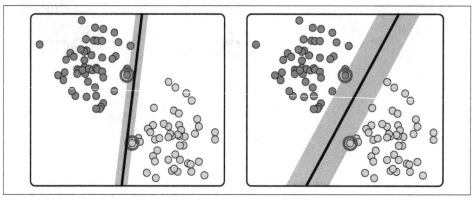

Figure 5-1. Maximum-margin classification

Of course, real data rarely lends itself to such clean separation. Overlap between classes inevitably prevents a perfect fit. To accommodate this, SVMs support a regularization parameter usually referred to as C that can be adjusted to loosen or tighten the fit. Lower values of C produce a wider margin with more errors on either side of the decision boundary, as shown in Figure 5-2. Higher values yield a tighter fit to the training data with a correspondingly thinner margin and fewer errors. If C is *too* high, the model might not generalize well. The optimum value varies by dataset. Data scientists typically try different values of C to determine which one performs the best against test data.

All of the aforementioned is true, but none of it explains why SVMs are so good at what they do. SVMs aren't the only models that mathematically look for boundaries separating the classes. What makes SVMs special are *kernels*, some of which add dimensions to data to find boundaries that don't exist at lower dimensions. Consider Figure 5-3. You can't draw a line that completely separates the red dots from the purple dots. But if you add a third dimension as shown on the right—a *z* dimension whose value is based on a point's distance from the center—then you can slide a plane between the purples and the reds and achieve 100% separation. In this example, data that isn't linearly separable in two dimensions is linearly separable in three dimensions. The principle at work is Cover's theorem (*https://oreil.ly/BAsz2*), which states that data that isn't linearly separable might be linearly separable if projected into higher-dimensional space using a nonlinear transform.

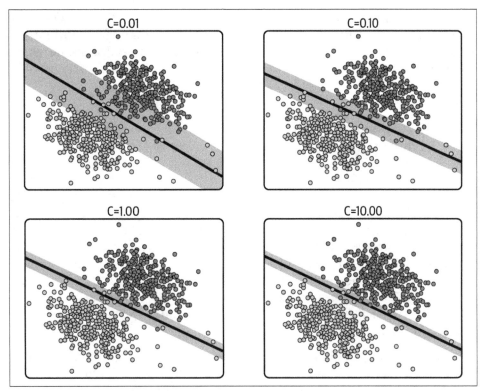

Figure 5-2. Effect of C on margin width

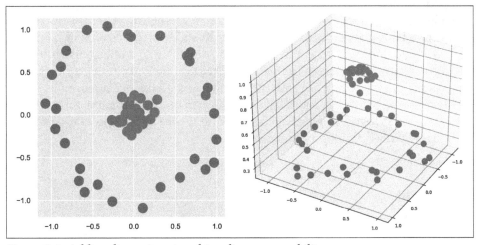

Figure 5-3. Adding dimensions to achieve linear separability

The kernel transformation used in this example, which projects two-dimensional data to three dimensions by adding a z to every x and y, works well with this particular dataset. But for SVMs to be broadly useful, you need a kernel that isn't tied to the shape of a specific dataset.

Kernels

Scikit-Learn has several general-purpose kernels built in, including the linear kernel, the RBF kernel,[1] the polynomial kernel, and the sigmoid kernel. The linear kernel doesn't add dimensions. It works well with data that is linearly separable out of the box, but it doesn't perform very well with data that isn't. Applying it to the problem in Figure 5-3 produces the decision boundary on the left in Figure 5-4. Applying the RBF kernel to the same data produces the decision boundary on the right. The RBF kernel projects the x and y values into a higher-dimensional space and finds a hyperplane that cleanly separates the purples from the reds. When projected back to two dimensions, the decision boundary roughly forms a circle. Similar results can be achieved on this dataset with a properly tuned polynomial kernel, but generally speaking, the RBF kernel can find decision boundaries in nonlinear data that the polynomial kernel cannot. That's why RBF is the default kernel type in Scikit if you don't specify otherwise.

A logical question to ask is, did the RBF kernel add a z to every x and y? The short answer is no. It effectively projected the data points into a space with an *infinite* number of dimensions. The key word is *effectively*. Kernels use mathematical shortcuts called *kernel tricks* to measure the effect of adding new dimensions without actually computing values for them. This is where the math for SVMs gets hairy. Kernels are carefully designed to compute the dot product (*https://oreil.ly/yGUYS*) between two n-dimensional vectors in m-dimensional space (where m is greater than n and can even be infinite) without generating all those new dimensions, and ultimately, the dot products are all an SVM needs to compute a decision boundary. It's the mathematical equivalent of having your cake and eating it too, and it's the secret sauce that makes SVMs awesome. SVMs can take a long time to train on large datasets, but one of the benefits of an SVM is that it tends to do better on smaller datasets with fewer rows or samples than other learning algorithms.

---

1 RBF is short for radial basis function (*https://oreil.ly/IRswE*).

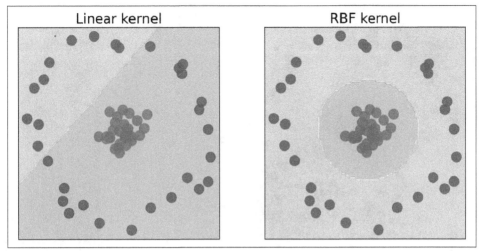

Figure 5-4. Linear kernel versus RBF kernel

Kernel Tricks

Want to see an example of how kernel tricks are used to compute dot products in high-dimensional spaces without computing values for the new dimensions? The following explanation is completely optional. But if you, like me, learn better from concrete examples, then you might find this section helpful.

Let's start with the two-dimensional circular dataset presented earlier, but this time let's project it into three-dimensional space with the following equations:

$$x' = x^2$$
$$y' = y^2$$
$$z = x \cdot y \cdot \sqrt{2}$$

In other words, we'll compute x and y in three-dimensional space (x' and y') by squaring x and y in two-dimensional space, and we'll add a z that's the product of the original x and y and the square root of 2. Projecting the data this way produces a clean separation between purples and reds, as shown in Figure 5-5.

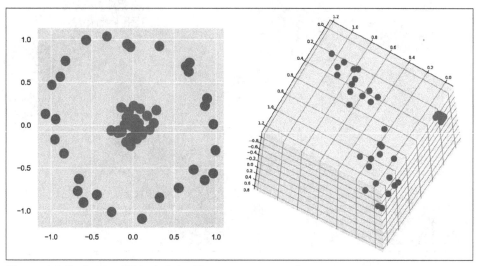

Figure 5-5. Projecting 2D points to 3D to separate two classes

The efficacy of SVMs depends on their ability to compute the dot product of two vectors (or points, which can be treated as vectors) in higher-dimensional space without projecting them into that space—that is, using only the values in the original space. Let's manufacture a couple of points to work with:

$$a = (3, 7)$$
$$b = (2, 4)$$

We can compute the dot product of these two points this way:

$$(3 \cdot 2) + (7 \cdot 4) = 34$$

Of course, the dot product in two dimensions isn't very helpful. An SVM needs the dot product of these points in 3D space. Let's use the preceding equations to project a and b to 3D, and then compute the dot product of the result:

$$a = \left(3^2, 7^2, 3 \cdot 7 \cdot \sqrt{2}\right) = (9, 49, 29.6984)$$
$$b = \left(2^2, 4^2, 2 \cdot 4 \cdot \sqrt{2}\right) = (4, 16, 11.3137)$$
$$(9 \cdot 4) + (49 \cdot 16) + (29.6984 \cdot 11.3137) = 1{,}156$$

We now have the dot product of a pair of 2D points in 3D space, but we had to generate coordinates in 3D space to get it. Here's where it gets interesting. The following function, or *kernel trick*, produces the same result using *only the values in the original 2D space*:

$$K(a, b) = \langle a, b \rangle^2$$

$\langle a, b \rangle$ is simply the dot product of a and b, so $\langle a, b \rangle^2$ is the square of the dot product of a and b. We already know how to compute the dot product of a and b. Therefore:

$$K((3, 7), (2, 4)) = ((3 \cdot 2) + (7 \cdot 4))^2 = 34^2 = 1,156$$

This agrees with the result computed by explicitly projecting the points, but with *no projection required*. That's the kernel trick in a nutshell. It saves time and memory when going from two dimensions to three. Just imagine the savings when projecting to an infinite number of dimensions—which, you'll recall, is exactly what the RBF kernel does.

The kernel trick used here wasn't manufactured from thin air. It happens to be the one used by a degree-2 polynomial kernel. With Scikit, you can fit an SVM classifier with a degree-2 polynomial kernel to a dataset this way:

```
model = SVC(kernel='poly', degree=2)
model.fit(x, y)
```

If you apply this to the preceding circular dataset and plot the decision boundary (Figure 5-6, right), the result is almost identical to the one generated by the RBF kernel. Interestingly, a degree-1 polynomial kernel (Figure 5-6, left) produces the same decision boundary as the linear kernel since a line is just a first-degree polynomial.

Kernel tricks are special. Each one is designed to simulate a specific projection into higher dimensions. Scikit gives you a handful of kernels to work with, but there are others that Scikit doesn't build in. You can extend Scikit with kernels of your own, but the ones that it provides are sufficient for the vast majority of use cases.

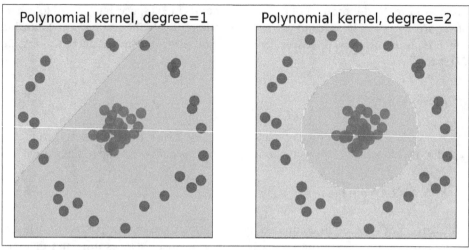

Figure 5-6. Degree-1 versus degree-2 polynomial kernel

Hyperparameter Tuning

At the outset, it's difficult to know which of the built-in kernels will produce the most accurate model. It's also difficult to know what the right value of C is—that is, the value that provides the best balance between underfitting and overfitting the training data and yields the best results when the model is run with test data. For the RBF and polynomial kernels, there's a third value called gamma that affects accuracy. And for polynomial kernels, the degree parameter impacts the model's ability to learn from the training data.

The C parameter controls how aggressively the model fits to the training data. The higher the value, the tighter the fit and the higher the risk of overfitting. Figure 5-7 shows how the RBF kernel fits a model to a set of training data containing three classes with different values of C. The default is C=1 in Scikit, but you can specify a different value to adjust the fit. You can see the danger of overfitting in the lower-right diagram. A point that lies to the extreme right would be classified as a blue, even though it probably belongs to the yellow or brown class. Underfitting is a problem too. In the upper-left example, virtually any data point that isn't a brown will be classified as a blue.

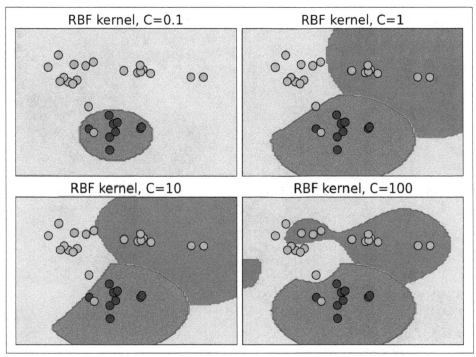

Figure 5-7. Effect of C on the RBF kernel

An SVM that uses the RBF kernel isn't properly tuned until you have the right value for gamma too. gamma controls how far the influence of a single data point reaches in computing decision boundaries. Lower values use more points and produce smoother decision boundaries; higher values involve fewer points and fit more tightly to the training data. This is illustrated in Figure 5-8, where increasing gamma while holding C constant closes the decision boundary more tightly around clusters of classes. gamma can be any nonzero positive value, but values between 0 and 1 are the most common. Rather than hardcode a default value for gamma, Scikit picks a default value algorithmically if you don't specify one.

In practice, data scientists experiment with different kernels and different parameter values to find the combination that produces the most accurate model, a process known as *hyperparameter tuning*. The usefulness of hyperparameter tuning isn't unique to SVMs, but you can almost always make an SVM more accurate by finding the optimum combination of kernel type, C, and gamma (and for polynomial kernels, degree).

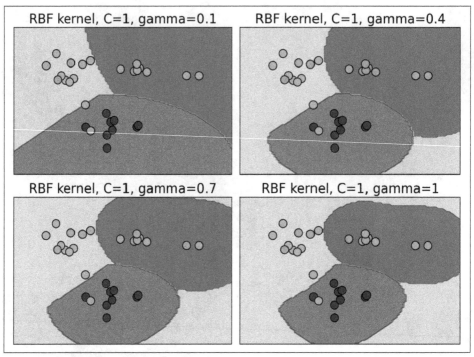

Figure 5-8. Effect of gamma on the RBF kernel

To aid in the process of hyperparameter tuning, Scikit provides a family of optimizers (*https://oreil.ly/IFpOA*) that includes `GridSearchCV` (*https://oreil.ly/n32OC*), which tries all combinations of a specified set of parameter values with built-in cross-validation to determine which combination produces the most accurate model. These optimizers prevent you from having to write code to do a brute-force search using all the unique combinations of parameter values. To be clear, they do brute-force searches themselves by training the model multiple times, each time with a different combination of values. At the end, you can retrieve the most accurate model from the `best_estimator_` attribute, the parameter values that produced the most accurate model from the `best_params_` attribute, and the best score from the `best_score_` attribute.

Here's an example that uses Scikit's `SVC` class to implement an SVM classifier. For starters, you can create an SVM classifier that uses default parameter values and fit it to a dataset with two lines of code:

```
model = SVC()
model.fit(x, y)
```

This uses the RBF kernel with C=1. You can specify the kernel type and values for C and gamma this way:

```
model = SVC(kernel='poly', C=10, gamma=0.1)
model.fit(x, y)
```

Suppose you wanted to try two different kernels and five values each for C and gamma to see which combination produces the best results. Rather than write a nested for loop, you could do this:

```
model = SVC()

grid = {
    'C': [0.01, 0.1, 1, 10, 100],
    'gamma': [0.01, 0.25, 0.5, 0.75, 1.0],
    'kernel': ['rbf', 'poly']
}

grid_search = GridSearchCV(estimator=model, param_grid=grid, cv=5, verbose=2)
grid_search.fit(x, y) # Train the model with different parameter combinations
```

The call to fit won't return for a while. It trains the model *250 times* since there are 50 different combinations of kernel, C, and gamma, and cv=5 says to use fivefold cross-validation to assess the results. Once training is complete, you retrieve the best model this way:

```
best_model = grid_search.best_estimator_
```

It is not uncommon to run a search regimen such as this one multiple times—the first time with course parameter values, and each time thereafter with narrower ranges of values centered on the values obtained from best_params_. More training time up front is the price you pay for an accurate model. To reiterate, you can almost always make an SVM more accurate by finding the optimum combination of parameters. And for better or worse, brute force is the most effective way to identify the best combination.

One nuance to be aware of regarding the SVC class is that it doesn't compute probabilities by default. If you want to call predict_proba on an SVC instance, you must set probability to True when creating the instance:

```
model = SVC(probability=True)
```

The model will train more slowly, but you'll be able to retrieve probabilities as well as predictions. Furthermore, the Scikit documentation warns that "predict_proba may be inconsistent with predict." For more information, see Section 1.4.1.2 in the documentation (*https://oreil.ly/Jg8X0*).

Data Normalization

In Chapter 2, I noted that some learning algorithms work better with normalized data. Unnormalized data contains columns of numbers with vastly different ranges— for example, values from 0 to 1 in one column and from 0 to 1,000,000 in another. SVM is a parametric learning algorithm. Training with normalized data is important because SVMs use distances to compute margins. If one dimension spans *much* larger distances than another, the internal algorithm used to find the maximum margins might have trouble converging on a solution.

The importance of training machine learning models with normalized data isn't limited to SVMs. Decision trees and learning algorithms such as random forests and gradient-boosted decision trees that rely on decision trees are nonparametric, so they work equally well with normalized and unnormalized data. They are the exception, however. Most other learning algorithms benefit to one degree or another from normalized data. That includes *k*-nearest neighbors, which although nonparametric uses distance-based calculations internally to discriminate between classes.

Scikit offers several classes for normalizing data. The most commonly used are MinMaxScaler (*https://oreil.ly/nz2wR*) and StandardScaler (*https://oreil.ly/OTTrm*). The former normalizes data by proportionally reducing the values in each column to values from 0.0 to 1.0. Mathematically, it's simple. For each column in a dataset, Min MaxScaler subtracts the minimum value in that column from all the column's values, then it divides each value by the difference between the minimum and maximum values. In the resulting column, the minimum value is 0.0 and the maximum is 1.0.

To demonstrate, I extracted subsets of two columns with vastly different ranges from the breast cancer dataset (*https://oreil.ly/IQAC9*) built into Scikit. Each column contains 100 values. Here are the first 10 rows:

```
[[1.001e+03 3.001e-01]
 [1.326e+03 8.690e-02]
 [1.203e+03 1.974e-01]
 [3.861e+02 2.414e-01]
 [1.297e+03 1.980e-01]
 [4.771e+02 1.578e-01]
 [1.040e+03 1.127e-01]
 [5.779e+02 9.366e-02]
 [5.198e+02 1.859e-01]
 [4.759e+02 2.273e-01]]
```

The values in the first column range from 201.9 to 1,878.0; the values in the second column range from 0.000692 to 0.3754. Figure 5-9 shows how the data looks if plotted with the x- and y-axis equally scaled. Because the values in the first column are much larger than the values in the second, the data points appear to form a line. If you adjust the scale of the axes to match the ranges of values in each column, you get a completely different picture (Figure 5-10).

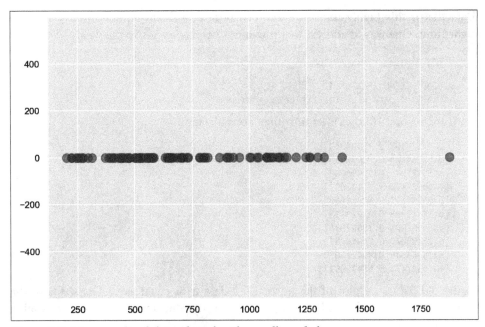

Figure 5-9. Unnormalized data plotted with equally scaled axes

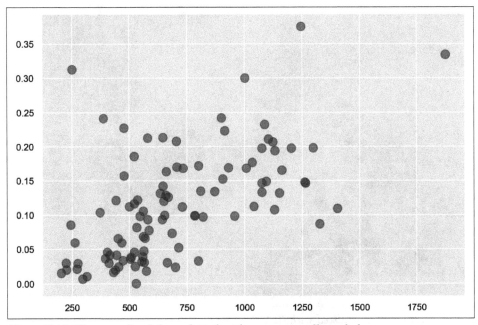

Figure 5-10. Unnormalized data plotted with proportionally scaled axes

Data that is this highly unnormalized can pose a problem for parametric learning algorithms. One way to address that is to apply `MinMaxScaler` to the data:

```
from sklearn.preprocessing import MinMaxScaler

scaler = MinMaxScaler()
normalized_data = scaler.fit_transform(data)
```

Here are the first 10 rows after min-max normalization:

```
[[0.47676153 0.79904352]
 [0.67066404 0.23006715]
 [0.5972794  0.52496344]
 [0.10989798 0.64238821]
 [0.65336197 0.52656469]
 [0.16419068 0.41928115]
 [0.50002983 0.29892076]
 [0.22433029 0.24810786]
 [0.18966649 0.49427287]
 [0.16347473 0.60475891]]]
```

Figure 5-11 shows a plot of the normalized data with equal axes. The shape of the data didn't change. What did change is that both columns now contain values ranging from 0.0 to 1.0.

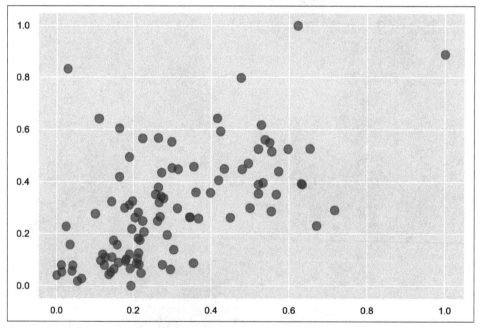

Figure 5-11. Data normalized with MinMaxScaler

SVMs almost always train better with normalized data, but the simple normalization performed by `MinMaxScaler` sometimes isn't enough. SVMs tend to respond better to data that is normalized to *unit variance* using a technique called *standardization* (*https://oreil.ly/UrsYP*) or *Z-score normalization*. Unit variance is achieved by doing the following to each column in a dataset:

- Computing the mean and standard deviations of all the values in the column
- Subtracting the mean from each value in the column
- Dividing each value in the column by the standard deviation

This is precisely the transform that Scikit's `StandardScaler` class performs on a dataset. Applying unit variance to a dataset is as simple as this:

```
from sklearn.preprocessing import StandardScaler

scaler = StandardScaler()
normalized_data = scaler.fit_transform(data)
```

The values in the original dataset may vary wildly from one column to the next, but the transformed dataset will contain columns of numbers anchored around 0 with ranges that are proportional to each column's standard deviation. Applying `StandardScaler` to the dataset produces the following values in the first 10 rows:

```
[[ 0.93457642  2.36212718]
 [ 1.95483237 -0.35495682]
 [ 1.56870474  1.05328794]
 [-0.99574783  1.61403698]
 [ 1.86379415  1.06093451]
 [-0.71007617  0.5486138 ]
 [ 1.05700714 -0.02615397]
 [-0.39363986 -0.26880538]
 [-0.57603023  0.90672853]
 [-0.71384326  1.4343424 ]]
```

And it produces the distribution shown in Figure 5-12. Once more, the shape of the data didn't change, but the values that *define* that shape changed substantially.

SVMs typically perform best when trained with standardized data, even if all the columns have similar ranges. (The same is true of neural networks, by the way.) The classic case in which columns have similar ranges but benefit from normalization anyway is image data, where each column holds pixel values from 0 to 255. There are exceptions, but it is usually a mistake to throw a bunch of data at an SVM without understanding the distribution of the data—specifically, whether it has unit variance.

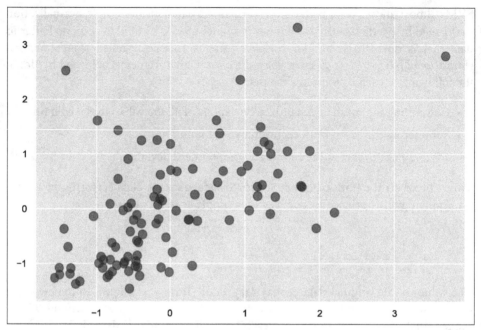

Figure 5-12. Data normalized with `StandardScaler`

Pipelining

If you normalize or standardize the values used to train a machine learning model, you must apply the same transform to values input to the model's `predict` method. In other words, if you train a model this way:

```
model = SVC()
scaler = StandardScaler()
x = scaler.fit_transform(x)
model.fit(x, y)
```

you make predictions with it this way:

```
input = [0, 1, 2, 3, 4]
model.predict([scaler.transform([input])
```

Otherwise, you'll get nonsensical predictions.

To simplify your code and make it harder to forget to transform training data and prediction data the same way, Scikit offers the `make_pipeline` function (*https://oreil.ly/HHN2p*). `make_pipeline` lets you combine predictive models—what Scikit calls *estimators*, or instances of classes such as `SVC`—with transforms applied to data input to those models. Here's how you use `make_pipeline` to ensure that any data input to the model is transformed with `StandardScaler`:

```
# Train the model
pipe = make_pipeline(StandardScaler(), SVC())
pipe.fit(x, y)

# Make a prediction with the model
input = [0, 1, 2, 3, 4]
pipe.predict([input])
```

Now data used to train the model has StandardScaler applied to it, and data input to make predictions is transformed the same way.

What if you wanted to use GridSearchCV to find the optimum set of parameters for a pipeline that combines a data transform and estimator? It's not hard, but there's a trick you need to know about. It involves using class names prefaced with double underscores in the param_grid dictionary passed to GridSearchCV. Here's an example:

```
pipe = make_pipeline(StandardScaler(), SVC())

grid = {
    'svc__C': [0.01, 0.1, 1, 10, 100],
    'svc__gamma': [0.01, 0.25, 0.5, 0.75, 1.0],
    'svc__kernel': ['rbf', 'poly']
}

grid_search = GridSearchCV(estimator=pipe, param_grid=grid, cv=5, verbose=2)
grid_search.fit(x, y) # Train the model with different parameter combinations
```

This example trains the model 250 times to find the best combination of kernel, C, and gamma for the SVC instance in the pipeline. Note the "svc__" nomenclature, which maps to the SVC instance passed to the make_pipeline function.

Using SVMs for Facial Recognition

Modern facial recognition is often accomplished with neural networks, but support vector machines can do a credible job too. Let's demonstrate by building a model that recognizes faces. The dataset we'll use is the Labeled Faces in the Wild (LFW) dataset (*https://oreil.ly/xG3LG*), which contains more than 13,000 facial images of famous people collected from around the web and is built into Scikit as a sample dataset. Of the more than 5,000 people represented in the dataset, 1,680 have two or more facial images, while only five have 100 or more. We'll set the minimum number of faces per person to 100, which means that five sets of faces corresponding to five famous people will be imported. Each facial image is labeled with the name of the person the face belongs to.

Start by creating a new Jupyter notebook and using the following statements to load the dataset and crop the facial images:

```python
import numpy as np
import pandas as pd
from sklearn.datasets import fetch_lfw_people

faces = fetch_lfw_people(min_faces_per_person=100, slice_=None)
faces.images = faces.images[:, 35:97, 39:86]
faces.data = faces.images.reshape(faces.images.shape[0], faces.images.shape[1] *
                                  faces.images.shape[2])
print(faces.target_names)
print(faces.images.shape)
```

In total, 1,140 facial images were loaded. Each image measures 47×62 pixels for a total of 2,914 pixels per image. That means the dataset contains 2,914 features. Use the following code to show the first 24 images in the dataset and the people to whom the faces belong:

```python
%matplotlib inline
import matplotlib.pyplot as plt

fig, ax = plt.subplots(3, 8, figsize=(18, 10))
for i, axi in enumerate(ax.flat):
    axi.imshow(faces.images[i], cmap='gist_gray')
    axi.set(xticks=[], yticks=[], xlabel=faces.target_names[faces.target[i]])
```

Here is the output:

Check the balance in the dataset by generating a histogram showing how many facial images were imported for each person:

```
import seaborn as sns
sns.set()

from collections import Counter
counts = Counter(faces.target)
names = {}

for key in counts.keys():
    names[faces.target_names[key]] = counts[key]

df = pd.DataFrame.from_dict(names, orient='index')
df.plot(kind='bar')
```

The output reveals that there are far more images of George W. Bush than of anyone else in the dataset:

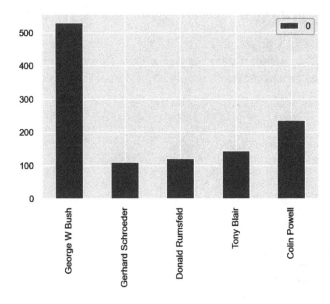

Classification models are best trained with balanced datasets. Use the following code to reduce the dataset to 100 images of each person:

```
mask = np.zeros(faces.target.shape, dtype=bool)

for target in np.unique(faces.target):
    mask[np.where(faces.target == target)[0][:100]] = 1

x = faces.data[mask]
y = faces.target[mask]
x.shape
```

Note that *x* contains 500 facial images and *y* contains the labels that go with them: 0 for Colin Powell, 1 for Donald Rumsfeld, and so on. Now let's see if an SVM can make sense of the data. We'll train three different models: one that uses a linear kernel, one that uses a polynomial kernel, and one that uses an RBF kernel. In each case, we'll use GridSearchCV to optimize hyperparameters. Start with a linear model and four different values of C:

```
from sklearn.svm import SVC
from sklearn.model_selection import GridSearchCV

svc = SVC(kernel='linear')

grid = {
    'C': [0.1, 1, 10, 100]
}

grid_search = GridSearchCV(estimator=svc, param_grid=grid, cv=5, verbose=2)
grid_search.fit(x, y) # Train the model with different parameters
grid_search.best_score_
```

This model achieves a cross-validated accuracy of 84.2%. It's possible that accuracy can be improved by standardizing the image data. Run the same grid search again, but this time use StandardScaler to apply unit variance to all the pixel values:

```
from sklearn.pipeline import make_pipeline
from sklearn.preprocessing import StandardScaler

scaler = StandardScaler()
svc = SVC(kernel='linear')
pipe = make_pipeline(scaler, svc)

grid = {
    'svc__C': [0.1, 1, 10, 100]
}

grid_search = GridSearchCV(estimator=pipe, param_grid=grid, cv=5, verbose=2)
grid_search.fit(x, y)
grid_search.best_score_
```

Standardizing the data produced an incremental improvement in accuracy. What value of C produced that accuracy?

```
grid_search.best_params_
```

Is it possible that a polynomial kernel could outperform a linear kernel? There's an easy way to find out. Note the introduction of the gamma and degree parameters to the parameter grid. These parameters, along with C, can greatly influence a polynomial kernel's ability to fit to the training data:

```
scaler = StandardScaler()
svc = SVC(kernel='poly')
pipe = make_pipeline(scaler, svc)

grid = {
    'svc__C': [0.1, 1, 10, 100],
    'svc__gamma': [0.01, 0.25, 0.5, 0.75, 1],
    'svc__degree': [1, 2, 3, 4, 5]
}

grid_search = GridSearchCV(estimator=pipe, param_grid=grid, cv=5, verbose=2)
grid_search.fit(x, y) # Train the model with different parameter combinations
grid_search.best_score_
```

The polynomial kernel achieved the same accuracy as the linear kernel. What parameter values led to this result?

```
grid_search.best_params_
```

The best_params_ attribute reveals that the optimum value of degree was 1, which means the polynomial kernel acted like a linear kernel. It's not surprising, then, that it achieved the same accuracy. Could an RBF kernel do better?

```
scaler = StandardScaler()
svc = SVC(kernel='rbf')
pipe = make_pipeline(scaler, svc)

grid = {
    'svc__C': [0.1, 1, 10, 100],
    'svc__gamma': [0.01, 0.25, 0.5, 0.75, 1.0]
}

grid_search = GridSearchCV(estimator=pipe, param_grid=grid, cv=5, verbose=2)
grid_search.fit(x, y)
grid_search.best_score_
```

The RBF kernel didn't perform as well as the linear and polynomial kernels. There's a lesson here. The RBF kernel often fits to nonlinear data better than other kernels, but it doesn't always fit better. That's why the best strategy with an SVM is to try different kernels with different parameter values. The best combination will vary from dataset to dataset. For the LFW dataset, it seems that a linear kernel is best. That's convenient, because the linear kernel is the fastest of all the kernels Scikit provides.

In addition to the SVC class, Scikit-Learn includes SVM classifiers named LinearSVC and NuSVC. The latter supports the same assortment of kernels as the SVC class, but it replaces C with a regularization parameter called nu that controls tightness of fit differently. NuSVC doesn't scale as well as SVC to large datasets, and in my experience it is rarely used. LinearSVC implements the linear kernel only, but it uses a different optimization algorithm that trains faster. If training is slow with SVC and you determine that a linear kernel yields the best model, consider swapping SVC for LinearSVC. Faster training times make a difference even for modestly sized datasets if you're using GridSearchCV to train a model hundreds of times. For a great summary of the functional differences between the two classes, see the article "SVM with Scikit-Learn: What You Should Know" (*https://oreil.ly/OQl1k*) by Angela Shi.

Confusion matrices are a great way to visualize a model's accuracy. Let's split the dataset, train an optimized linear model with 80% of the images, test it with the remaining 20%, and show the results in a confusion matrix.

The first step is to split the dataset. Note the stratify=y parameter, which ensures that the training dataset and the test dataset have the same proportion of samples of each class as the original dataset. In this example, the training dataset will contain 20 samples of each of the five people:

```
from sklearn.model_selection import train_test_split

x_train, x_test, y_train, y_test = train_test_split(x, y, train_size=0.8,
                                         stratify=y, random_state=0)
```

Now train a linear SVM with the optimum C value revealed by the grid search:

```
scaler = StandardScaler()
svc = SVC(kernel='linear', C=0.1)
pipe = make_pipeline(scaler, svc)
pipe.fit(x_train, y_train)
```

Cross-validate the model to confirm its accuracy:

```
from sklearn.model_selection import cross_val_score

cross_val_score(pipe, x, y, cv=5).mean()
```

Use a confusion matrix to see how the model performs against the test data:

```
from sklearn.metrics import ConfusionMatrixDisplay as cmd

fig, ax = plt.subplots(figsize=(6, 6))
ax.grid(False)
cmd.from_estimator(pipe, x_test, y_test, display_labels=faces.target_names,
                cmap='Blues', xticks_rotation='vertical', ax=ax)
```

Here is the output:

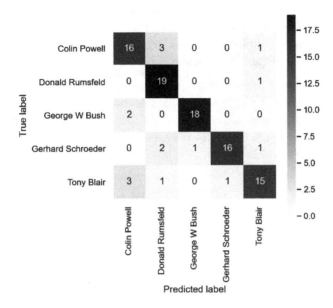

The model correctly identified Colin Powell 16 times out of 20, Donald Rumsfeld 19 times out of 20, and so on. That's not bad. And it's a great example of support vector machines at work.

Summary

Support vector machines, or SVMs, frequently fit to datasets better than other learning algorithms. SVMs are maximum-margin classifiers that use kernel tricks to simulate adding dimensions to data. The theory is that data that isn't linearly separable in m dimensions might be separable in n dimensions if n is higher than m. SVMs are most often used for classification, but they can perform regression too. As an experiment, try replacing `GradientBoostingRegressor` in the taxi-fare example in Chapter 2 with `SVR` and using `GridSearchCV` to optimize the model's hyperparameters. Which model produces the highest cross-validated coefficient of determination?

SVMs usually train better with data that is normalized to unit variance. That's true even if the values in all the columns have similar ranges, but it's especially true if they *don't* have similar ranges. Scikit's `StandardScaler` class applies unit variance to data. Unit variance is achieved by dividing the values in a column by the mean of all the values in the column and dividing by the standard deviation. Scikit's `make_pipeline` function enables you to combine transformers such as `StandardScaler` and classifiers such as `SVC` into one logical unit to ensure that data passed to `fit`, `predict`, and `predict_proba` undergoes the same transformations.

SVMs require tuning in order to achieve optimum accuracy. Tuning means finding the right values for parameters such as C, gamma, and kernel, and it entails trying different parameter combinations and assessing the results. Scikit provides classes such as GridSearchCV to help, but they increase training time by training the model once for each unique combination of parameter values.

SVMs can seem magical in their ability to fit mathematical models to complex datasets. But in my view, that magic takes a back seat to the numerical gymnastics performed by principal component analysis (PCA), which solves a variety of problems routinely encountered in machine learning. I often introduce PCA by telling audiences that it's the best-kept secret in machine learning. After Chapter 6, it will be a secret no longer.

Principal Component Analysis

Principal component analysis, or PCA, is one of the minor miracles of machine learning. It's a dimensionality reduction technique that reduces the number of dimensions in a dataset without sacrificing a commensurate amount of information. While that might seem underwhelming on the face of it, it has profound implications for engineers and software developers working to build predictive models from their data.

What if I told you that you could take a dataset with 1,000 columns, use PCA to reduce it to 100 columns, and retain 90% or more of the information in the original dataset? That's relatively common, believe it or not. And it lends itself to a variety of practical uses, including:

- Reducing high-dimensional data to two or three dimensions so that it can be plotted and explored

- Reducing the number of dimensions in a dataset and then restoring the original number of dimensions, which finds application in anomaly detection and noise filtering

- Anonymizing datasets so that they can be shared with others without revealing the nature or meaning of the data

And that's not all. A side effect of applying PCA to a dataset is that less important features—columns of data that have less relevance to the outcome of a predictive model—are removed, while dependencies between columns is eliminated. And in datasets with a low ratio of samples (rows) to features (columns), PCA can be used to increase that ratio. As a rule of thumb, you typically want a dataset used for machine learning to have *at least* five times as many rows as it has columns. If you can't add rows, an alternative is to use PCA to shave columns.

Once you learn about PCA, you'll wonder how you lived without it. Let's take a few moments to understand what it is and how it works. Then we'll look at some examples demonstrating why it's such an indispensable tool.

Understanding Principal Component Analysis

One way to wrap your head around PCA is to see how it reduces a two-dimensional dataset to one dimension. Figure 6-1 depicts a 2D dataset comprising a somewhat random collection of x and y values. If you reduced this dataset to a single dimension by simply dropping the x column or the y column, you'd be left with a horizontal or vertical line that bears little resemblance to the original dataset.

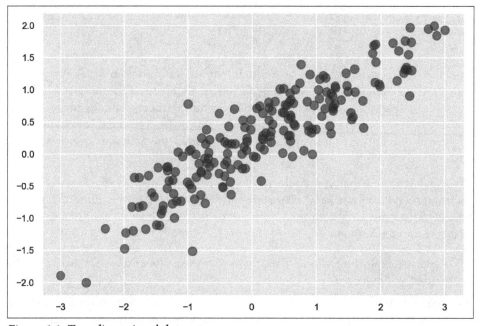

Figure 6-1. Two-dimensional dataset

Figure 6-2 adds arrows representing the dataset's two principal components. Essentially, the coordinate system has been transformed so that one axis (the longer of the two arrows) captures most of the variance in the dataset. This is the dataset's *primary principal component*. The other axis contains a narrower range of values and represents the secondary principal component. The number of principal components equals the number of dimensions in a dataset, so in this example, there are two principal components.

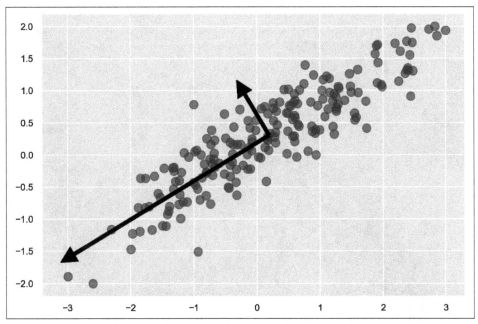

Figure 6-2. Arrows depicting the principal components of a two-dimensional dataset

To reduce a two-dimensional dataset to one dimension, PCA finds the two principal components and eliminates the one with less variance. This effectively projects the data points onto the primary principal component axis, as shown in Figure 6-3. The red data points don't retain all of the information in the original dataset, but they contain most of it. In this example, the PCAed dataset retains more than 95% of the information in the original. PCA reduced the number of dimensions by 50%, but it sacrificed less than 5% of the meaningful information in the dataset. That's the gist of PCA: reducing the number of dimensions without incurring a commensurate loss of information.

Under the hood, PCA works its magic by building a covariance matrix (*https:// oreil.ly/Rz9oI*) that quantifies the variance of each dimension with respect to the others, and from the matrix computing eigenvectors and eigenvalues (*https://oreil.ly/ LGSv1*) that identify the dataset's principal components. If you'd like to dig deeper, I suggest reading "A Step-by-Step Explanation of Principal Component Analysis (PCA)" (*https://oreil.ly/zLREl*) by Zakaria Jaadi. The good news is that you don't have to understand the math to make PCA work, because Scikit-Learn's PCA class (*https:// oreil.ly/8Ld8y*) does the math for you. The following statements reduce the dataset *x* to five dimensions, regardless of the number of dimensions it originally contains:

```
pca = PCA(n_components=5)
x = pca.fit_transform(x)
```

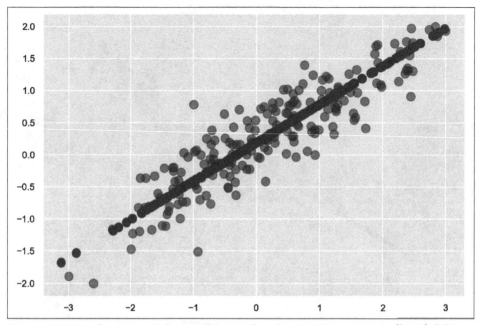

Figure 6-3. Two-dimensional dataset (blue) reduced to one dimension (red) with PCA

You can also invert a PCA transform to restore the original number of dimensions:

```
x = pca.inverse_transform(x)
```

The `inverse_transform` method restores the dataset to its original number of dimensions, but it doesn't restore the original dataset. The information that was discarded when the PCA transform was applied will be missing from the restored dataset.

You can visualize the loss of information when a PCA transform is applied and then inverted using the Labeled Faces in the Wild (LFW) dataset (*https://oreil.ly/USXRc*) introduced in Chapter 5. To demonstrate, fire up a Jupyter notebook and run the following code:

```
%matplotlib inline
import matplotlib.pyplot as plt
from sklearn.datasets import fetch_lfw_people

faces = fetch_lfw_people(min_faces_per_person=100, slice_=None)
faces.images = faces.images[:, 35:97, 39:86]
faces.data = faces.images.reshape(faces.images.shape[0], faces.images.shape[1] *
                                   faces.images.shape[2])
fig, ax = plt.subplots(3, 8, figsize=(18, 10))

for i, axi in enumerate(ax.flat):
    axi.imshow(faces.images[i], cmap='gist_gray')
    axi.set(xticks=[], yticks=[], xlabel=faces.target_names[faces.target[i]])
```

The output shows the first 24 images in the dataset:

Each image measures 47 × 62 pixels, for a total of 2,914 pixels per image. That means the dataset has 2,914 dimensions. Now use the following code to reduce the number of dimensions to 150 (roughly 5% of the original number), restore the original 2,914 dimensions, and plot the restored images:

```python
from sklearn.decomposition import PCA

pca = PCA(n_components=150, random_state=0)
pca_faces = pca.fit_transform(faces.data)
unpca_faces = pca.inverse_transform(pca_faces).reshape(1140, 62, 47)

fig, ax = plt.subplots(3, 8, figsize=(18, 10))

for i, axi in enumerate(ax.flat):
    axi.imshow(unpca_faces[i], cmap='gist_gray')
    axi.set(xticks=[], yticks=[], xlabel=faces.target_names[faces.target[i]])
```

Even though you removed almost 95% of the dimensions in the dataset, little meaningful information was discarded. The restored images are slightly blurrier than the originals, but the faces are still recognizable:

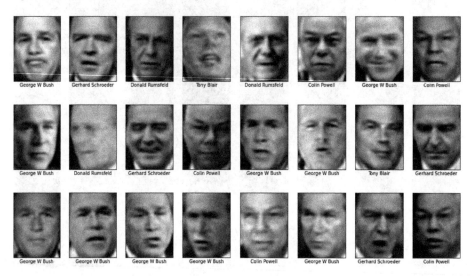

To reiterate, you reduced the number of dimensions from 2,914 to 150, but because PCA found 2,914 principal components and removed the ones that are least important (the ones with the least variance), you retained the bulk of the information in the original dataset. Which begs a question: precisely how much of the original information was retained?

After a PCA object is fit to a dataset, you can find out how much variance is encoded in each principal component from the explained_variance_ratio_ attribute. It's an array with one element for each principal component in the transformed dataset. Here's how it looks after the LFW dataset is reduced to 150 dimensions:

```
array([0.20098166, 0.1436709 , 0.0694095 , 0.0554688 , 0.04888214,
       0.02838693, 0.02344352, 0.02056908, 0.01904505, 0.01790946,
       0.01446775, 0.0141357 , 0.01173403, 0.01033751, 0.00927581,
       0.00900304, 0.00895557, 0.00830898, 0.00770731, 0.00712525,
       0.0064077 , 0.00619189, 0.00582111, 0.00557892, 0.00535471,
       0.00494034, 0.00482188, 0.00446195, 0.00443723, 0.00404091,
       0.00382839, 0.00370596, 0.00363874, 0.00352478, 0.00335927,
       0.00328615, 0.00315452, 0.00309412, 0.00290268, 0.00284517,
       0.00278296, 0.00267253, 0.00258336, 0.00249151, 0.00243766,
       0.00239778, 0.00237984, 0.00231506, 0.00223432, 0.00220306,
       0.00208555, 0.00207567, 0.00204288, 0.00196099, 0.00192303,
       0.00189352, 0.0018381 , 0.00180081, 0.00178862, 0.00174389,
       0.00168321, 0.00165759, 0.00162565, 0.00159976, 0.00153559,
```

```
       0.00152782, 0.00150262, 0.0014841 , 0.00147757, 0.00144323,
       0.00140246, 0.00138122, 0.00136053, 0.00132581, 0.00130121,
       0.00128062, 0.00126851, 0.00123904, 0.00123427, 0.00120644,
       0.00118998, 0.00117278, 0.00116551, 0.00115161, 0.00111428,
       0.00108951, 0.00107443, 0.00105793, 0.00104903, 0.00104119,
       0.00099986, 0.00098006, 0.00097077, 0.00095622, 0.00093874,
       0.00092516, 0.00091716, 0.00091061, 0.00090051, 0.00087887,
       0.00086778, 0.0008543 , 0.00084502, 0.00082587, 0.00081203,
       0.00080346, 0.00079375, 0.00077893, 0.00077295, 0.00077045,
       0.00075456, 0.00073704, 0.00073038, 0.00072013, 0.0007093 ,
       0.00070115, 0.00069389, 0.00067964, 0.00067382, 0.00065503,
       0.0006506 , 0.00063969, 0.00063328, 0.00062684, 0.00062352,
       0.0006103 , 0.00060463, 0.00059769, 0.00058182, 0.00057901,
       0.00056648, 0.00056551, 0.00054979, 0.00054543, 0.00053753,
       0.0005361 , 0.00053067, 0.00051841, 0.00051382, 0.00050711,
       0.00049933, 0.0004919 , 0.00048888, 0.00047992, 0.00047919,
       0.00046916, 0.00046408, 0.00046142, 0.00045397, 0.0004432 ],
      dtype=float32)
```

This reveals that 20% of the variance in the dataset is explained by the primary principal component, 14% is explained by the secondary principal component, and so on. Observe that the numbers decrease as the index increases. By definition, each principal component in a PCAed dataset contains more information than the principal component after it. In this example, the 2,764 principal components that were discarded contained so little information that their loss was barely noticeable when the transform was inverted. In fact, the sum of the 150 numbers in the preceding example is 0.938. This means reducing the dataset from 2,914 dimensions to 150 retained 93.8% of the information in the original dataset. In other words, you reduced the number of dimensions by almost 95%, and yet you retained almost 94% of the information in the dataset. If that's not awesome, I don't know what is.

A logical question to ask is, what is the "right" number of components? In other words, what number of components strikes the best balance between reducing the number of dimensions in the dataset and retaining most of the information? One way to find that number is with a *scree plot*, which charts the proportion of explained variance for each dimension. The following code produces a scree plot for the PCA transform used on the facial images:

```python
import seaborn as sns
sns.set()

plt.plot(pca.explained_variance_ratio_)
plt.xlabel('Principal Component')
plt.ylabel('Explained Variance')
```

Here is the output:

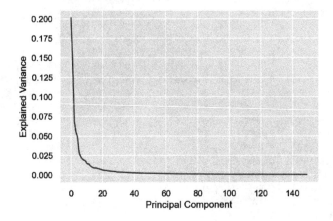

Another way to look at it is to plot the cumulative sum of the variances as a function of component count:

```
import numpy as np

plt.plot(np.cumsum(pca.explained_variance_ratio_))
plt.xlabel('Number of Components')
plt.ylabel('Cumulative Explained Variance');
```

Here is the output:

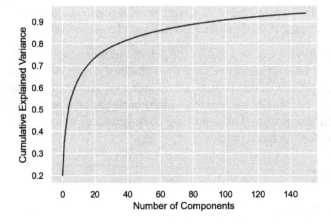

Either way you look at it, the bulk of the information is contained in the first 50 to 100 dimensions. Based on these plots, if you reduced the number of dimensions to 50 instead of 150, would you expect the restored facial images to look substantially different? If you're not sure, try it and see.

Filtering Noise

One very practical use for PCA is to filter noise from data. *Noise (https://oreil.ly/yjLYD)* is data that is random, corrupt, or otherwise meaningless, and it's particularly likely to occur when the data comes from physical devices such as pressure sensors or accelerometers. The basic approach to using PCA for noise reduction is to PCA-transform the data and then invert the transform, reducing the dataset from m dimensions to n and then restoring it to m. Because PCA discards the least important information when reducing dimensions and noise tends to have little or no informational value, this ideally eliminates much of the noise while retaining most of the meaningful data.

You can test this supposition with the LFW dataset. Use the following statements to add noise to the facial images using a random-number generator and plot the first 24 images:

```
%matplotlib inline
import matplotlib.pyplot as plt
from sklearn.datasets import fetch_lfw_people
import numpy as np

faces = fetch_lfw_people(min_faces_per_person=100, slice_=None)
faces.images = faces.images[:, 35:97, 39:86]
faces.data = faces.images.reshape(faces.images.shape[0], faces.images.shape[1] *
                                  faces.images.shape[2])

np.random.seed(0)
noisy_faces = np.random.normal(faces.data, 0.0765)

fig, ax = plt.subplots(3, 8, figsize=(18, 10))

for i, axi in enumerate(ax.flat):
    axi.imshow(noisy_faces[i].reshape(62, 47), cmap='gist_gray')
    axi.set(xticks=[], yticks=[], xlabel=faces.target_names[faces.target[i]])
```

The resulting facial images resemble a staticky 1960s TV screen:

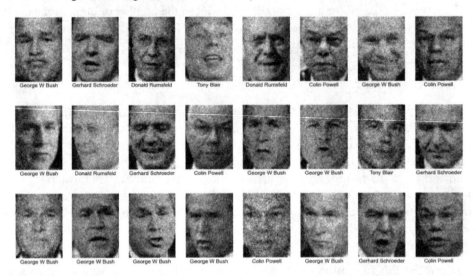

Now use PCA to reduce the number of dimensions. Rather than specify the number of dimensions (components), we'll specify that we want to reduce the amount of information in the dataset to 80%. We'll let Scikit decide how many dimensions will remain, and then show the count:

```
from sklearn.decomposition import PCA

pca = PCA(0.8, random_state=0)
pca_faces = pca.fit_transform(noisy_faces)
pca.n_components_
```

PCA reduced the number of dimensions from 2,914 to 179, but the remaining dimensions contain 80% of the information in the original 2,914. Now reconstruct the facial images from the PCAed faces and show the results:

```
unpca_faces = pca.inverse_transform(pca_faces)

fig, ax = plt.subplots(3, 8, figsize=(18, 10))

for i, axi in enumerate(ax.flat):
    axi.imshow(unpca_faces[i].reshape(62, 47), cmap='gist_gray')
    axi.set(xticks=[], yticks=[], xlabel=faces.target_names[faces.target[i]])
```

Here is the output:

The reconstructed dataset isn't quite as clean as the original, but it's clean enough that you can make out the faces in the photos.

Anonymizing Data

Chapter 3 demonstrated how to use various learning algorithms to build a binary classification model that detects credit card fraud. The dataset used in the example contained real credit card data that had been anonymized to protect the card holders (and the credit card company's intellectual property). The first 10 rows of that dataset are pictured in Figure 6-4.

	Time	V1	V2	V3	...	V27	V28	Amount	Class
0	0.0	-1.359807	-0.072781	2.536347	...	0.133558	-0.021053	149.62	0
1	0.0	1.191857	0.266151	0.166480	...	-0.008983	0.014724	2.69	0
2	1.0	-1.358354	-1.340163	1.773209	...	-0.055353	-0.059752	378.66	0
3	1.0	-0.966272	-0.185226	1.792993	...	0.062723	0.061458	123.50	0
4	2.0	-1.158233	0.877737	1.548718	...	0.219422	0.215153	69.99	0
5	2.0	-0.425966	0.960523	1.141109	...	0.253844	0.081080	3.67	0
6	4.0	1.229658	0.141004	0.045371	...	0.034507	0.005168	4.99	0
7	7.0	-0.644269	1.417964	1.074380	...	-1.206921	-1.085339	40.80	0
8	7.0	-0.894286	0.286157	-0.113192	...	0.011747	0.142404	93.20	0
9	9.0	-0.338262	1.119593	1.044367	...	0.246219	0.083076	3.68	0

Figure 6-4. Anonymized fraud detection dataset

Another practical use for PCA is to anonymize data in this manner. It's generally a two-step process:

1. Use PCA to "reduce" the dataset from *m* dimensions to *m*, where *m* is the original number of dimensions (as well as the number of dimensions after "reduction").

2. Normalize the data so that it has unit variance.

The second step isn't required, but it does make the ranges of values more uniform. Data anonymized this way can still be used to train a machine learning model, but its original meaning can't be inferred.

Try it with a dataset of your own. First, use the following code to load Scikit's breast cancer dataset (*https://oreil.ly/siksH*) and display the first five rows:

```
import pandas as pd
from sklearn.datasets import load_breast_cancer

data = load_breast_cancer()
df = pd.DataFrame(data=data.data, columns=data.feature_names)
pd.set_option('display.max_columns', 6)
df.head()
```

The output is as follows:

	mean radius	mean texture	mean perimeter	...	worst concave points	worst symmetry	worst fractal dimension
0	17.99	10.38	122.80	...	0.2654	0.4601	0.11890
1	20.57	17.77	132.90	...	0.1860	0.2750	0.08902
2	19.69	21.25	130.00	...	0.2430	0.3613	0.08758
3	11.42	20.38	77.58	...	0.2575	0.6638	0.17300
4	20.29	14.34	135.10	...	0.1625	0.2364	0.07678

5 rows × 30 columns

The dataset contains 30 columns, not counting the label column. Now use the following statements to find the 30 principal components and apply StandardScaler to the transformed data:

```
from sklearn.decomposition import PCA
from sklearn.preprocessing import StandardScaler

pca = PCA(n_components=30, random_state=0)
pca_data = pca.fit_transform(df)

scaler = StandardScaler()
anon_df = pd.DataFrame(scaler.fit_transform(pca_data))
```

```
pd.set_option('display.max_columns', 8)
anon_df.head()
```

The result is as follows:

	0	1	2	3	...	26	27	28	29
0	1.743043	-3.440692	1.832695	-1.179529	...	-1.033900	0.767070	1.406020	0.841434
1	1.906779	0.182972	-1.335313	2.418269	...	-0.043492	-0.798802	0.484854	-1.267746
2	1.496120	0.458381	-0.064503	0.568556	...	0.092680	0.010964	-0.547972	0.484234
3	-0.611764	-0.788775	0.327197	-1.592188	...	0.008095	0.811865	-1.511794	-1.978890
4	1.397781	2.216483	0.051866	1.150718	...	1.716566	0.161769	1.260500	0.390467

5 rows × 30 columns

The dataset is unrecognizable after the PCA transform. Without the transform, it's impossible to work backward and reconstruct the original data. Yet the sum of the `explained_variance_ratio_` values is 1.0, which means no information was lost. You can prove it this way:

```
import numpy as np

np.sum(pca.explained_variance_ratio_)
```

The PCAed dataset is just as useful for machine learning as the original. Furthermore, if you want to share the dataset with others so that they can train models of their own, there is no risk of divulging sensitive or proprietary information.

Visualizing High-Dimensional Data

Yet another use for PCA is to reduce a dataset to two or three dimensions so that it can be plotted with libraries such as Matplotlib (*https://oreil.ly/p7u82*). You can't plot a dataset that has 1,000 columns. You *can* plot a dataset that has two or three columns. The fact that PCA can reduce high-dimensional data to two or three dimensions while retaining much of the original information makes it a great tool for exploring data and visualizing relationships between classes.

Suppose you're building a classification model and want to assess up front whether there is sufficient separation between classes to support such a model. Take the Optical Recognition of Handwritten Digits dataset (*https://oreil.ly/RuXLJ*) built into Scikit, for example. Each digit in the dataset is represented by an 8 × 8 array of pixel values, meaning the dataset has 64 dimensions. If you could plot a 64-dimensional diagram, you might be able to inspect the dataset and look for separation between classes. But 64 dimensions is 61 too many for most humans.

Enter PCA. The following code loads the dataset, uses PCA to reduce it to two dimensions, and plots the result, with different colors representing different classes (digits):

```
from sklearn.decomposition import PCA
from sklearn.datasets import load_digits
import matplotlib.pyplot as plt
%matplotlib inline

digits = load_digits()
pca = PCA(n_components=2, random_state=0)
pca_digits = pca.fit_transform(digits.data)

plt.figure(figsize=(12, 8))
plt.scatter(pca_digits[:, 0], pca_digits[:, 1], c=digits.target,
            cmap=plt.cm.get_cmap('Paired', 10))
plt.colorbar(ticks=range(10))
plt.clim(-0.5, 9.5)
```

The resulting plot provides an encouraging sign that you might be able to train a classifier with the data. While there is clearly some overlap between classes, the different classes form rather distinct clusters. There is significant overlap between red (the digit 4) and light purple (the digit 6), indicating that a model might have some difficulty distinguishing between 4s and 6s. However, 0s and 1s lie at the top and bottom, while 3s and 4s fall on the far left and far right. A model would presumably be proficient at telling these digits apart:

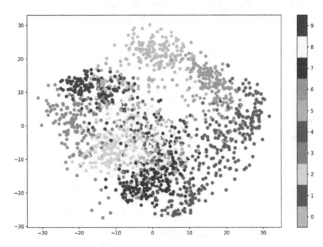

You can better visualize relationships between classes with a 3D plot. The following code uses PCA to reduce the dataset to three dimensions and Mplot3D (*https://oreil.ly/sNdl9*) to produce an interactive plot. Note that if you run this code in Jupyter Lab, you'll probably have to change the first line to %matplotlib widget:

```
%matplotlib notebook
from mpl_toolkits.mplot3d import Axes3D

digits = load_digits()
pca = PCA(n_components=3, random_state=0)
pca_digits = pca.fit_transform(digits.data)

ax = plt.figure(figsize=(12, 8)).add_subplot(111, projection='3d')
ax.scatter(xs = pca_digits[:, 0], ys = pca_digits[:, 1], zs = pca_digits[:, 2],
           c=digits.target, cmap=plt.cm.get_cmap('Paired', 10))
```

You can rotate the resulting plot in 3D and look at it from different angles. Here we can see that there is more separation between 4s and 6s than was evident in two dimensions:

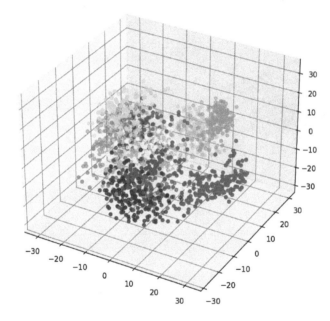

PCA isn't the only way to reduce a dataset to two or three dimensions for plotting. You can also use Scikit's Isomap class (*https://oreil.ly/f9TL6*) or its TSNE class (*https://oreil.ly/vtIcf*). TSNE implements *t-distributed stochastic neighbor embedding* (*https://oreil.ly/ncz7O*), or *t*-SNE for short. *t*-SNE is a dimensionality reduction algorithm that is used almost exclusively for visualizing high-dimensional data. Whereas PCA uses a linear function to transform data, *t*-SNE uses a nonlinear transform that tends to heighten the separation between classes by keeping similar data points close together in low-dimensional space. (PCA, by contrast, focuses on keeping dissimilar points far apart.) Here's an example that plots the Digits dataset in two dimensions after reducing it with *t*-SNE:

```
%matplotlib inline
from sklearn.manifold import TSNE

digits = load_digits()
tsne = TSNE(n_components=2, init='pca', learning_rate='auto',
            random_state=0)
tsne_digits = tsne.fit_transform(digits.data)

plt.figure(figsize=(12, 8))
plt.scatter(tsne_digits[:, 0], tsne_digits[:, 1], c=digits.target,
            cmap=plt.cm.get_cmap('Paired', 10))
plt.colorbar(ticks=range(10))
plt.clim(-0.5, 9.5)
```

And here is the output:

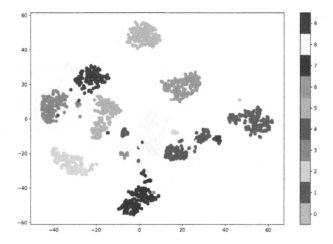

t-SNE does a better job of separating groups of digits into clusters, indicating there are patterns in the data that machine learning can exploit. The chief drawback is that *t*-SNE is compute intensive, which means it can take a prohibitively long time to run on large datasets. One way to mitigate that is to run *t*-SNE on a subset of rows rather than the entire dataset. Another strategy is to use PCA to reduce the number of dimensions, and then subject the PCAed dataset to *t*-SNE.

Anomaly Detection

Anomaly detection is a branch of machine learning that seeks to identify anomalies in datasets or data streams. Airbus uses it to predict failures in jet engines (*https://oreil.ly/Aafuz*) and detect anomalies in telemetry data (*https://oreil.ly/OiEKG*) beamed down from the International Space Station. Credit card companies use it to detect credit card fraud (*https://oreil.ly/tjAtA*). The goal of anomaly detection is to identify outliers in data—samples that aren't "normal" when compared to others. In the case of credit card fraud, the assumption is that if transactions are subjected to an anomaly detection algorithm, fraudulent transactions will show up as anomalous, while legitimate transactions will not.

There are many ways to perform anomaly detection. They go by names such as isolation forests (*https://oreil.ly/PtlRg*), one-class SVMs (*https://oreil.ly/KFoT4*), and local outlier factor (LOF) (*https://oreil.ly/LumWj*). Most rely on unsupervised learning methods and therefore do not require labeled data. They simply look at a collection of samples and determine which ones are anomalous. Unsupervised anomaly detection is particularly interesting because it doesn't require a priori knowledge of what constitutes an anomaly, nor does it require an unlabeled dataset to be meticulously labeled.

One of the most popular forms of anomaly detection relies on principal component analysis. You already know that PCA can be used to reduce data from m dimensions to n, and that a PCA transform can be inverted to restore the original m dimensions. You also know that inverting the transform doesn't recover the data that was lost when the transform was applied. The gist of PCA-based anomaly detection is that an anomalous sample should exhibit more loss or *reconstruction error* (the difference between the original data and the same data after a PCA transform is applied and inverted) than a normal one. In other words, the loss incurred when an anomalous sample is PCAed and un-PCAed should be higher than the loss incurred when the same operation is applied to a normal sample. Let's see if this assumption holds up in the real world.

Using PCA to Detect Credit Card Fraud

Supervised learning isn't the only option for detecting credit card fraud. Here's an alternative approach that uses PCA-based anomaly detection to identify fraudulent transactions. Begin by loading the dataset, separating the samples by class into one dataset representing legitimate transactions and another representing fraudulent transactions, and dropping the Time and Class columns. If you didn't download the dataset in Chapter 3, you can get it now from the ZIP file (*https://oreil.ly/RTNYn*).

```
import pandas as pd

df = pd.read_csv('Data/creditcard.csv')
df.head()

# Separate the samples by class
legit = df[df['Class'] == 0]
fraud = df[df['Class'] == 1]

# Drop the "Time" and "Class" columns
legit = legit.drop(['Time', 'Class'], axis=1)
fraud = fraud.drop(['Time', 'Class'], axis=1)
```

Use PCA to reduce the two datasets from 29 to 26 dimensions, and then invert the transform to restore each dataset to 29 dimensions. The transform is fitted to legitimate transactions only because we need a baseline value for reconstruction error that allows us to discriminate between legitimate and fraudulent transactions. It is applied, however, to both datasets:

```python
from sklearn.decomposition import PCA

pca = PCA(n_components=26, random_state=0)
legit_pca = pd.DataFrame(pca.fit_transform(legit), index=legit.index)
fraud_pca = pd.DataFrame(pca.transform(fraud), index=fraud.index)

legit_restored = pd.DataFrame(pca.inverse_transform(legit_pca),
                              index=legit_pca.index)

fraud_restored = pd.DataFrame(pca.inverse_transform(fraud_pca),
                              index=fraud_pca.index)
```

Some information was lost in the transition. Hopefully, the fraudulent transactions incurred more loss than the legitimate ones, and we can use that to differentiate between them. The next step is to compute the loss for each row in the two datasets by summing the squares of the differences between the values in the original rows and the restored rows:

```python
import numpy as np

def get_anomaly_scores(df_original, df_restored):
    loss = np.sum((np.array(df_original) - np.array(df_restored)) ** 2, axis=1)
    loss = pd.Series(data=loss, index=df_original.index)
    return loss

legit_scores = get_anomaly_scores(legit, legit_restored)
fraud_scores = get_anomaly_scores(fraud, fraud_restored)
```

Now plot the losses incurred when the legitimate transactions were transformed and restored:

```python
%matplotlib inline
import matplotlib.pyplot as plt
import seaborn as sns
sns.set()

legit_scores.plot(figsize = (12, 6))
```

Here is the result:

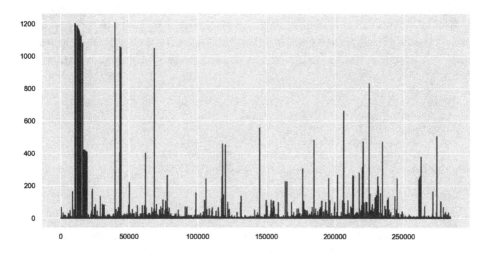

Next, plot the losses for the fraudulent transactions:

```
fraud_scores.plot(figsize = (12, 6))
```

Here is the result:

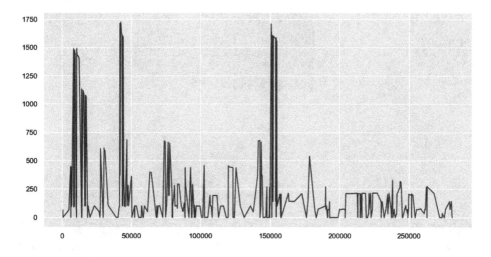

The plots reveal that most of the rows in the dataset representing legitimate transactions incurred a loss of less than 200, while many of the rows in the dataset representing fraudulent transactions incurred a loss greater than 200. Separate the rows on this basis—classifying transactions with a loss of less than 200 as legitimate and transactions with a higher loss as fraudulent—and use a confusion matrix to visualize the results:

```
threshold = 200

true_neg = legit_scores[legit_scores < threshold].count()
false_pos = legit_scores[legit_scores >= threshold].count()
true_pos = fraud_scores[fraud_scores >= threshold].count()
false_neg = fraud_scores[fraud_scores < threshold].count()

labels = ['Legitimate', 'Fraudulent']
mat = [[true_neg, false_pos], [false_neg, true_pos]]

sns.heatmap(mat, square=True, annot=True, fmt='d', cbar=False, cmap='Blues',
            xticklabels=labels, yticklabels=labels)

plt.xlabel('Predicted label')
plt.ylabel('True label')
```

Here is the result:

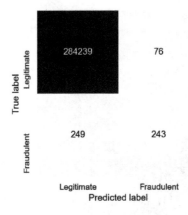

The results aren't quite as good as they were with the random forest, but the model still caught about 50% of the fraudulent transactions while mislabeling just 76 out of 284,315 legitimate transactions. That's an error rate of less than 0.03% for legitimate transactions, compared to 0.007% for the supervised learning model.

Two parameters in this model drive the error rate: the number of dimensions the datasets were reduced to (26), and the threshold chosen to distinguish between legitimate and fraudulent transactions (200). You can tweak the accuracy by experimenting with different values. I did some informal testing and concluded that this was a reasonable combination. Picking a lower threshold improves the model's ability to identify fraudulent transactions, but at the cost of misclassifying more legitimate transactions. In the end, you have to decide what error rate you're willing to live with, keeping in mind that declining a legitimate credit card purchase is likely to anger a customer.

Using PCA to Predict Bearing Failure

One of the classic uses for anomaly detection is to predict failures in rotating machinery. Let's apply PCA-based anomaly detection to a subset of a dataset published by NASA (*https://oreil.ly/B9jOD*) to predict failures in bearings. The dataset contains vibration data for four bearings supporting a rotating shaft with a radial load of 6,000 pounds applied to it. The bearings were run to failure, and vibration data was captured by high-sensitivity quartz accelerometers at regular intervals until failure occurred.

First, download the CSV file (*https://oreil.ly/vOUCc*) containing the subset that I culled from the larger NASA dataset. Then create a Jupyter notebook and load the data:

```
import pandas as pd

df = pd.read_csv('Data/bearings.csv', index_col=0, parse_dates=[0])
df.head()
```

Here are the first five rows in the dataset:

	Bearing 1	Bearing 2	Bearing 3	Bearing 4
2004-02-12 10:32:39	0.058333	0.071832	0.083242	0.043067
2004-02-12 10:42:39	0.058995	0.074006	0.084435	0.044541
2004-02-12 10:52:39	0.060236	0.074227	0.083926	0.044443
2004-02-12 11:02:39	0.061455	0.073844	0.084457	0.045081
2004-02-12 11:12:39	0.061361	0.075609	0.082837	0.045118

The dataset contains 984 samples. Each sample contains vibration data for four bearings, and the samples were taken 10 minutes apart. Plot the vibration data for all four bearings as a time series:

```
%matplotlib inline
import matplotlib.pyplot as plt
import seaborn as sns
sns.set()

df.plot(figsize = (12, 6))
```

Here is the output:

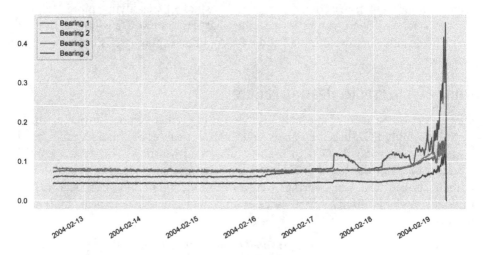

About four days into the test, vibrations in bearing 1 began increasing. They spiked a day later, and about two days after that, bearing 1 suffered a catastrophic failure. Our goal is to build a model that recognizes increased vibration in any bearing as a sign of impending failure, and to do it *without* a labeled dataset.

The next step is to extract samples representing "normal" operation from the dataset (x_train in the following code) and reduce four dimensions to one using PCA— essentially combining the data from all four bearings. Then apply the same PCA transform to the remainder of the dataset (x_test), combine the two partial datasets, and plot the result:

```
from sklearn.decomposition import PCA

x_train = df['2004-02-12 10:32:39':'2004-02-13 23:42:39']
x_test = df['2004-02-13 23:52:39':]

pca = PCA(n_components=1, random_state=0)
x_train_pca = pd.DataFrame(pca.fit_transform(x_train))
x_train_pca.index = x_train.index

x_test_pca = pd.DataFrame(pca.transform(x_test))
x_test_pca.index = x_test.index

df_pca = pd.concat([x_train_pca, x_test_pca])
df_pca.plot(figsize = (12, 6))
plt.legend().remove()
```

The output is shown here:

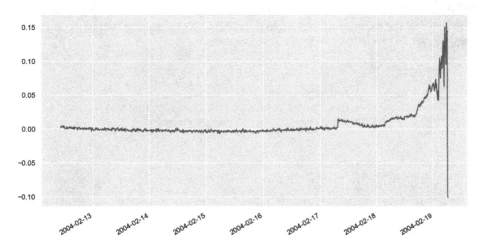

Now invert the PCA transform and plot the "restored" dataset:

```
df_restored = pd.DataFrame(pca.inverse_transform(df_pca), index=df_pca.index)
df_restored.plot(figsize = (12, 6))
```

The results are as follows:

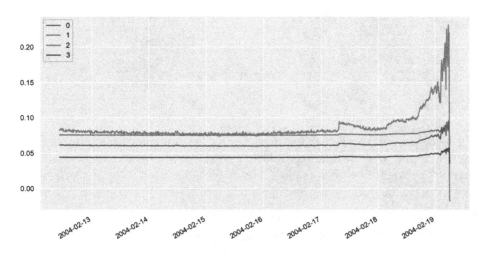

It is obvious that a loss was incurred by applying and inverting the transform. Let's define a function that computes the loss in a range of samples, then apply that function to all of the samples in the original dataset and the restored dataset and plot the differences over time:

```python
import numpy as np

def get_anomaly_scores(df_original, df_restored):
    loss = np.sum((np.array(df_original) - np.array(df_restored)) ** 2, axis=1)
    loss = pd.Series(data=loss, index=df_original.index)
    return loss

scores = get_anomaly_scores(df, df_restored)
scores.plot(figsize = (12, 6))
```

Here is the output:

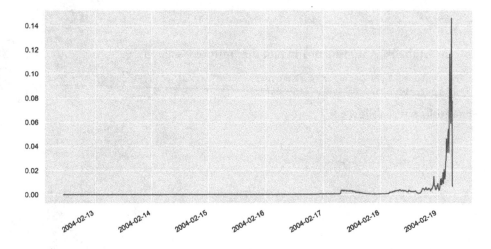

The loss is very small when all four bearings are operating normally, but it begins to rise when one or more bearings exhibit greater-than-normal vibration. From the chart, it's apparent that when the loss rises above a threshold value of approximately 0.002, that's an indication a bearing might fail.

Now that you've selected a tentative loss threshold, you can use it to detect anomalous behavior in the bearings. Begin by defining a function that takes a sample and returns True or False indicating whether the sample is anomalous by applying and inverting a PCA transform, measuring the loss for each bearing, and comparing it to a specified loss threshold:

```python
def is_anomaly(row, pca, threshold):
    pca_row = pca.transform(row)
    restored_row = pca.inverse_transform(pca_row)
    losses = np.sum((row - restored_row) ** 2)
```

```
    for loss in losses:
        if loss > threshold:
            return True;

    return False
```

Apply the function to a row early in the time series that represents normal behavior and confirm that it returns `False`:

```
x = df.loc[['2004-02-16 22:52:39']]
is_anomaly(x, pca, 0.002)
```

Apply the function to a row later in the time series that represents anomalous behavior and confirm that it returns `True`:

```
x = df.loc[['2004-02-18 22:52:39']]
is_anomaly(x, pca, 0.002)
```

Now apply the function to all the samples in the dataset and shade anomalous samples red in order to visualize when anomalous behavior is detected:

```
df.plot(figsize = (12, 6))

for index, row in df.iterrows():
    if is_anomaly(pd.DataFrame([row]), pca, 0.002):
        plt.axvline(row.name, color='r', alpha=0.2)
```

Here is the output:

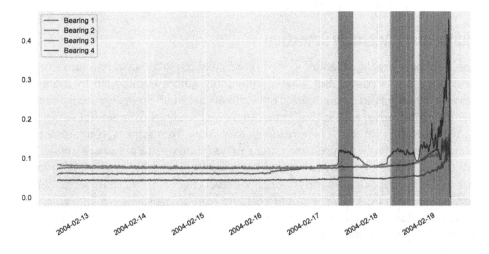

Repeat this procedure, but this time use a loss threshold of 0.0002 rather than 0.002:

```
df.plot(figsize = (12, 6))

for index, row in df.iterrows():
```

```
if is_anomaly(pd.DataFrame([row]), pca, 0.0002):
    plt.axvline(row.name, color='r', alpha=0.2)
```

Here is the output:

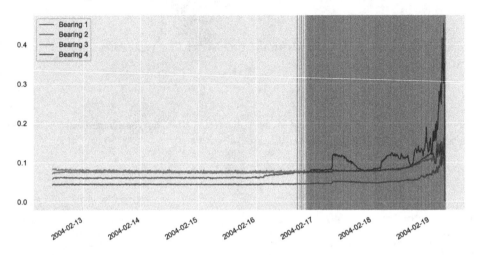

You can adjust the sensitivity of the model by adjusting the threshold value used to detect anomalies. Using a loss threshold of 0.002 predicts bearing failure about two days before it occurs, while a loss threshold of 0.0002 predicts the failure about three days before. You typically want to choose a loss threshold that predicts failure as early as possible without raising false alarms.

Multivariate Anomaly Detection

Could we have predicted failure in the preceding example by simply monitoring individual bearings? Perhaps. But what if impending failure is indicated by marginally elevated vibrations in *two* bearings rather than just one? Engineers frequently find that it isn't individual sensors but a combination of readings from several sensors that signal impending trouble. These readings may come from sensors of different types: temperature sensors and pressure gauges in automotive and aerospace applications (*https://oreil.ly/18v1V*), for example, or heart monitors and blood pressure monitors in health-care applications (*https://oreil.ly/UoeIc*). Reducing the number of dimensions to one with PCA is an attempt to capture relationships between data emanating from individual sensors and treat the readings systemically, a technique known as *multivariate anomaly detection*.

One limitation of using PCA to detect anomalies in multivariate systems is that because it uses linear transforms, PCA is better at modeling linear relationships between variables than nonlinear relationships. Neural networks, by contrast, excel at modeling nonlinear data. That's the primary reason why state-of-the-art multivariate anomaly detection today commonly relies on deep learning.

As the number of variables increases, so too does the challenge of modeling the interdependencies between them. It is not uncommon for overall system health to be determined by dozens of otherwise independent variables. In September 2020, a team of researchers at Microsoft and Peking University published a paper titled "Multivariate Time-series Anomaly Detection via Graph Attention Network" (*https://oreil.ly/ T5qMi*) that proposed a novel architecture for multivariate anomaly detection. It combines two deep-learning models: one that relies on prediction error and another that relies on reconstruction error. Microsoft uses this architecture in its Azure Multivariate Anomaly Detector (*https://oreil.ly/5ixTv*) service, which can model dependencies between up to 300 independent data sources and is used by companies such as Airbus and Siemens to detect irregularities in space-station telemetry (*https://oreil.ly/ xO1nI*) and to test medical devices before they're sent to market. The Azure Multivariate Anomaly Detector service is part of Azure Cognitive Services (*https://oreil.ly/ rwRSv*), which is covered in Chapter 14.

Summary

Principal component analysis is a technique for reducing the number of dimensions in a dataset without incurring a commensurate loss of information. It enjoys a number of uses in machine learning, including visualizing high-dimensional data, anonymizing data, reducing noise, and increasing the ratio of rows to columns by reducing the number of dimensions. It can also be used to perform anomaly detection by measuring the loss incurred when a PCA transform is applied and then inverted. Anomalous samples tend to incur more loss.

When I teach classes, I often introduce PCA as "the best-kept secret in machine learning." It shouldn't remain a secret, because it's an indispensable tool in the hands of machine learning engineers. Now that you know about it, I can just about guarantee that you'll find ways to put it to work.

## CHAPTER 7

Operationalizing Machine Learning Models

All of the machine learning models presented so far in this book are written in Python. Models don't *have* to be written in Python, but many are, thanks in part to the numerous world-class Python libraries available, including Pandas (*https://oreil.ly/lSq91*) and Scikit-Learn (*https://oreil.ly/rg9GI*). ML models written in Python are easily consumed in Python apps. Calling them from other languages such as C++, Java, and C# requires a little more work. You can't simply call a Python function from C++ as if it were a C++ function. So how do you invoke models written in Python from apps written in other languages? Put another way, how do you *operationalize* Python models such that they are usable in any app on any platform written in any programming language?

The diagram on the left in Figure 7-1 shows one strategy: wrap the model in a web service and expose its functionality through a REST API (*https://oreil.ly/CEvxK*). Then any client that can generate an HTTP(S) request can invoke the model. It's relatively easy to do the wrapping with help from Python frameworks such as Flask (*https://oreil.ly/x5vWh*). The web service can be hosted locally or in the cloud, and it can be containerized for easy deployment using tools such as Docker (*https://oreil.ly/wVDsr*).

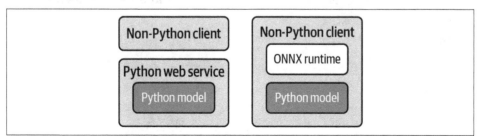

Figure 7-1. Architectures for consuming Python models in other languages

The diagram on the right shows another strategy—one that's relatively new but rapidly growing in popularity. It involves exporting a Python model to a platform-agnostic format called ONNX, short for Open Neural Network Exchange (*https://oreil.ly/qdwhs*), and then using an ONNX runtime to load the model in Java, C++, C#, and other programming languages. Once loaded, the model can be called via the ONNX runtime.

Of course, if the client app and the model are written in the same language, you need neither a web service nor ONNX. In this chapter, I'll walk you through several scenarios:

- How to save a trained Python model and invoke it from a Python client
- How to invoke a Python model from a non-Python client using a web service
- How to containerize a Python model (and web service) for easy deployment
- How to use ONNX to invoke a Python model from other programming languages
- How to write machine learning models in C# rather than Python

I'll finish up by demonstrating a novel way to operationalize a machine learning model by exposing its functionality through Microsoft Excel. There's a lot to cover, so let's get started.

Consuming a Python Model from a Python Client

Ostensibly, invoking a Python model from a Python client is simple: just call `predict` (or, for a classifier, `predict_proba`) on the model. Of course, you don't want to have to retrain the model every time you use it. You want to train it once, and then empower a client app to re-create the model in its trained state. For that, Python programmers use the Python pickle module (*https://oreil.ly/DgFfH*).

To demonstrate, the following code builds and trains the Titanic model featured in Chapter 3. Rather than use the model to make predictions, however, it saves the model to a *.pkl* file (it "pickles" the model) with a call to `pickle.dump` on the final line:

```
import pickle
import pandas as pd
from sklearn.linear_model import LogisticRegression

df = pd.read_csv('Data/titanic.csv')
df = df[['Survived', 'Age', 'Sex', 'Pclass']]
df = pd.get_dummies(df, columns=['Sex', 'Pclass'])
df.dropna(inplace=True)

x = df.drop('Survived', axis=1)
```

```
y = df['Survived']

model = LogisticRegression(random_state=0)
model.fit(x, y)

pickle.dump(model, open('titanic.pkl', 'wb'))
```

To invoke the model, a Python client uses `pickle.load` to deserialize the model from the *.pkl* file, *re-creating the model in its trained state*, and calls `predict_proba` to compute the odds of a passenger's survival:

```
import pickle
import pandas as pd

model = pickle.load(open('titanic.pkl', 'rb'))

female = pd.DataFrame({ 'Age': [30], 'Sex_female': [1], 'Sex_male': [0],
                       'Pclass_1': [1], 'Pclass_2': [0], 'Pclass_3': [0] })

probability = model.predict_proba(female)[0][1]
print(f'Probability of survival: {probability:.1%}')
```

Now the client can use the model to make a prediction without retraining it. And once the model is loaded, it can persist for the lifetime of the client and be called upon for predictions whenever needed.

Chapter 5 introduced Scikit's `make_pipeline` function (*https://oreil.ly/wiBqe*), which allows estimators (objects that make predictions) and transformers (objects that transform data input to a model) to be combined into a single unit, or *pipeline*. The `pickle` module can be used to serialize and deserialize pipelines too. Example 7-1 recasts the sentiment analysis model featured in Chapter 4 to use `make_pipeline` to combine a `CountVectorizer` (*https://oreil.ly/zZcAM*) for vectorizing text with a `LogisticRegression` object (*https://oreil.ly/bqimV*) for classifying text. The call to `pickle.dump` saves the model, `CountVectorizer` and all.

Example 7-1. Training and saving a sentiment analysis pipeline

```
import pickle
import pandas as pd
from sklearn.feature_extraction.text import CountVectorizer
from sklearn.linear_model import LogisticRegression
from sklearn.pipeline import make_pipeline

df = pd.read_csv('Data/reviews.csv', encoding="ISO-8859-1")
df = df.drop_duplicates()

x = df['Text']
y = df['Sentiment']

vectorizer = CountVectorizer(ngram_range=(1, 2), stop_words='english',
```

```
                            min_df=20)

    model = LogisticRegression(max_iter=1000, random_state=0)
    pipe = make_pipeline(vectorizer, model)
    pipe.fit(x, y)

    pickle.dump(pipe, open('sentiment.pkl', 'wb'))
```

A Python client can deserialize the pipeline and call `predict_proba` to score a line of text for sentiment with a few simple lines of code:

```
import pickle

pipe = pickle.load(open('sentiment.pkl', 'rb'))
score = pipe.predict_proba(['Great food and excellent service!'])[0][1]
print(score)
```

Pickling in this manner works not just with `CountVectorizer` but with other transformers as well, such as `StandardScaler`.

Pickling a pipeline containing a `CountVectorizer` can produce a large *.pkl* file. In this example, *sentiment.pkl* is 50 MB in length because it contains the entire vocabulary built by `CountVectorizer` from the training text. Remove the `min_df=20` parameter and the file swells to nearly 90 MB.

The solution is to replace `CountVectorizer` with `Hashing Vectorizer`, which doesn't create a vocabulary but instead uses word hashes to index a table of word frequencies. That reduces *sentiment.pkl* to 8 MB—without the `min_df` parameter, which `HashingVectorizer` doesn't support anyway.

If you'd like to write a standalone Python client that performs sentiment analysis, run the code in Example 7-1 in a Jupyter notebook to generate *sentiment.pkl*. Optionally, you can change `CountVectorizer` to `HashingVectorizer` and remove the `min_df` parameter to reduce the size of the *.pkl* file. Then create a Python script named *sentiment.py* containing the following code:

```
import pickle, sys

# Get the text to analyze
if len(sys.argv) > 1:
    text = sys.argv[1]
else:
    text = input('Text to analyze: ')

# Load the pipeline containing the model and the vectorizer
pipe = pickle.load(open('sentiment.pkl', 'rb'))
```

```
# Pass the input text to the pipeline and print the result
score = pipe.predict_proba([text])[0][1]
print(score)
```

Copy *sentiment.pkl* into the same directory as *sentiment.py*, and then pop out to the command line and run the script:

```
python sentiment.py "Great food and excellent service!"
```

The output should look something like this, which is proof that you succeeded in re-creating the model in its trained state and invoking it to analyze the input text for sentiment:

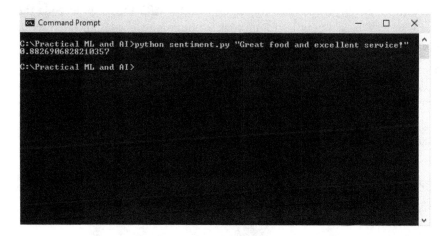

Note that the sentiment score will differ slightly if you replaced CountVectorizer with HashingVectorizer, in part due to the omission of the min_df parameter.

Versioning Pickle Files

Generally speaking, a model pickled (saved) with one version of Scikit can't be unpickled with another version. Sometimes the result is warning messages; other times, it doesn't work at all. Be sure to save models with the same version of Scikit that you use to consume them. This requires a bit of planning from an engineering perspective, because if you store serialized models in a centralized repository and update the version of Scikit used in your apps, you'll need to update the saved models too.

Which prompts some interesting questions: How do you set up a repository for machine learning models? How do you deploy models from the repository to the devices that host them? For that matter, how do you *version* those models as well as the datasets you train them with?

The answers come from a nascent field known as *MLOps*, which is short for *ML operations*. I don't cover MLOps in this book because it's a rich subject that is a book unto itself. If you want to learn more, I recommend reading *Practical MLOps: Operationalizing Machine Learning Models* by Noah Gift and Alfredo Deza (O'Reilly).

Consuming a Python Model from a C# Client

Suppose you wanted to invoke the sentiment analysis model in the previous section from an app written in another language—say, C#. You can't directly call a Python function from C#, but you *can* wrap a Python model in a web service and expose its `predict` (or `predict_proba`) method using a REST API. One way to code the web service is to use Flask, a popular framework for building websites and web services in Python.

To see for yourself, make sure Flask is installed on your computer. Then create a file named *app.py* and paste in the following code. This code uses Flask to implement a Python web service that listens on port 5000:

```python
import pickle
from flask import Flask, request

app = Flask(__name__)
pipe = pickle.load(open('sentiment.pkl', 'rb'))

@app.route('/analyze', methods=['GET'])
def analyze():
    if 'text' in request.args:
        text = request.args.get('text')
    else:
        return 'No string to analyze'

    score = pipe.predict_proba([text])[0][1]
    return str(score)

if __name__ == '__main__':
    app.run(debug=True, port=5000, host='0.0.0.0')
```

At startup, the service deserializes the pipeline comprising the sentiment analysis model in *sentiment.pkl*. The `@app.route` statement decorating the `analyze` function tells Flask to call the function when the service's `analyze` method is called. If the service is hosted locally, the following request invokes `analyze` and returns a string containing a sentiment score for the text in the query string:

```
http://localhost:5000/analyze?text=Great food and excellent service!
```

To demonstrate, go to the directory where *app.py* is located (make sure *sentiment.pkl* is there too) and start Flask by typing:

```
flask run
```

Then go to a separate command prompt and use a `curl` command to fire off a request to the URL:

```
curl -G -w "\n" http://localhost:5000/analyze --data-urlencode "text=Great food
and excellent service!"
```

Here's the output:

If you have Visual Studio or Visual Studio Code installed on your computer and are set up to compile and run C# apps, you can use the following code in a C# console app to invoke the web service and score a text string for sentiment. Of course, you're not limited to invoking the web service (and by extension, the model) from C#. Any language will do, because virtually all modern programming languages provide a means for sending HTTP requests.

```csharp
using System;
using System.Net.Http;
using System.Threading.Tasks;

class Program
{
    static async Task Main(string[] args)
    {
        string text;

        // Get the text to analyze
        if (args.Length > 0)
        {
            text = args[0];
        }
        else
        {
            Console.Write("Text to analyze: ");
            text = Console.ReadLine();
```

```
    }

        // Pass the text to the web service
        var client = new HttpClient();
        var url = $"http://localhost:5000/analyze?text={text}";
        var response = await client.GetAsync(url);
        var score = await response.Content.ReadAsStringAsync();

        // Show the sentiment score
        Console.WriteLine(score);
    }
}
```

This web service is a simple one that reads input from a query string and returns a
string. For more complex input and output, you can serialize the input into JSON and
transmit it in the body of an HTTP POST, and you can return a JSON payload in the
response. For a tutorial demonstrating how to do it in Python, see the article "Python
Post JSON Using Requests Library" (*https://oreil.ly/rOJcF*).

Containerizing a Machine Learning Model

One downside to wrapping a machine learning model in a web service and running it
locally is that the client computer must have Python installed, as well as all the packages that the model and web service require. An alternative is to host the web service
in the cloud where it can be called via the internet. It's not hard to go out to Azure or
AWS, spin up a virtual machine (VM), and install the software there. But there's a
better way. That better way is containers.

Containers (*https://oreil.ly/7AKWf*) have revolutionized the way software is built and
deployed. A container includes an app and everything the app needs to run, including a runtime (for example, Python), the packages the app relies on, and even a virtual filesystem. If you're not familiar with containers, think of them as lightweight
VMs that start quickly and consume far less memory. Docker (*https://oreil.ly/
w2mU6*) is the world's most popular container platform, although it is rapidly being
supplanted by Kubernetes (*https://oreil.ly/IRpur*).

Containers are created from *container images*, which serve as blueprints for containers in the same way that classes in object-oriented programming languages constitute
blueprints for objects. The first step in creating a Docker container image containing
the sentiment analysis model and web service is to create a file named *Dockerfile* (no
filename extension) in the same directory as *app.py* and *sentiment.pkl* and then paste
the following statements into it:

```
FROM python:3.8
RUN pip install flask numpy scipy scikit-learn && \
    mkdir /app
COPY app.py /app
COPY sentiment.pkl /app
```

```
WORKDIR /app
EXPOSE 5000
ENTRYPOINT ["python"]
CMD ["app.py"]
```

A *Dockerfile* contains instructions for building a container image. This one creates a container image that includes a Python runtime, several Python packages such as Flask and Scikit-Learn, and *app.py* and *sentiment.pkl*. It also instructs the Docker runtime that hosts the container to open port 5000 for HTTP requests and to execute *app.py* when the container starts.

There are several ways to build a container image from a *Dockerfile*. If Docker is installed on your computer, you can use a `docker build` command:

```
docker build -t sentiment-server .
```

Alternatively, you can upload the *Dockerfile* to a cloud platform such as Microsoft Azure and build it there. This prevents you from having to have Docker installed on the local machine, and it makes it easy to store the resulting container image in the cloud. Container images are stored in *container registries*, and modern cloud platforms host container registries as well as containers. If you launch a container instance in Azure, the web service in the container can be invoked with a URL similar to this one:

```
http://wintellect.northcentralus.azurecontainer.io:5000/analyze?text=Great food
and excellent service!
```

One of the benefits of hosting the container instance in the cloud is that it can be reached from any client app running on any machine and any operating system, and the computer that hosts the app needs nothing special installed. Containers can be beneficial even if you host the web service locally rather than in the cloud. As long as you deploy a container stack such as the Docker runtime to the local machine, you don't have to install Python and all the packages that the web service requires. You just launch a container instance and direct HTTP requests to it via localhost.

Using ONNX to Bridge the Language Gap

Is it possible to bridge the gap between a C# client and a Python ML model without using a web service as a middleman? In a word, yes! The solution lies in a four-letter acronym: *ONNX*. As mentioned earlier, ONNX stands for Open Neural Network Exchange, and it was originally devised to allow deep-learning models written with one framework—for example, TensorFlow—to be used with other frameworks such as PyTorch. But today it can be used with Scikit models too. I'll say more about ONNX in Chapter 12, but for now, let's use it to call a Python model *directly* from an app written in C#—no web service required.

The first step in using ONNX to bridge the language gap is to install Skl2onnx (*https://oreil.ly/LIOiY*) in your Python environment. Then use that package's `convert_sklearn` method to save a trained Scikit model to a *.onnx* file. Here's a short code snippet that saves the sentiment analysis model in Example 7-1 to a file named *sentiment.onnx*. The `initial_types` parameter specifies that the model expects one input value: a string containing the text to score for sentiment:

```
from skl2onnx import convert_sklearn
from skl2onnx.common.data_types import StringTensorType

initial_type = [('string_input', StringTensorType([None, 1]))]
onnx = convert_sklearn(pipe, initial_types=initial_type)

with open('sentiment.onnx', 'wb') as f:
    f.write(onnx.SerializeToString())
```

If you wish to consume the model from Python, you first install a Python package named Onnxruntime (*https://oreil.ly/aEax9*) containing the ONNX runtime, also known as the ORT. This provides support for loading and running ONNX models. Then you call the runtime's `InferenceSession` method with a path to a *.onnx* file to deserialize the model, and call `run` on the returned session object to call the model's `predict` or `predict_proba` method. Here's how it looks in code:

```
import numpy as np
import onnxruntime as rt

session = rt.InferenceSession('sentiment.onnx')
input_name = session.get_inputs()[0].name
label_name = session.get_outputs()[1].name # 0 = predict, 1 = predict_proba

input = np.array('Great food and excellent service!').reshape(1, -1)
score = session.run([label_name], { input_name: input })[0][0][1]
print(score)
```

Note that the string input to the model is passed as a NumPy array. The value returned from the `run` method reveals the sentiment score returned by `predict_proba`. If you'd rather call `predict` to predict a class, change `session.get_outputs()[1].name` to `session.get_outputs()[0].name`.

> In real life, you wouldn't load the model every time you call it. You'd load it once, allow it to persist for the lifetime of the client, and call it whenever you want to make a prediction.

Of course, the whole point of this discussion is to call the model from C# instead of Python. That's not difficult either, thanks to a NuGet package from Microsoft called Microsoft.ML.OnnxRuntime (*https://oreil.ly/UYHqr*). You can install it from the command line or using an integrated development environment such as Visual Studio. Then it's a relatively simple matter to write a C# console app that re-creates the trained sentiment analysis model from the *.onnx* file and calls it to score a text string:

```csharp
using System;
using Microsoft.ML.OnnxRuntime;
using Microsoft.ML.OnnxRuntime.Tensors;
using System.Collections.Generic;
using System.Linq;

class Program
{
    static void Main(string[] args)
    {
        string text;

        // Get the text to analyze
        if (args.Length > 0)
        {
            text = args[0];
        }
        else
        {
            Console.Write("Text to analyze: ");
            text = Console.ReadLine();
        }


        // Create the model and pass the text to it
        var tensor = new DenseTensor<string>(new string[]
            { text }, new int[] { 1, 1 });

        var input = new List<NamedOnnxValue>
        {
            NamedOnnxValue.CreateFromTensor<string>("string_input", tensor)
        };

        var session = new InferenceSession("sentiment.onnx");
        var output = session.Run(input)
            .ToList().Last().AsEnumerable<NamedOnnxValue>();

        var score = output.First().AsDictionary<Int64, float>()[1];

        // Show the sentiment score
        Console.WriteLine(score);
    }
}
```

This code assumes that *sentiment.onnx* is present in the current directory. Instantiating the `InferenceSession` object creates the model, and calling `Run` on the object invokes the model. Under the hood, `Run` calls the model's predict and `predict_proba` methods and makes the results of both available. There's a little work required to convert C# types into ONNX types and vice versa, but once you get the hang of it, it's pretty remarkable that you can call a machine learning model written in Python directly from C#.

Here's another example. Suppose you wanted to consume Chapter 2's taxi-fare regression model in C#. Recall that the model accepts three floating-point values as input—the day of the week (0–6), the hour of day (0–23), and the distance to travel in miles—and returns a predicted taxi fare. The following Python code saves the trained model in an ONNX file:

```python
from skl2onnx import convert_sklearn
from skl2onnx.common.data_types import FloatTensorType

initial_type = [('float_input', FloatTensorType([None, 3]))]
onnx = convert_sklearn(model, initial_types=initial_type)

with open('taxi.onnx', 'wb') as f:
    f.write(onnx.SerializeToString())
```

The following C# code loads the model and makes a prediction. The big difference between this and the previous example is that you pass the model an array of three floating-point values rather than a string:

```csharp
// Package the input
var input = new float[] {
    4.0f,  // Day of week
    17.0f, // Pickup time (hour of day)
    2.0f   // Distance to travel
};

var tensor = new DenseTensor<float>(input, new int[] { 1, 3 });

// Create the model and pass the input to it
var session = new InferenceSession("taxi.onnx");

var output = session.Run(new List<NamedOnnxValue>
{
    NamedOnnxValue.CreateFromTensor("float_input", tensor)
});

var score = output.First().AsTensor<float>().First();

// Show the predicted fare
Console.WriteLine($"{score:#.00}");
```

So which is faster? Calling a machine learning model wrapped in a web service, or calling the same model using ONNX? I wrote a simple test harness to answer that question. On my computer, calling the sentiment analysis model in a Flask web service running locally required slightly more than 2 seconds per round trip. Calling the same model through the Python ONNX runtime took 0.001 seconds on average. That's a difference of more than three orders of magnitude. And you would incur additional latency if the web service was hosted on a remote server.

Significantly, ONNX isn't limited to C# and Python. ONNX runtimes are available for Python, C, C++, C#, Java, JavaScript, and Objective-C, and they run on Windows, Linux, macOS, Android, and iOS. When it comes to projecting machine learning models written in Python to other platforms and languages, ONNX is a game changer. For more information, check out the ONNX runtime website (*https://oreil.ly/mnSO5*).

Building ML Models in C# with ML.NET

Scikit-Learn is arguably the world's most popular machine learning framework. The efficacy of the library, the documentation that accompanies it, and the mindshare that surrounds it are the primary reasons more ML models are written in Python than any other language. But Scikit isn't the only machine learning framework. Others exist for other languages, and if you can code an ML model in the same programming language as the client that consumes it, you can avoid jumping through hoops to operationalize the model.

ML.NET (*https://oreil.ly/YhIiZ*) is Microsoft's free, open source, cross-platform machine learning library for .NET developers. It does most of what Scikit does and a few things that Scikit doesn't do. And when it comes to writing ML/AI solutions in C#, there is no better tool for the job.

ML.NET derives from an internal library that was developed by Microsoft—and has been used in Microsoft products—for more than a decade. The ML algorithms that it implements have been tried and tested in the real world and tuned to optimize performance and accuracy. Because ML.NET is consumed from C#, you get all the benefits of a compiled programming language, including type safety and fast execution.

A paper published by the ML.NET team (*https://oreil.ly/ixrrY*) at Microsoft in 2019 discusses the motivations and design goals behind ML.NET. It also compares ML.NET's accuracy and performance to that of Scikit-Learn and another machine learning framework named H2O (*https://oreil.ly/vDbtD*). Using a 9 GB Amazon review dataset, ML.NET trained a sentiment analysis model to 95% accuracy. Neither Scikit nor H2O could process the dataset due to its size. When all three frameworks were trained on 10% of the dataset, ML.NET achieved the highest accuracy, and trained six times faster than Scikit and almost 10 times faster than H2O.

ML.NET is compatible with Windows, Linux, and macOS. Thanks to an innovation called IDataView, it can handle datasets of virtually unlimited size. While it can't be used to build neural networks from scratch, it does have the ability to load existing neural networks and use a technique called *transfer learning* to repurpose those networks to solve domain-specific problems. (Transfer learning will be covered in Chapter 10.) It also builds in ONNX support that allows it to load sophisticated models, including deep-learning models, stored in ONNX format. Finally, it can be consumed in Python and even combined with Scikit-Learn using a set of Python bindings called NimbusML (*https://oreil.ly/CsG6N*).

If you're a .NET developer who is interested in machine learning, there has never been a better time to get acquainted with ML.NET. This section isn't meant to provide an exhaustive treatment of ML.NET but to introduce it, show the basics of building ML models with it, and hopefully whet your appetite enough to motive you to learn more. There are plenty of great resources available online, including the official ML.NET documentation (*https://oreil.ly/ay04E*), a GitHub repo containing ML.NET samples (*https://oreil.ly/IjAw7*), and the ML.NET cookbook (*https://oreil.ly/06YMb*).

Sentiment Analysis with ML.NET

The following C# code uses ML.NET to build and train a sentiment analysis model. It's equivalent to the Python implementation in Example 7-1:

```
var context = new MLContext(seed: 0);

// Load the data
var data = context.Data.LoadFromTextFile<Input>("reviews.csv",
    hasHeader: true, separatorChar: ',', allowQuoting: true);

// Split the data into a training set and a test set
var trainTestData = context.Data.TrainTestSplit(data,
    testFraction: 0.2, seed: 0);
var trainData = trainTestData.TrainSet;
var testData = trainTestData.TestSet;

// Build and train the model
var pipeline = context.Transforms.Text.FeaturizeText
    (outputColumnName: "Features", inputColumnName: "Text")
    .Append(context.BinaryClassification.Trainers.SdcaLogisticRegression());

var model = pipeline.Fit(trainData);

// Evaluate the model
var predictions = model.Transform(testData);
var metrics = context.BinaryClassification.Evaluate(predictions);
Console.WriteLine($"AUC: {metrics.AreaUnderPrecisionRecallCurve:P2}");

// Score a line of text for sentiment
var predictor = context.Model.CreatePredictionEngine<Input, Output>(model);
var input = new Input { Text = "Among the best movies I have ever seen"};
var prediction = predictor.Predict(input);

// Show the score
Console.WriteLine($"Sentiment score: {prediction.Probability}");
```

Every ML.NET app begins by creating an instance of the MLContext class (*https://oreil.ly/1qkYh*). The optional seed parameter (*https://oreil.ly/CsKBJ*) initializes the random-number generator used by ML.NET so that you get repeatable results from one run to the next. MLContext exposes a number of properties through which large parts of the ML.NET API are accessed. One example of this is the call to LoadFrom TextFile (*https://oreil.ly/zzzIB*), which is a DataOperationsCatalog method (*https://oreil.ly/EkbKU*) accessed through MLContext's Data property (*https://oreil.ly/rsAOo*).

LoadFromTextFile is one of several methods ML.NET provides for loading data from text files, databases, and other data sources. It returns a *data view*, which is an object that implements the IDataView interface (*https://oreil.ly/Q0k2S*). Data views in ML.NET are similar to DataFrames in Pandas, with one important difference: whereas DataFrames have to fit in memory, data views do not. Internally, data views use a SQL-like cursor to access data. This means they can wrap a theoretically unlimited amount of data. That's why ML.NET was able to process the entire 9 GB Amazon dataset, while Scikit and H2O were not.

After loading the data and splitting it for training and testing, the preceding code creates a pipeline containing a TextFeaturizingEstimator object (*https://oreil.ly/D2Eql*)—created with the FeaturizeText method (*https://oreil.ly/pnKic*)—and an SdcaLogisticRegressionBinaryTrainer object (*https://oreil.ly/6yMxg*)—created with the SdcaLogisticRegression method (*https://oreil.ly/4qn9u*). This is analogous in Scikit to creating a pipeline containing a CountVectorizer object for vectorizing input text and a LogisticRegression object for fitting a model to the data. Calling Fit on the pipeline trains the model, just like calling fit in Scikit. It's no coincidence that ML.NET employs some of the same patterns as Scikit. This was done intentionally to impart a sense of familiarity to programmers who use Scikit.

After evaluating the model's accuracy by computing the area under the precision-recall curve, a call to ModelOperationsCatalog.CreatePredictionEngine (*https://oreil.ly/nFQoT*) creates a prediction engine whose Predict method (*https://oreil.ly/HOGQp*) makes predictions. Unlike Scikit, which has you call predict on the estimator itself, ML.NET encapsulates prediction capability in a separate object, in part so that multiple prediction engines can be created to achieve scalability in high-traffic scenarios.

In this example, Predict accepts an Input object as input and returns an Output object. One of the benefits of building models with ML.NET is strong typing. Load FromTextFile is a generic method that accepts a class name as a type parameter—in this case, Input. Similarly, CreatePredictionEngine uses type parameters to specify schemas for input and output. The Input and Output classes are application specific and in this instance are defined as follows:

```
public class Input
{
    [LoadColumn(0)]
    public string Text;

    [LoadColumn(1), ColumnName("Label")]
    public bool Sentiment;
}

public class Output
{
    [ColumnName("PredictedLabel")]
    public bool Prediction { get; set; }

    public float Probability { get; set; }
}
```

The LoadColumn attributes (*https://oreil.ly/d32Bc*) map columns in the data file to properties in the Input class. Here, they tell ML.NET that values for the Text field come from column 0 in the input file, and values for Sentiment (the 1s and 0s indicating whether the sentiment expressed in the text is positive or negative) come from column 1. The ColumnName("Label") attribute identifies the second column as the label column—the one containing the values that the model will attempt to predict.

The Output class defines the output schema. In this example, it contains properties named Prediction and Probability, which, following a prediction, hold the predicted label (0 or 1) and the probability that the sample belongs to the positive class, which doubles as a sentiment score. The ColumnName("PredictedLabel") attribute maps the value returned by Predict to the Output object's Prediction property.

> There is nothing magic about the class names Input and Output. You could name them SentimentData and SentimentPrediction and the code would work just the same.

Saving and Loading ML.NET Models

Earlier, you learned how to use Python's pickle module to save and load trained models. You do the same in ML.NET by calling ModelOperationsCatalog.Save (*https://oreil.ly/dIGXI*) and ModelOperationsCatalog.Load (*https://oreil.ly/PKo3g*) through the MLContext object's Model property (*https://oreil.ly/yp9JL*):

```
// Save a trained model to a zip file
context.Model.Save(model, data.Schema, "model.zip");

// Load a trained model from a zip file
var model = context.Model.Load("model.zip", out DataViewSchema schema);
```

This enables clients to re-create a model in its trained state and use it to make predictions without having to train the model repeatedly.

Adding Machine Learning Capabilities to Excel

Want to see a novel way to operationalize a machine learning model? Imagine that you're a software developer at an internet vacation rentals firm. The company's communications department has asked you to create a spreadsheet that lets them analyze text for sentiment. The idea is that if sentiment toward the company turns negative on social media, the communications team can get out in front of it.

You already know how to build, train, and save a sentiment analysis model in Python using Scikit. Excel supports user-defined functions (UDFs) (*https://oreil.ly/8Pv4R*), which enable users to write custom functions that are called just like SUM(), AVG(), and other functions built into Excel. But UDFs are written in Visual Basic for Applications (VBA) (*https://oreil.ly/cJt6d*), not Python. To marry Scikit with Excel, you need to write UDFs in Python.

Fortunately, there are libraries that let you do just that. One of them is Xlwings (*https://oreil.ly/uSmSL*), an open source library that combines the power of Excel with the versatility of Python. With it, you can write Python code that loads or creates Excel spreadsheets and manipulates their content, write Python macros triggered by button clicks in Excel, access Excel spreadsheets from Jupyter notebooks, and more. You can also use Xlwings to write Python UDFs for Excel for Windows.

The first step in building the spreadsheet the communications team wants is to configure Excel to trust VBA add-ins. Launch Microsoft Excel and use the File → Options command to open the Excel Options dialog. Click Trust Center in the menu on the left, and then click the Trust Center Settings button. Click Macro Settings on the left, and check the "Trust access to the VBA project object model" box, as shown in Figure 7-2. Then click OK to dismiss the Trust Center dialog, followed by OK to dismiss the Excel Options dialog.

The next step is to install Xlwings on your computer using a pip install xlwings command or equivalent for your Python environment. Afterward, go to the command line and use the following command to install the Xlwings add-in in Excel:

```
xlwings addin install
```

Now create a project directory on your computer and cd to it. Then execute the following command:

```
xlwings quickstart sentiment
```

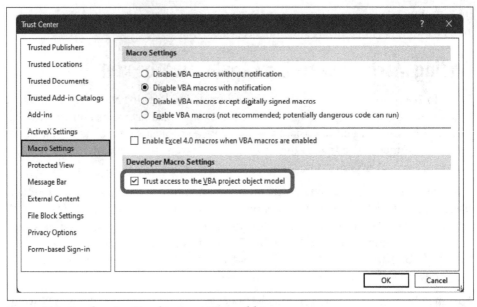

Figure 7-2. Configuring Excel to trust VBA add-ins

This command creates a subdirectory named *sentiment* in the current directory and initializes it with a pair of files: a spreadsheet named *sentiment.xlsm* and a Python file named *sentiment.py*. It is the latter of these in which you will write a UDF that analyzes text for sentiment.

Next, copy *sentiment.pkl* into the *sentiment* directory. (If you didn't generate that file earlier, run the code in Example 7-1 in a Jupyter notebook to generate it now.) Then open *sentiment.py* in your favorite text editor and replace its contents with the following code. This code loads the sentiment analysis model from the *.pkl* file and stores a reference to the model in the variable named model. Then it defines a UDF named analyze_text that can be called from Excel. analyze_text vectorizes the text passed to it and inputs it to the model's predict_proba method. That method returns a number from 0.0 to 1.0, with 0.0 representing negative sentiment and 1.0 representing positive sentiment. When you're done, save your changes to *sentiment.py*. The UDF is written. Now it's time to call it from Excel.

```
import pickle, os
import xlwings as xw

# Load the model and the vocabulary and create a CountVectorizer
model_path = os.path.abspath(os.path.join(os.path.dirname(__file__),
                             'sentiment.pkl'))

model = pickle.load(open(model_path, 'rb'))
```

```
@xw.func
def analyze_text(text):
    score = model.predict_proba([text])[0][1]
    return score
```

The elegance of Xlwings is that once you've written a UDF such as `analyze_text`, you can call it the same way you call functions built into Excel. But first you must import the UDF. To do that, open *sentiment.xlsm* in Excel. Go to the "xlwings" tab and click Import Functions to import the `analyze_text` function, as shown in Figure 7-3. Excel doesn't tell you if the import was successful, but it does let you know if the import failed—if, for example, you forgot to copy *sentiment.pkl* into the directory where *sentiment.py* is stored.

Figure 7-3. Importing the `analyze_text` *function*

Type **Great food and excellent service** into cell A1 (Figure 7-4).

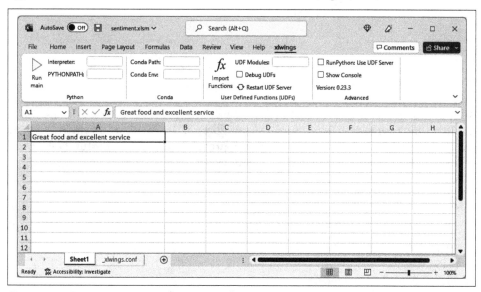

Figure 7-4. Entering a string of text to analyze

Type the following expression into cell B1 to pass the text in cell A1 to the analyze_text function imported from *sentiment.py*:

```
=analyze_text(A1)
```

Confirm that a number from 0.0 to 1.0 appears in cell B1, as shown in Figure 7-5. This is the score that the machine learning model assigned to the text "Great food and excellent service." Do you agree with the score? Finish up by entering some text strings of your own to see how they score for sentiment.

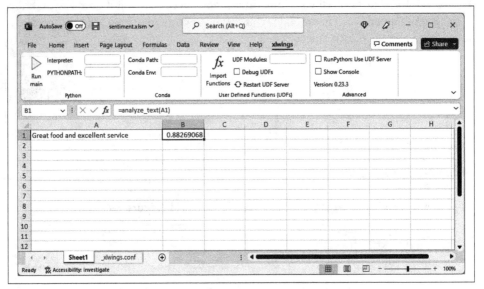

Figure 7-5. Sentiment analysis in Excel

UDFs written in Python present Excel users with a new whole new world of possibilities thanks to the rich ecosystem of Python libraries available for machine learning, statistical analysis, and other tasks. And they provide a valuable opportunity for Excel users to operationalize machine learning models written in Python.

Summary

Writing a Python client that invokes a Python machine learning model requires little more than an extra line of code to deserialize the model from a *.pkl* file. One way for a non-Python client to invoke a Python model is to wrap the model in a Python web service and invoke the model using REST APIs. The web service can be hosted locally or in the cloud, and containerizing the web service (and the model) simplifies deployment and makes the software more portable.

An alternative approach is to use ONNX to bridge the language gap. With ONNX, you can save a Scikit model to a *.onnx* file and load the model from a variety of programming languages, including C, C++, C#, Java, and JavaScript. Once loaded, the model can be invoked just as if it were called from Python.

Another option for invoking machine learning models from non-Python clients is to write the model in the same language as the client. Microsoft's ML.NET, which is free, cross-platform, and open source, is a great option for C# developers. Other libraries include Java-ML (*https://oreil.ly/7NDrq*) for Java developers and Caret (*https://oreil.ly/SB6Hn*) and Tidymodels (*https://oreil.ly/KcMVW*) for R developers. The APIs supported by these libraries are different from the APIs in Scikit, but the principles embodied in them are the same.

Deep Learning with Keras and TensorFlow

## CHAPTER 8

Deep Learning

Every model in Part I of this book employed classic machine learning algorithms that form the core of ML itself: logistic regression, random forests, and so on. Such models are often referred to as *traditional* machine learning models to differentiate them from deep-learning models. Recall from Chapter 1 that deep learning is a subset of machine learning that relies primarily on neural networks, and that most of what's considered AI today is accomplished with deep learning. From recognizing objects in photos to real-time speech translation to using computers to generate art, music, poetry, and photorealistic faces (*https://oreil.ly/2cDLY*), deep learning allows computers to perform feats that traditional machine learning does not.

I frequently introduce deep learning to software developers by challenging them to devise an algorithmic means for determining whether a photo contains a dog. If they offer a solution, I'll counter with a dog picture that foils the algorithm. Traditional ML models can partially solve the problem, but when it comes to recognizing objects in images, deep learning represents the state of the art. It's not terribly difficult to train a neural network to recognize dog pictures, sometimes more accurately than humans. Once you learn how to do that, it's a small step forward to recognizing defective parts coming off an assembly line or bicycles passing in front of a self-driving car.

Neural networks have been around for decades, but it's only in the past 10 years or so that sufficient compute power has been available to train sophisticated networks. Cutting-edge neural networks are trained on graphics processing units (GPUs) (*https://oreil.ly/25jZk*) and tensor processing units (TPUs) (*https://oreil.ly/f9pV6*), often attached to high-performance computing clusters. GPUs are great for gaming because they deliver high-performance graphics. They are also efficient parallel processing machines that allow data scientists to train neural networks in a fraction of the time required on ordinary CPUs. Today, any researcher with a credit card can purchase an NVIDIA GPU or spin up GPUs in Azure or AWS and have access to compute power that researchers 20 years ago could only have dreamed of. This, more than anything else, has driven AI's resurgence and precipitated continual advances in the state of the art.

This chapter is the first of several focused on deep learning. In it, you'll learn:

- What a neural network is and where the "deep" in deep learning comes from
- How a neural network transforms input into output using simple mathematical operations
- What happens when a neural network is trained, as well as the challenges that training entails

You won't start building and training neural networks just yet; that begins in Chapter 9. Before you build a house, you need a foundation to build upon. That foundation begins right now.

Understanding Neural Networks

Neural networks come in many varieties. Convolutional neural networks (CNNs) (*https://oreil.ly/Kdts5*), for example, excel at computer-vision tasks such as classifying images. Recurrent neural networks (RNNs) (*https://oreil.ly/0nUec*) find application in handwriting recognition and natural language processing (NLP), while generative adversarial networks (*https://oreil.ly/LL9Wt*), or GANs, enable computers to create art, music, and other content. But the first step in wrapping your head around deep learning is to understand what a neural network is and how it works.

The simplest type of neural network is the *multilayer perceptron* (*https://oreil.ly/KqFX1*). It consists of nodes or *neurons* arranged in layers. The depth of the network is the number of layers; the width is the number of neurons in each layer, which can be different for every layer. State-of-the-art neural networks sometimes contain 100 or more layers and thousands of neurons in individual layers. A *deep neural network* is one that contains many layers, and it's where the term *deep learning* is derived from.

The multilayer perceptron in Figure 8-1 contains three layers: an input layer with two neurons, a middle layer (also known as a *hidden layer*) with three neurons, and an output layer with one neuron. Because the input layer is often ignored when counting layers, some would argue that this network contains *two* layers, not three. Regardless, the network's job is to take two floating-point values as input and produce a single floating-point number as output. Neural networks work with floating-point numbers. They *only* work with floating-point numbers. As with traditional machine learning models, a neural network can only process non-numeric data—for example, text strings—if the data is first converted to numbers.

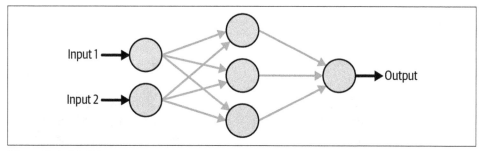

Figure 8-1. Multilayer perceptron

The orange arrows in Figure 8-1 represent connections between neurons. Each neuron in each layer is connected to each neuron in the next layer, giving rise to the term *fully connected layers*. Each connection is assigned a *weight*, which is typically a small floating-point number. In addition, each neuron outside the input layer is assigned a *bias*, which is also a small floating-point number. Figure 8-2 shows a set of weights and biases that enable the network to sum two inputs (for example, to add 2 and 2). The blocks labeled "ReLU" represent *activation functions* (*https://oreil.ly/Q4wqT*), which apply simple nonlinear transforms to values propagated through the network. The most commonly used activation function is the *rectified linear units* (ReLU) function (*https://oreil.ly/KdPE0*), which passes positive numbers through unchanged while converting negative numbers to 0s. Without activation functions, neural networks would struggle to model nonlinear data. And it's no secret that real-world data tends to be nonlinear.

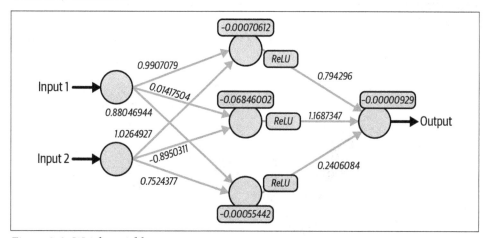

Figure 8-2. Weights and biases

Neurons perform simple linear transformations on data input to them. For a neuron with a single input x, the neuron's value y is computed by multiplying x by the weight m assigned to the input and adding b, the neuron's bias:

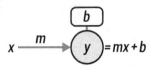

$$x \xrightarrow{m} \left(y\right) = mx + b$$

Look familiar? That's the equation for linear regression. Scikit-Learn has a `Perceptron` class (*https://oreil.ly/HLCh9*) that models this behavior and can be used to build neural linear regression models. It even offers classes named `MLPRegressor` (*https://oreil.ly/vJUCI*) and `MLPClassifier` (*https://oreil.ly/XY5GE*) for building simple multilayer perceptrons. Scikit is not, however, a deep-learning library. Real deep-learning libraries do more to support advanced neural networks.

 The combination of neurons that perform linear transformations and activation functions that apply nonlinear transforms is an embodiment of the universal approximation theorem (*https://oreil.ly/xfDMb*), which states that you can approximate any function f by summing the output from linear functions and transforming it with a nonlinear function. Textbooks often say that activation functions "add nonlinearity" to neural networks. Now you know why.

To turn inputs into outputs, a neural network assigns the input values to the neurons in the input layer. Then it multiplies the values of the input neurons by the weights connecting them to the neurons in the next layer, sums the inputs for each neuron, and adds the biases. It repeats this process to propagate values from left to right all the way to the output layer. Figure 8-3 shows what happens in the first two layers when the network in Figure 8-2 adds 2 and 2.

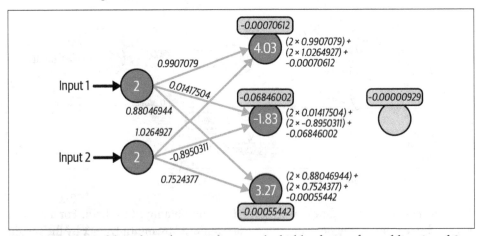

Figure 8-3. Flow of data from the input layer to the hidden layer when adding 2 and 2

Values propagate from the hidden layer to the output layer the same way, with one exception: they are transformed by an activation function before they're multiplied by weights. Remember that the ReLU activation function turns negative numbers into 0s. In Figure 8-4, the −1.83 calculated for the middle neuron in the hidden layer is converted to 0 when forwarded to the output layer, effectively eliminating that neuron's contribution to the output.

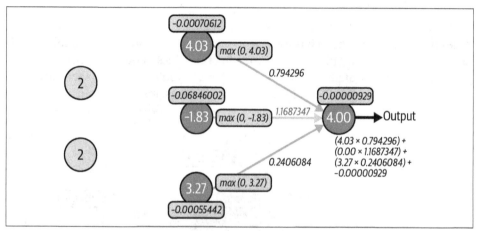

Figure 8-4. Flow of data from the hidden layer to the output layer when adding 2 and 2

Given a set of weights and biases, it isn't difficult to code a neural network by hand. The following Python code models the network in Figure 8-2:

```python
# Weights
w0 = 0.9907079
w1 = 1.0264927
w2 = 0.01417504
w3 = -0.8950311
w4 = 0.88046944
w5 = 0.7524377
w6 = 0.794296
w7 = 1.1687347
w8 = 0.2406084

# Biases
b0 = -0.00070612
b1 = -0.06846002
b2 = -0.00055442
b3 = -0.00000929

def relu(x):
    return max(0, x)

def predict(x1, x2):
    h1 = (x1 * w0) + (x2 * w1) + b0
```

```
h2 = (x1 * w2) + (x2 * w3) + b1
h3 = (x1 * w4) + (x2 * w5) + b2
y = (relu(h1) * w6) + (relu(h2) * w7) + (relu(h3) * w8) + b3
return y
```

If you'd like to see for yourself, paste the code into a Jupyter notebook and call the predict function with the inputs 2 and 2. The answer should be very close to the actual sum of 2 and 2.

For a given problem, there is an *infinite combination of weights and biases* that produces the desired outcome. Figure 8-5 shows the same network with a completely different set of weights and biases. Yet, if you plug the values into the preceding code (or propagate values through the network by hand), you'll find that the network is equally capable of adding 2 and 2—or other small values, for that matter.

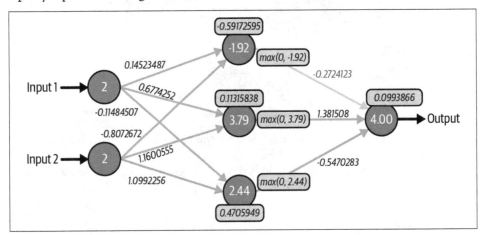

Figure 8-5. Adding 2 and 2 with a different set of weights and biases

Given a set of weights and biases, using a neural network to make predictions is simplicity itself. It's little more than multiplication and addition. But coming up with a set of weights and biases to begin with is a challenge. It's why neural networks must be *trained*.

Training Neural Networks

Training a traditional machine learning model fits it to a dataset. Neural networks require training too, and it is during training that weights and biases are calculated. Weights are typically initialized with small random numbers. Biases are usually initialized with 0s. In its untrained state, a neural network can do little more than generate random outputs. Once training is complete, the weights and biases enable the network to distinguish dogs from cats, translate a book review to another language, or do whatever else it was designed to do.

What happens when a neural network is trained? At a high level, training samples are fed through the network, the error (the difference between the computed output and the correct output) is computed using a *loss function*, and a *backpropagation algorithm* (*https://oreil.ly/ScACh*) goes backward through the network adjusting the weights and biases (Figure 8-6). This is done repeatedly until the error is sufficiently small. With each iteration, the weights and biases become incrementally more refined and the error commensurately smaller.

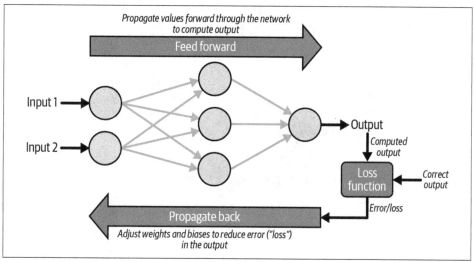

Figure 8-6. Adjusting weights and biases during training

The most critical component of the backpropagation regimen is the *optimizer*, which on each backward pass decides how much and in which direction, positive or negative, to adjust the weights and biases. Data scientists work constantly to find better and more efficient optimizers to train networks more accurately and in less time.

Do a search on "neural networks" and you'll turn up lots of articles with lots of complex math. Most of the math is related to optimization. An optimizer can't just *guess* how to adjust the weights and biases due to their sheer numbers. A neural network containing two hidden layers with 1,000 neurons each has 1,000,000 connections between layers, and therefore 1,000,000 weights to adjust. Training would take forever if the optimization strategy were simply randomly guessing. An optimizer must be intelligent enough to make adjustments that reduce the error in each successive iteration.

Data scientists use plots like the one in Figure 8-7 to visualize what optimizers do. The plot is called a *loss landscape*. It has been reduced to three dimensions for visualization purposes, but in reality, it contains *many* dimensions—sometimes millions of them. The multicolored contour charts the error for different combinations of weights and biases. The optimizer's goal is to navigate the contour and find the combination that produces the least error, which corresponds to the lowest point, or *global minimum*, in the loss landscape.

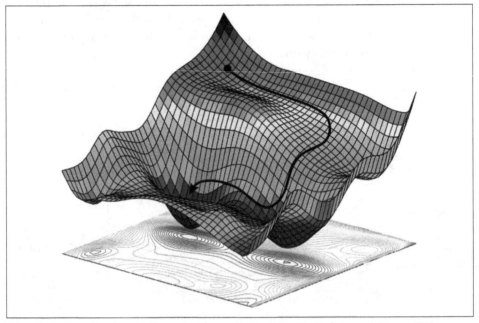

Figure 8-7. Loss landscape (Source: Alexander Amini, Ava Soleimany, Sertac Karaman, and Daniela Rus, "Spatial Uncertainty Sampling for End-to-End Control," NeurIPS Bayesian Deep Learning [2017], https://arxiv.org/pdf/1805.04829)

The optimizer's job isn't an easy one. It involves partial derivatives (calculating the slope of the contour with respect to each weight and bias), gradient descent (adjusting the weights and biases to go *down* the slope rather than up it or sideways), and *learning rates*, which drive the fractional adjustments made to the weights and biases in each backpropagation pass. If the learning rate is too great, the optimizer might miss the global minimum. If it's too small, the network will take a long time to train. Modern optimizers use *adaptive learning rates* that take smaller steps as they approach a minimum, where the slope of the contour is 0. To complicate matters, the optimizer must avoid getting trapped in local minima so that it can continue traversing the contour toward the global minimum where the error is the smallest. It also has to be wary of "saddle points" where the slope increases in one direction but falls off in a perpendicular direction.

If you research gradient descent, you'll encounter terms such as *stochastic gradient descent* (SGD) (*https://oreil.ly/Cawlk*) and *mini-batch gradient descent* (MBGD) (*https://oreil.ly/TLiCP*). Optimization via gradient descent is an iterative process in which samples are fed forward through the network, gradients are computed, and the gradients are combined with the learning rate to update weights and biases. Updating the weights and biases after every sample is fed forward through the network is computationally expensive, so training typically involves running batches of perhaps 30 to 40 samples through the network, averaging the error, and *then* performing a backpropagation pass. That's MBGD. It speeds training and helps the optimizer bypass local minima. For more information, and for a very readable introduction to the challenges inherent to training neural networks, see the article "How Neural Networks Are Trained" (*https://oreil.ly/HXX57*).

Neural networks are fundamentally simple. Training them is mathematically complex. Fortunately, you don't have to understand everything that happens during training in order to build them. Deep-learning libraries such as Keras (*https://oreil.ly/7K0Ko*) and TensorFlow (*https://oreil.ly/dWamB*) insulate you from the math and provide cutting-edge optimizers to do the heavy lifting. But now when you use one of these libraries and it asks you to pick a loss function and an optimizer, you'll understand what it's asking for and why.

Summary

Deep learning is a subset of machine learning that relies on deep neural networks, and it is the root of modern AI. It's how computers identify objects in images, translate text and speech into other languages, generate artwork and music, and perform other tasks that were virtually impossible a few years ago.

The multilayer perceptron is a simple neural network comprising layers of neurons. Each neuron turns input into output using a simple mathematical formula. Activation functions further transform the data as it passes between layers by introducing nonlinearities, enabling neural networks to fit to a variety of datasets. Hidden layers between the input layer and the output layer perform the bulk of the computational work, and a multilayer perceptron with many hidden layers is referred to as a deep neural network.

Training a neural network fits it to a dataset by iteratively adjusting weights and biases—the weights connecting neurons in adjacent layers and the biases assigned to the neurons themselves—to produce the desired outcome. The backpropagation passes that adjust the weights and biases are the heart of the training regimen. The component responsible for making adjustments is the optimizer; its ultimate goal is to find the optimum combination of weights and biases with as few backpropagation passes as possible.

Now that you understand how neural networks work, the next step is to learn how to build and train them. For that, data scientists rely on frameworks such as Keras and TensorFlow. Chapter 9 begins a deep dive into both.

Neural Networks

Machine learning isn't hard when you have a properly engineered dataset to work with. The *reason* it's not hard is libraries such as Scikit-Learn and ML.NET, which reduce complex learning algorithms to a few lines of code. Deep learning isn't difficult either, thanks to libraries such as the Microsoft Cognitive Toolkit (CNTK) (*https://oreil.ly/3jPye*), Theano (*https://oreil.ly/l4V0X*), and PyTorch (*https://oreil.ly/grwxl*). But the library that most of the world has settled on for building neural networks is TensorFlow (*https://oreil.ly/xuxYG*), an open source framework created by Google that was released under the Apache License 2.0 in 2015.

TensorFlow isn't limited to building neural networks. It is a framework for performing fast mathematical operations at scale using *tensors*, which are generalized arrays. Tensors can represent scalar values (0-dimensional tensors), vectors (1D tensors), matrices (2D tensors), and so on. A neural network is basically a workflow for transforming tensors. The three-layer perceptron featured in Chapter 8 takes a 1D tensor containing two values as input, transforms it into a 1D tensor containing three values, and produces a 0D tensor as output. TensorFlow lets you define directed graphs that in turn define how tensors are computed. And unlike Scikit, it supports GPUs.

The learning curve for TensorFlow is rather steep. Another library, named Keras (*https://oreil.ly/goUK8*), provides a simplified Python interface to TensorFlow and has emerged as the Scikit of deep learning. Keras is all about neural networks. It began life as a standalone project in 2015 but was integrated into TensorFlow in 2019. Any code that you write using TensorFlow's built-in Keras module ultimately executes in (and is optimized for) TensorFlow. Even Google recommends using the Keras API.

Keras offers two APIs for building neural networks: a sequential API (*https://oreil.ly/6C1Mp*) and a functional API (*https://oreil.ly/2vOt2*). The former is simpler and is sufficient for most neural networks. The latter is useful in more advanced scenarios such as networks with multiple inputs or outputs—for example, a classification output and a regression output, which is common in neural networks that perform object detection—or shared layers. Most of the examples in this book use the sequential API. If curiosity compels you to learn more about the functional API, see "How to Use the Keras Functional API for Deep Learning" (*https://oreil.ly/QQJ5M*) by Jason Brownlee for a very readable introduction.

Building Neural Networks with Keras and TensorFlow

Creating a neural network using Keras's sequential API is simple. You first create an instance of the Sequential class (*https://oreil.ly/23D4a*). Then you call add (*https://oreil.ly/Zx0me*) on the Sequential object to add layers. The layers are instances of classes such as Dense (*https://oreil.ly/lTzcp*), which represents a fully connected layer with a specified number of neurons. The following statements create the three-layer network featured in Chapter 8:

```
from tensorflow.keras.models import Sequential
from tensorflow.keras.layers import Dense

model = Sequential()
model.add(Dense(3, activation='relu', input_dim=2))
model.add(Dense(1))
```

This network contains an input layer with two neurons, a hidden layer with three neurons, and an output layer with one neuron. Values passed from the hidden layer to the output layer are transformed by the rectified linear units (ReLU) activation function, which, you'll recall, turns negative numbers into 0s and helps the model fit to nonlinear datasets. Observe that you don't have to add the input layer explicitly. The input_dim=2 parameter in the first hidden layer implicitly creates an input layer with two neurons.

 relu (*https://oreil.ly/dQUt0*) is one of several activation functions (*https://oreil.ly/uAIZt*) included in Keras. Others include tanh (*https://oreil.ly/IGwZp*), sigmoid (*https://oreil.ly/WXtNw*), and soft max (*https://oreil.ly/QlGs9*). You will rarely if ever use anything other than relu in the hidden layers. Later in this chapter, you'll see why functions such as sigmoid and softmax are useful in the output layers of networks that perform classification rather than regression.

Once all the layers are added, the next step is to call compile (*https://oreil.ly/YZvZy*) and specify important attributes such as which optimizer and loss function to use during training. Here's an example:

```
model.compile(optimizer='adam', loss='mae', metrics=['mae'])
```

Let's walk through the parameters one at a time:

`optimizer='adam'`

Tells Keras to use the `Adam` optimizer (*https://oreil.ly/ebDXm*) to adjust weights and biases in each backpropagation pass during training. `Adam` is one of eight optimizers (*https://oreil.ly/XpozA*) built into Keras, and it is among the most advanced. It employs an adaptive learning rate and is always the one I start with in the absence of a compelling reason to do otherwise.

`loss='mae'`

Tells Keras to use mean absolute error (MAE) to measure loss. This is common for neural networks intended to solve regression problems. Another frequently used option for regression models is `loss='mse'` for mean squared error (MSE).

`metrics=['mae']`

Tells Keras to capture MAE values as the network is trained. This information is used after training is complete to judge the efficacy of the training.

String values such as `'adam'` and `'mae'` are shortcuts for functions built into Keras. For example, `optimizer='adam'` is equivalent to `optimizer=Adam()`. The longhand form is useful for calling the function with nondefault parameter values—for example, `optimizer=Adam(learning_rate=2e-5)` to create an `Adam` optimizer with a custom learning rate. You'll see an example of this in Chapter 13 when we fine-tune a model by training it with a low learning rate.

Inside the `compile` method, Keras creates a TensorFlow object graph to speed execution. Once the network is compiled, you train it by calling `fit` (*https://oreil.ly/tXh6S*):

```
hist = model.fit(x, y, epochs=100, batch_size=100, validation_split=0.2)
```

The `fit` method accepts many parameters. Here are the ones used in this example:

`x`

The dataset's feature columns.

`y`

The dataset's label column—the one containing the values the network will attempt to predict.

`epochs=100`

Tells Keras to train the network for 100 iterations, or *epochs*. In each epoch, all of the training data passes through the network one time.

`batch_size=100`

Tells Keras to pass 100 training samples through the network before making a backpropagation pass to adjust the weights and biases. Training takes less time if the batch size is large, but accuracy could suffer. You typically experiment with different batch sizes to find the right balance between training time and accuracy. Do *not* assume that lowering the batch size will improve accuracy. It frequently does, but sometimes does not.

`validation_split=0.2`

Tells Keras that in each epoch, it should train with 80% of the rows in the dataset and validate the network's accuracy with the remaining 20%. If you prefer, you can split the dataset yourself and use the `validation_data` parameter to pass the validation data to `fit`. Keras doesn't offer an explicit function for splitting a dataset, but you can use Scikit's `train_test_split` function (*https://oreil.ly/ObHAM*) to do it. One difference between `train_test_split` and `validation_split` is that the former splits the data randomly and includes an option for performing a stratified split. `validation_split`, by contrast, simply divides the dataset into two partitions and does not attempt to shuffle or stratify. Don't use `validation_split` on ordered data without shuffling the data first.

It might surprise you to learn that if you train the same network on the same dataset several times, the results will be different each time. By default, weights are initialized with random values, and different starting points produce different outcomes. Additional randomness baked into the training process means the network will train differently even if it's initialized with the same random weights. Rather than fight it, data scientists learn to "embrace the randomness." If you work the tutorial in the next section, your results will differ from mine. They shouldn't differ by a lot, but they *will* differ.

The random weights assigned to the connections between neurons aren't perfectly random. Keras includes about a dozen initializers (*https://oreil.ly/SPGKn*), each of which initializes parameters in a different way. By default, `Dense` layers use the `Zeroes` initializer (*https://oreil.ly/a4mXG*) to initialize biases and the `GlorotUniform` initializer (*https://oreil.ly/FZTzQ*) to initialize weights. The latter generates random numbers that fall within a uniform distribution whose limits are computed from the network topology.

You judge the efficacy of training by examining information returned by the fit method. fit returns a history object containing the training and validation metrics specified in the metrics parameter passed to the compile method. For example, metrics=['mae'] captures MAE at the end of each epoch. Charting these metrics lets you determine whether you trained for the right number of epochs. It also lets you know if the network is underfitting or overfitting. Figure 9-1 plots MAE over the course of 30 training epochs.

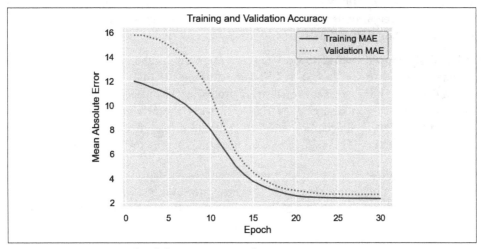

Figure 9-1. Training and validation accuracy during training

The blue curve in Figure 9-1 reveals how the network fit to the training data. The orange curve shows how it tested against the validation data. Most of the learning was done in the first 20 epochs, but MAE continued to drop as training progressed. The validation MAE nearly matched the training MAE at the end, which is an indication that the network isn't overfitting. You typically don't care how well the network fits to the training data. You care about the fit to the validation data because that indicates how the network performs with data it hasn't seen before. The greater the gap between the training and validation accuracy, the greater the likelihood that the network is overfitting.

Once a neural network is trained, you call its predict method (*https://oreil.ly/UO1uW*) to make a prediction:

```
prediction = model.predict(np.array([[2, 2]]))
```

In this example, the network accepts two floating-point values as input and returns a single floating-point value as output. The value returned by predict is that output.

Sizing a Neural Network

A neural network is characterized by the number of layers (the *depth* of the network), the number of neurons in each layer (the *widths* of the layers), the types of layers (in this example, Dense layers of fully connected neurons), and the activation functions used. There are other layer types (*https://oreil.ly/7xehq*), many of which I will introduce in later chapters. Dropout layers (*https://oreil.ly/HsN2s*), for example, can increase a network's ability to generalize by randomly dropping connections between layers during weight updates, while Conv2D layers (*https://oreil.ly/avOeo*) enable us to build convolutional neural networks (CNNs) that excel at image processing.

When designing a network, how do you pick the right number of layers and the right number of neurons for each layer? The short answer is that the "right" width and depth depends on the problem you're trying to solve, the dataset you're training with, and the accuracy you desire. As a rule, you want the *minimum width and depth required* to achieve that accuracy, and you get there using a combination of intuition and experimentation. That said, here are a few guidelines to keep in mind:

- Greater widths and depths give the network more capacity to "learn" by fitting more tightly to the training data. They also increase the likelihood of overfitting. It's the validation results that matter, and sometimes loosening the fit to the training data allows the network to generalize better. The simplest way to loosen the fit is to reduce the number of neurons.

- Generally speaking, you prefer greater width to greater depth in part to avoid the vanishing gradient problem (*https://oreil.ly/9q1xK*), which diminishes the impact of added layers. The ReLU activation function provides some protection against vanishing gradients, but that protection isn't absolute. For an explanation, see "How to Fix the Vanishing Gradients Problem Using the ReLU" (*https://oreil.ly/yHIqy*). In addition, a network with, say, 100 neurons in one layer trains faster than a network with five layers of 20 neurons each because the former has fewer weights. Think about it: there are no connections between neurons in one layer, but there are 1,600 connections ($20^2 \times 4$) between five layers containing 20 neurons each.

- Fewer neurons means less training time. State-of-the-art neural networks trained with large datasets sometimes take days or weeks to train on high-end GPUs, so training time is important.

In real life, data scientists experiment with various widths and depths to find the right balance between training time, accuracy, and the network's ability to generalize. For a multilayer perceptron, you rarely ever need more than two hidden layers, and one is often sufficient. A network with one or two hidden layers has the capacity to solve even complex nonlinear problems. Two layers with 128 neurons each, for example, gives you 16,384 (128^2) weights that can be adjusted, plus 256 biases. That's a lot of fitting power. I frequently start with one or, at most, two layers of 512 neurons each and halve the width or depth until the validation accuracy drops below an acceptable threshold.

Using a Neural Network to Predict Taxi Fares

Let's put this knowledge to work building and training a neural network. The problem that we'll solve is the same one presented in Chapter 2: using data from the New York City Taxi and Limousine Commission to predict taxi fares. We'll use a neural network as a regression model to make the predictions.

Download the CSV file containing the dataset (*https://oreil.ly/LZ4OO*) if you didn't download it in Chapter 2 and copy it into the *Data* directory where your Jupyter notebooks are hosted. Then use the following code to load the dataset and show the first five rows. It contains about 55,000 rows and is a subset of a much larger dataset that was recently used in Kaggle's New York City Taxi Fare Prediction competition (*https://oreil.ly/1vIVL*):

```python
import pandas as pd

df = pd.read_csv('Data/taxi-fares.csv', parse_dates=['pickup_datetime'])
df.head()
```

The data requires a fair amount of prep work before it's useful—something that's quite common in machine learning and in deep learning too. Use the following statements to transform the raw dataset into one suitable for training, and refer to the taxi-fare example in Chapter 2 for a step-by-step explanation of the transformations applied:

```python
from math import sqrt

df = df[df['passenger_count'] == 1]
df = df.drop(['key', 'passenger_count'], axis=1)

for i, row in df.iterrows():
    dt = row['pickup_datetime']
    df.at[i, 'day_of_week'] = dt.weekday()
    df.at[i, 'pickup_time'] = dt.hour
    x = (row['dropoff_longitude'] - row['pickup_longitude']) * 54.6
    y = (row['dropoff_latitude'] - row['pickup_latitude']) * 69.0
    distance = sqrt(x**2 + y**2)
    df.at[i, 'distance'] = distance

df.drop(['pickup_datetime', 'pickup_longitude', 'pickup_latitude',
         'dropoff_longitude', 'dropoff_latitude'], axis=1, inplace=True)

df = df[(df['distance'] > 1.0) & (df['distance'] < 10.0)]
df = df[(df['fare_amount'] > 0.0) & (df['fare_amount'] < 50.0)]
df.head()
```

The resulting dataset contains columns for the day of the week (0–6, where 0 corresponds to Monday), the hour of the day (0–23), and the distance traveled in miles, and from which outliers have been removed:

	fare_amount	day_of_week	pickup_time	distance
2	6.1	0.0	15.0	1.038136
4	10.5	5.0	10.0	2.924341
5	15.3	4.0	20.0	4.862893
8	7.7	5.0	1.0	2.603493
9	8.9	3.0	16.0	1.365739

The next step is to create the neural network. Use the following statements to create a network with an input layer that accepts three values (day, time, and distance), two hidden layers with 512 neurons each, and an output layer with a single neuron (the predicted fare amount):

```
from tensorflow.keras.models import Sequential
from tensorflow.keras.layers import Dense

model = Sequential()
model.add(Dense(512, activation='relu', input_dim=3))
model.add(Dense(512, activation='relu'))
model.add(Dense(1))
model.compile(optimizer='adam', loss='mae', metrics=['mae'])
model.summary()
```

The call to summary (*https://oreil.ly/ZAqMT*) in the last statement produces a concise summary of the network topology, including the number of *trainable parameters*—weights and biases that can be adjusted to fit the network to a dataset (Figure 9-2). For a given layer, the parameter count is the product of the number of neurons in that layer and the previous layer (the number of weights connecting the neurons in the two layers) plus the number of neurons in the layer (the biases associated with those neurons). This network is a relatively simple one, and yet it features more than a quarter million knobs and dials that can be adjusted to fit it to a dataset.

Now separate the feature columns from the label column and use them to train the network. Set validation_split to 0.2 to validate the network using 20% of the training data. Train for 100 epochs and use a batch size of 100. Given that the dataset contains more than 38,000 samples, this means that about 380 backpropagation passes will be performed in each epoch:

```
x = df.drop('fare_amount', axis=1)
y = df['fare_amount']

hist = model.fit(x, y, validation_split=0.2, epochs=100, batch_size=100)
```

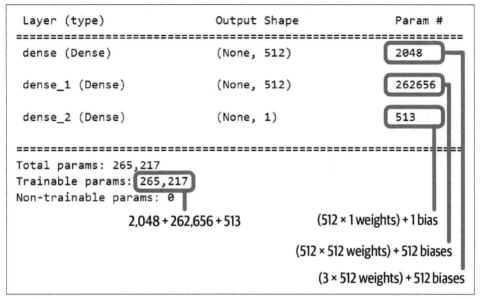

```
Layer (type)                 Output Shape              Param #
=================================================================
 dense (Dense)               (None, 512)                2048

 dense_1 (Dense)             (None, 512)                262656

 dense_2 (Dense)             (None, 1)                  513


=================================================================
Total params: 265,217
Trainable params: 265,217
Non-trainable params: 0
```

2,048 + 262,656 + 513

(512 × 1 weights) + 1 bias

(512 × 512 weights) + 512 biases

(3 × 512 weights) + 512 biases

Figure 9-2. Trainable parameters in a simple neural network

Use the history object returned by `fit` to plot the training and validation accuracy for each epoch:

```
%matplotlib inline
import matplotlib.pyplot as plt
import seaborn as sns
sns.set()

err = hist.history['mae']
val_err = hist.history['val_mae']
epochs = range(1, len(err) + 1)

plt.plot(epochs, err, '-', label='Training MAE')
plt.plot(epochs, val_err, ':', label='Validation MAE')
plt.title('Training and Validation Accuracy')
plt.xlabel('Epoch')
plt.ylabel('Mean Absolute Error')
plt.legend(loc='upper right')
plt.plot()
```

Your results will be slightly different from mine, but they should look something like this:

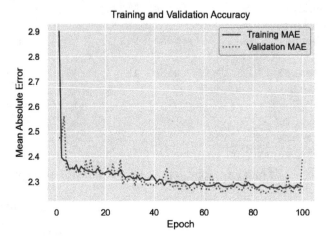

The final validation MAE was about 2.25, which means that on average, a taxi fare predicted by this network should be accurate to within about $2.25.

Recall from Chapter 2 that a common accuracy measure for regression models is the coefficient of determination, or R^2 score. Keras doesn't have a function for computing R^2 scores, but Scikit does. To that end, use the following statements to compute R^2 for the network:

```
from sklearn.metrics import r2_score

r2_score(y, model.predict(x))
```

Again, your results will differ from mine but will probably land at around 0.75.

Finish up by using the model to predict what it will cost to hire a taxi for a 2-mile trip at 5:00 p.m. on Friday:

```
import numpy as np

model.predict(np.array([[4, 17, 2.0]]))
```

Now predict the fare amount for a 2-mile trip taken at 5:00 p.m. one day later (on Saturday):

```
model.predict(np.array([[5, 17, 2.0]]))
```

Does the model predict a higher or lower fare amount for the same trip on Saturday afternoon? Do the results make sense given that the data comes from New York City cabs?

Before you close out this notebook, use it as a basis for further experimentation. Here are a few things you can try in order to gain further insights into neural networks:

- Run the notebook from start to finish a few times and note the differences in R^2 scores as well as the MAE curves. Remember that neural networks are initialized with random weights each time they're created, and additional randomness during the training process further ensures that the results will vary from run to run.

- Vary the width of the hidden layers. I used 512 neurons in each layer and found that doing so produced acceptable results. Would 128, 256, or 1,024 neurons per layer improve the accuracy? Try it and find out. Since the results will vary slightly from one run to the next, it might be useful to train the network several times in each configuration and average the results.

- Vary the batch size. What effect does that have on training time, and why? How about the effect on accuracy?

Finally, try reducing the network to *one hidden layer* containing just 16 neurons. Train it again and check the R^2 score. Does the result surprise you? How many trainable parameters does this network contain?

Binary Classification with Neural Networks

One of the common uses for machine learning is binary classification, which looks at an input and predicts which of two possible classes it belongs to. Practical uses include sentiment analysis, spam filtering, and fraud detection. Such models are trained with datasets labeled with 1s and 0s representing the two classes, employ popular learning algorithms such as logistic regression (*https://oreil.ly/EGn82*) and Naive Bayes (*https://oreil.ly/1pyWP*), and are frequently built with libraries such as Scikit-Learn.

Deep learning can be used for binary classification too. In fact, building a neural network that acts as a binary classifier is not much different than building one that acts as a regressor. In the previous section, you built a neural network that solved a regression problem. That network had an input layer that accepted three values—distance to travel, hour of the day, and day of the week—and output a predicted taxi fare. Building a neural network that performs binary classification involves making two simple changes:

- Add an activation function—specifically, the `sigmoid` activation function—to the output layer. `sigmoid` produces a value from 0.0 to 1.0 representing the probability that the input belongs to the positive class. For a reminder of what a sigmoid function does, refer to the discussion of logistic regression in Chapter 3.

- Change the loss function to `binary_crossentropy` (*https://oreil.ly/Kid2S*), which is purpose-built for binary classifiers. Accordingly, change `metrics` to `'[accuracy]'` so that accuracies computed by the loss function are captured in the history object returned by `fit`.

Here's a network designed to perform binary classification rather than regression:

```
from tensorflow.keras.models import Sequential
from tensorflow.keras.layers import Dense

model = Sequential()
model.add(Dense(512, activation='relu', input_dim=3))
model.add(Dense(512, activation='relu'))
model.add(Dense(1, activation='sigmoid'))
model.compile(optimizer='adam', loss='binary_crossentropy', metrics=['accuracy'])
```

That's it. That's all it takes to create a neural network that serves as a binary classifier. You still call `fit` to train the network, and you use the returned history object to plot the training and validation accuracy to determine whether you trained for a sufficient number of epochs and see how well the network fit to the data.

What is binary cross-entropy, and what does it do to help a binary classifier converge on a solution? During training, the cross-entropy loss function exponentially increases the penalty for wrong outputs to drive the weights and biases more aggressively in the right direction.

Let's say a sample belongs to the positive class (its label is 1), and the network predicts that the probability it's a 1 is 0.9. The cross-entropy loss, also known as *log loss*, is `-log(0.9)`, which is 0.04. But if the network outputs a probability of 0.1 for the same sample, the error is `-log(0.1)`, which equals 1. What's significant is that if the predicted probability is *really* wrong, the penalty is much higher. If the sample is a 1 and the network says the probability it's a 1 is a mere 0.0001, the cross-entropy loss is `-log(0.0001)`, or 4. Cross-entropy loss basically pats the optimizer on the back when it's close to the right answer and slaps it on the hand when it's not. The worse the prediction, the harder the slap.

To sum up, you build a neural network that performs binary classification by including a single neuron with `sigmoid` activation in the output layer and specifying `binary_crossentropy` as the loss function. The output from the network is a probability from 0.0 to 1.0 that the input belongs to the positive class. Doesn't get much simpler than that!

Making Predictions

One of the benefits of a neural network is that it can easily fit nonlinear datasets. You don't have to worry about trying different learning algorithms as you do with conventional machine learning models; the network *is* the learning algorithm. As an example, consider the dataset in Figure 9-3, in which each data point consists of an *x–y* coordinate pair and belongs to one of two classes.

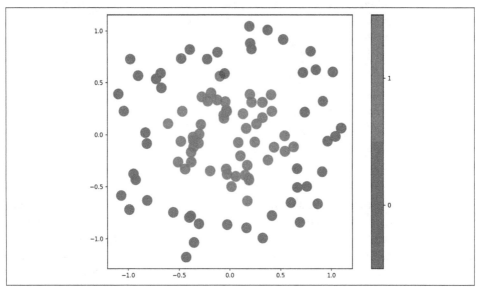

Figure 9-3. Nonlinear dataset containing two classes

The following code trains a neural network to predict a class based on a point's *x* and *y* coordinates:

```
from tensorflow.keras.models import Sequential
from tensorflow.keras.layers import Dense

model = Sequential()
model.add(Dense(128, activation='relu', input_dim=2))
model.add(Dense(1, activation='sigmoid'))
model.compile(optimizer='adam', loss='binary_crossentropy', metrics=['accuracy'])
hist = model.fit(x, y, epochs=40, batch_size=10, validation_split=0.2)
```

This network contains just one hidden layer with 128 neurons, and yet a plot of the training and validation accuracy reveals that it is remarkably successful in separating the classes:

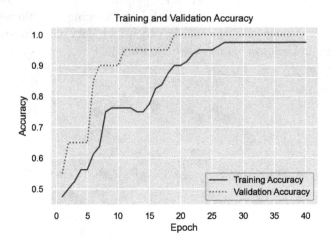

Once a binary classifier is trained, you make predictions by calling its `predict` method. Thanks to the `sigmoid` activation function, `predict` returns a number from 0.0 to 1.0 representing the probability that the input belongs to the positive class. In this example, purple data points represent the negative class (0), while red data points represent the positive class (1). Here the network is asked to predict the probability that a data point at (–0.5, 0.0) belongs to the red class:

```
model.predict(np.array([[-0.5, 0.0]]))
```

The answer is 0.57, which indicates that (–0.5, 0.0) is more likely to be red than purple. If you simply want to know which class the point belongs to, do it this way:

```
(model.predict(np.array([[-0.5, 0.0]])) > 0.5).astype('int32')
```

The answer is 1, which corresponds to red. Older versions of Keras included a `predict_classes` method that did the same without the `astype` cast, but that method was recently deprecated and removed.

Training a Neural Network to Detect Credit Card Fraud

Let's train a neural network to detect credit card fraud. Begin by downloading a ZIP file (*https://oreil.ly/uMFFJ*) containing the dataset if you haven't already and copying *creditcard.csv* from the ZIP file into your notebooks' *Data* subdirectory. It's the same one used in Chapters 3 and 6. It contains information about 284,808 credit card transactions, including the amount of each transaction and a label: 0 for legitimate transactions and 1 for fraudulent transactions. It also contains 28 columns named V1 through V28 whose meaning has been obfuscated with principal component analysis.

The dataset is highly imbalanced, containing fewer than 500 examples of fraudulent transactions.

Now load the dataset:

```python
import pandas as pd

df = pd.read_csv('Data/creditcard.csv')
df.head(10)
```

Use the following statements to drop the Time column, divide the dataset into features x and labels y, and split the dataset into two datasets: one for training and one for testing. Rather than allow Keras to do the split for us, we'll do it ourselves so that we can later run the test data through the network and use a confusion matrix to analyze the results:

```python
from sklearn.model_selection import train_test_split

x = df.drop(['Time', 'Class'], axis=1)
y = df['Class']

x_train, x_test, y_train, y_test = train_test_split(
    x, y, test_size=0.2, stratify=y, random_state=0)
```

Create a neural network for binary classification:

```python
from tensorflow.keras.models import Sequential
from tensorflow.keras.layers import Dense

model = Sequential()
model.add(Dense(128, activation='relu', input_dim=29))
model.add(Dense(1, activation='sigmoid'))
model.compile(loss='binary_crossentropy', optimizer='adam', metrics=['accuracy'])
model.summary()
```

The next step is to train the model. Notice the validation_data parameter passed to fit, which uses the test data split off from the larger dataset to assess the model's accuracy as training takes place:

```python
hist = model.fit(x_train, y_train, validation_data=(x_test, y_test),
                 epochs=10, batch_size=100)
```

Now plot the training and validation accuracy using the per-epoch values in the history object:

```python
%matplotlib inline
import matplotlib.pyplot as plt
import seaborn as sns
sns.set()

acc = hist.history['accuracy']
val = hist.history['val_accuracy']
epochs = range(1, len(acc) + 1)
```

```
plt.plot(epochs, acc, '-', label='Training accuracy')
plt.plot(epochs, val, ':', label='Validation accuracy')
plt.title('Training and Validation Accuracy')
plt.xlabel('Epoch')
plt.ylabel('Accuracy')
plt.legend(loc='lower right')
plt.plot()
```

The result looked like this for me. Remember that your results will be different thanks to the randomness inherent to training neural networks:

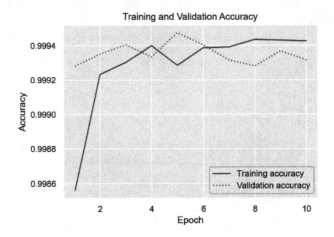

On the surface, the validation accuracy (around 0.9994) appears to be very high. But remember that the dataset is imbalanced. Fraudulent transactions represent less than 0.2% of all the samples, which means that the model could simply guess that every transaction is legitimate and get it right about 99.8% of the time. Use a confusion matrix to visualize how the model performs during testing with data it wasn't trained with:

```
from sklearn.metrics import ConfusionMatrixDisplay as cmd

sns.reset_orig()
y_predicted = model.predict(x_test) > 0.5
labels = ['Legitimate', 'Fraudulent']

cmd.from_predictions(y_test, y_predicted, display_labels=labels,
                     cmap='Blues', xticks_rotation='vertical')
```

Here's how it turned out for me:

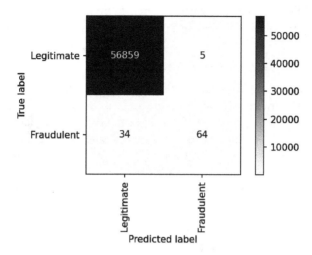

Your results will probably vary. Indeed, train the model several times and you'll get different results each time. In this run, the model correctly identified 56,858 transactions as legitimate while misclassifying legitimate transactions just six times. This means legitimate transactions are classified correctly about 99.99% of the time. Meanwhile, the model caught more than 73% of the fraudulent transactions. That's acceptable, because credit card companies would rather allow 100 fraudulent transactions to go through than decline one legitimate transaction.

Data scientists often use *three* datasets, not two, to train and assess the accuracy of a neural network: a training dataset for training, a validation dataset for validating the network (and scoring its progress) as training takes place, and a test dataset for evaluating the network's accuracy once training is complete.

The preceding example used the same dataset for validation and testing—the 20% split off from the original dataset with train_test_split. That's ostensibly fine because validation data is *not* used to adjust the network's weights and biases during training. However, if you really want to have confidence in the network's accuracy, it is never a bad idea to test it with a third dataset not used for training or validation. In the real world, the ultimate test of a deep-learning model's accuracy is how it performs against data that it has never seen before.

A final note regarding this example has to do with an extra parameter you can pass to the `fit` method that is particularly useful when dealing with imbalanced datasets. As an experiment, try replacing the call to `fit` with the following statement:

```
hist = model.fit(x_train, y_train, validation_data=(x_test, y_test), epochs=10,
                 batch_size=100, class_weight={ 0: 1.0, 1: 0.01 })
```

Then run the notebook again from start to finish. In all likelihood, the resulting confusion matrix will show zero (or at most, one or two) misclassified legitimate transactions, but the percentage of correctly identified fraudulent transactions will decrease too. The `class_weight` parameter in this example tells the model that you care a *lot* more about classifying legitimate samples correctly than correctly identifying fraudulent samples. You can experiment with different weights for the two classes and find the balance that best suits the business requirements that prompted you to build the model in the first place.

Multiclass Classification with Neural Networks

Here again is a simple binary classifier that accepts two inputs, has a hidden layer with 128 neurons, and outputs a value from 0.0 to 1.0 representing the probability that the input belongs to the positive class:

```
from tensorflow.keras.models import Sequential
from tensorflow.keras.layers import Dense

model = Sequential()
model.add(Dense(128, activation='relu', input_dim=2))
model.add(Dense(1, activation='sigmoid'))
model.compile(optimizer='adam', loss=' binary_crossentropy',
              metrics=['accuracy'])
```

Key elements include an output layer with one neuron assigned the `sigmoid` activation function, and `binary_crossentropy` as the loss function. Three simple modifications repurpose this network to do multiclass classification:

```
from tensorflow.keras.models import Sequential
from tensorflow.keras.layers import Dense

model = Sequential()
model.add(Dense(128, activation='relu', input_dim=2))
model.add(Dense(4, activation='softmax'))
model.compile(optimizer='adam', loss='sparse_categorical_crossentropy',
              metrics=['accuracy'])
```

The changes are as follows:

- The output layer contains *one neuron per class* rather than just one neuron. If the dataset contains four classes, then the output layer has four neurons. If the dataset contains 10 classes, then the output layer has 10 neurons. Each neuron corresponds to one class.

- The output layer uses the `softmax` activation function (*https://oreil.ly/sdtB9*) rather than the `sigmoid` activation function. Each neuron in the output layer yields a probability for the corresponding class, and thanks to the `softmax` function, the sum of all the probabilities is 1.0.

- The loss function is `sparse_categorical_crossentropy` (*https://oreil.ly/WpSp5*). During training, this loss function exponentially penalizes error in the probabilities predicted by a multiclass classifier, just as `binary_crossentropy` does for binary classifiers.

After defining the network, you call `fit` to train it and `predict` to make predictions. Since an example is worth a thousand words, let's fit a neural network to a two-dimensional dataset comprising four classes (Figure 9-4).

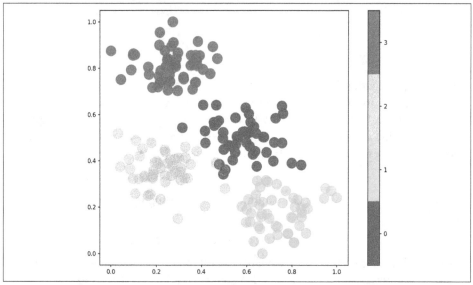

Figure 9-4. Nonlinear dataset containing four classes

The following code trains a neural network to predict a class based on a point's *x* and *y* coordinates:

```
from tensorflow.keras.models import Sequential
from tensorflow.keras.layers import Dense
```

```
model = Sequential()
model.add(Dense(128, activation='relu', input_dim=2))
model.add(Dense(4, activation='softmax'))
model.compile(optimizer='adam', loss='sparse_categorical_crossentropy',
              metrics=['accuracy'])

hist = model.fit(x, y, epochs=40, batch_size=10, validation_split=0.2)
```

A plot of the training and validation accuracy reveals that the network had little trouble separating the classes:

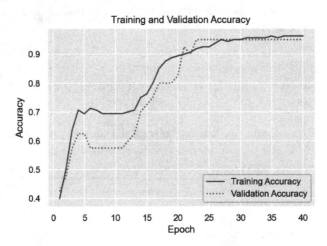

You make predictions by calling the classifier's `predict` method. For each input, `predict` returns an array of probabilities—one per class. The predicted class is the one assigned the highest probability. In this example, purple data points represent class 0, light blue represent class 1, taupe represent class 2, and red represent class 3. Here the network is asked to classify a point that lies at (0.2, 0.8):

```
model.predict(np.array([[0.2, 0.8]]))
```

The answer is an array of four probabilities corresponding to classes 0, 1, 2, and 3, in that order:

```
[2.1877741e-02, 5.3804164e-05, 5.0240371e-02, 9.2782807e-01]
```

The network predicted there's a 2% chance that (0.2, 0.8) corresponds to class 0, a 0% chance that it corresponds to class 1, a 5% chance that it corresponds to class 2, and a 93% chance that it corresponds to class 3. Looking at the plot, that seems like a reasonable answer.

If you simply want to know which class the point belongs to, you can do it this way:

```
np.argmax(model.predict(np.array([[0.2, 0.8]])), axis=1)
```

The answer is 3, which corresponds to red. Older versions of Keras included a `predict_classes` method that did the same without the call to `argmax`, but that method has since been deprecated and removed.

 Keras also includes a loss function named `categorical_cross entropy` (*https://oreil.ly/Ue2bC*) that is frequently used for multi-class classification. It works like `sparse_categorical_cross entropy`, but it requires labels to be one-hot-encoded. Rather than pass `fit` a label column containing values from 0 to 3, for example, you pass it four columns containing 0s and 1s. Keras provides a utility function named to_categorical (*https://oreil.ly/yt6hu*) to do the encoding. If you use `sparse_categorical_crossentropy`, however, you can use the label column as is.

Training a Neural Network to Recognize Faces

Chapter 5 documented the steps for training a support vector machine to recognize faces. Let's train a neural network to do the same. We'll use the same dataset as before: the Labeled Faces in the Wild (LFW) dataset (*https://oreil.ly/vkEpV*), which contains more than 13,000 facial images of famous people and is built into Scikit as a sample dataset. Recall that of the more than 5,000 people represented in the dataset, 1,680 have two or more facial images, while only 5 have 100 or more. We'll set the minimum number of faces per person to 100, which means that five sets of faces corresponding to five famous people will be imported.

Start by creating a new Jupyter notebook and using the following statements to load the dataset:

```
import pandas as pd
from sklearn.datasets import fetch_lfw_people

faces = fetch_lfw_people(min_faces_per_person=100, slice_=None)
faces.images = faces.images[:, 35:97, 39:86]
faces.data = faces.images.reshape(faces.images.shape[0], faces.images.shape[1] *
                                  faces.images.shape[2])
image_count = faces.images.shape[0]
image_height = faces.images.shape[1]
image_width = faces.images.shape[2]
class_count = len(faces.target_names)
```

In total, 1,140 facial images were loaded. After cropping, each measures 47 × 62 pixels. Use the following code to show the first 24 images in the dataset and the people to whom the faces belong:

```
%matplotlib inline
import matplotlib.pyplot as plt
```

```
fig, ax = plt.subplots(3, 8, figsize=(18, 10))
for i, axi in enumerate(ax.flat):
    axi.imshow(faces.images[i], cmap='gist_gray')
    axi.set(xticks=[], yticks=[], xlabel=faces.target_names[faces.target[i]])
```

Check the balance in the dataset by generating a histogram showing how many facial images were imported for each person:

```
from collections import Counter
import seaborn as sns
sns.set()

counts = Counter(faces.target)
names = {}

for key in counts.keys():
    names[faces.target_names[key]] = counts[key]

df = pd.DataFrame.from_dict(names, orient='index')
df.plot(kind='bar')
```

There are far more images of George W. Bush than of anyone else in the dataset. Classification models are best trained with balanced datasets. Use the following code to reduce the dataset to 100 images of each person:

```
import numpy as np

mask = np.zeros(faces.target.shape, dtype=bool)

for target in np.unique(faces.target):
    mask[np.where(faces.target == target)[0][:100]] = 1

x_faces = faces.data[mask]
y_faces = faces.target[mask]
x_faces.shape
```

x_faces contains 500 facial images, and y_faces contains the labels that go with them: 0 for Colin Powell, 1 for Donald Rumsfeld, and so on.

The next step is to split the data for training and testing. We'll set aside 20% of the data for testing, let Keras use it to validate the model during training, and later use it to assess the results with a confusion matrix:

```
from sklearn.model_selection import train_test_split

x_train, x_test, y_train, y_test = train_test_split(
    face_images, y_faces, train_size=0.8, stratify=y_faces, random_state=0)
```

Normally you'd divide all the pixel values by 255 because neural networks frequently train better with normalized data, and dividing by 255 is a simple way to normalize pixel values. It's not uncommon to use Scikit's StandardScaler class to apply unit variance instead. Dividing by 255 is unnecessary in this example because the pixels in the LFW dataset have already been normalized that way.

Create a neural network containing one hidden layer with 512 neurons. Use sparse_categorical_crossentropy as the loss function and softmax as the activation function in the output layer since this is a multiclass classification task:

```
from tensorflow.keras.layers import Dense
from tensorflow.keras.models import Sequential

model = Sequential()
model.add(Dense(512, activation='relu',
                input_shape=(image_width * image_height,)))
model.add(Dense(class_count, activation='softmax'))
model.compile(optimizer='adam', loss='sparse_categorical_crossentropy',
              metrics=['accuracy'])
model.summary()
```

Now train the network:

```
hist = model.fit(x_train, y_train, validation_data=(x_test, y_test),
                 epochs=100, batch_size=20)
```

Plot the training and validation accuracy:

```
acc = hist.history['accuracy']
val_acc = hist.history['val_accuracy']
epochs = range(1, len(acc) + 1)

plt.plot(epochs, acc, '-', label='Training Accuracy')
plt.plot(epochs, val_acc, ':', label='Validation Accuracy')
plt.title('Training and Validation Accuracy')
plt.xlabel('Epoch')
plt.ylabel('Accuracy')
plt.legend(loc='lower right')
plt.plot()
```

Finally, use a confusion matrix to visualize how the network performs against test data:

```
from sklearn.metrics import ConfusionMatrixDisplay as cmd

sns.reset_orig()
y_pred = model.predict(x_test)
fig, ax = plt.subplots(figsize=(5, 5))
ax.grid(False)
```

```
cmd.from_predictions(y_test, y_pred.argmax(axis=1),
                     display_labels=faces.target_names, colorbar=False,
                     cmap='Blues', xticks_rotation='vertical', ax=ax)
```

How many times did the model correctly identify George W. Bush? How many times did it identify him as someone else? Would the network be just as accurate with 128 neurons in the hidden layer as it is with 512?

Dropout

The goal of any machine learning model is to make accurate predictions. In a perfect world, the gap between training accuracy and validation accuracy would be close to 0 in the later stages of training a neural network. In the real world, it rarely happens that way. Training accuracy for the model in the previous section approached 100%, but validation accuracy probably peaked between 80% and 85%. This means the model isn't generalizing as well as you'd like. It learned the training data very well, but when presented with data it hadn't seen before (the validation data), it underperformed. This may be a sign that the model is overfitting. In the end, it's not training accuracy that matters; it's how accurately the model responds to new data.

One way to combat overfitting is to reduce the depth of the network, the width of individual layers, or both. Fewer neurons means fewer trainable parameters, and fewer parameters makes it harder for the network to fit too tightly to the training data.

Another way to guard against overfitting is to introduce *dropout* to the network. Dropout randomly drops connections between layers during training to prevent the network from learning the training data too well. It's like reading a book but skipping every other page in hopes that you'll learn high-level concepts without getting bogged down in the details. Dropout was introduced in a 2014 paper titled "Dropout: A Simple Way to Prevent Neural Networks from Overfitting" (*https://oreil.ly/yqLRk*).

Keras's Dropout class (*https://oreil.ly/O2crM*) makes adding dropout to a network dead-simple. To demonstrate, go back to the example in the previous section and redefine the network this way:

```
from tensorflow.keras.layers import Dense, Dropout
from tensorflow.keras.models import Sequential

model = Sequential()
model.add(Dense(512, activation='relu',
                input_shape=(image_width * image_height,)))
model.add(Dropout(0.2))
model.add(Dense(class_count, activation='softmax'))
model.compile(optimizer='adam', loss='sparse_categorical_crossentropy',
              metrics=['accuracy'])
```

Now train the network again and plot the training and validation accuracy. Here's how it turned out for me:

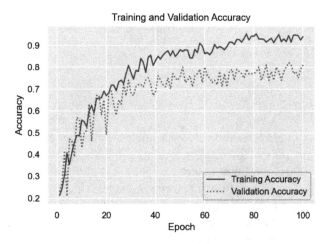

The gap didn't close much (if at all), but sometimes adding dropout in this manner *will* increase the validation accuracy. The key statement in this example is `model.add(Dropout(0.2))`, which adds a `Dropout` layer that randomly drops (ignores) 20% of the connections between the neurons in the hidden layer and the neurons in the output layer in each backpropagation pass. You can be more aggressive by dropping more connections—increasing 0.2 to 0.4, for example—but you can also reach a point of diminishing returns. Note that if you're super-aggressive with the dropout rate (for example, 0.5), you'll probably need to train the model for more epochs. Making it harder to learn means taking longer to learn too.

In practice, the only way to know whether dropout will improve a model's ability to generalize is to try it. In addition to trying different dropout percentages, you can try introducing dropout between two or more layers in hopes of finding a combination that works.

Saving and Loading Models

In Chapter 7, you learned how to serialize (save) a trained Scikit model and load it in a client app. The same requirement applies to neural networks: you need a way to save a trained network and load it later in order to operationalize it.

You can get the weights and biases from a model with Keras's `get_weights` method (*https://oreil.ly/7twtR*), and you can restore them with `set_weights` (*https://oreil.ly/JyNu9*). But saving a trained model so that you can re-create it later requires an additional step. Specifically, you must save the network architecture: the number of and types of layers, the number of neurons in each layer, the activation functions used in each layer, and so on. Fortunately, all that requires just one line of code. That line of code differs depending on which of two formats you want the model saved in:

```
model.save('my_model.h5') # Save the model in Keras's H5 format
model.save('my_model')    # Save the model in TensorFlow's native format
```

Loading a saved model is equally simple:

```
from tensorflow.keras.models import load_model

model = load_model('my_model.h5') # Load model saved in H5 format
model = load_model('my_model') # Load model saved in TensorFlow format
```

Saving the model in H5 format produces a single *.h5* file that encapsulates the entire model. Saving it in TensorFlow's native format, also known as the *SavedModel format* (*https://oreil.ly/hI3Rt*), produces a series of files and subdirectories containing the serialized model. Google recommends using the latter, although it's still common to see Keras's H5 format used. There is no functional difference between the two, but *.h5* files can only be read by Keras apps written in Python, while models saved in Saved-Model format can be loaded by other frameworks. Apps written in C# with Micro-soft's ML.NET, for example, can load models saved in SavedModel format regardless of the programming language in which the model was crafted.

The H5 format was originally devised so that Keras models could be saved in a manner independent of the deep-learning framework used as the backend. Keras is still available in a standalone version that supports backends other than TensorFlow (specifically, CNTK and Theano), but those frameworks have been deprecated—they are no longer being developed—and are rarely used today other than in legacy models. The version of Keras built into TensorFlow supports *only* TensorFlow backends.

Once a saved model is loaded, it acts identically to the original. The predictions that it makes, for example, are identical to the predictions made by the original model. You can even further train the model by running additional training samples through it. This highlights one of the major differences between neural networks and traditional machine learning models built with Scikit. Since the state of a network is defined by its weights and biases and loading a model restores the weights and biases, neural networks inherently support incremental training, also known as *continual learning*, so that they can become smarter over time. Most Scikit models do not because serial-izing the models doesn't save the internal state accumulated as training takes place.

To recap: you can run a million training samples through a neural network, save it, load it, and run another million training samples through it and the network picks up right where it left off. The results are identical to running 2 million training samples through the network to begin with save for minor differences that result from the randomness that is always inherent to training.

Keras Callbacks

As you train a neural network and it achieves peak validation accuracy, the peak is hard to capture. Rather than nicely level out in later epochs, the validation accuracy may go down or oscillate between peaks and valleys. Given the stochastic (random) nature of neural networks, if you mark the epoch that achieved maximum validation accuracy and train again for exactly that number of epochs, you won't get the same results the second time. How do you train for *exactly* the right number of epochs to produce the best (most accurate) network possible?

An elegant solution is Keras's *callbacks API* (*https://oreil.ly/RP1Ht*), which lets you write callback functions that are called at various points during training—for example, at the end of each epoch—and that have the ability to alter and even stop the training process. Here's an example that creates a child class named StopCallback that inherits from Keras's Callback class (*https://oreil.ly/CULyB*). The child class implements the on_epoch_end function that's called at the end of each training epoch and stops training if the validation accuracy reaches 95%:

```
from tensorflow.keras.callbacks import Callback

class StopCallback(Callback):
    accuracy_threshold = None

    def __init__(self, threshold):
        self.accuracy_threshold = threshold

    def on_epoch_end(self, epoch, logs=None):
        if (logs.get('val_accuracy') >= self.accuracy_threshold):
            self.model.stop_training = True

callback = StopCallback(0.95)
model.fit(x, y, validation_split=0.2, epochs=100, batch_size=20,
        callbacks=[callback])
model.save('best_model.h5')
```

Note the validation accuracy threshold (0.95) passed to StopCallback's constructor. The call to fit ostensibly trains the network for 100 epochs, but if the validation accuracy reaches 0.95 before that, training stops in its tracks. The final statement saves the model that achieved that accuracy.

on_epoch_end is one of several functions you can implement in classes that inherit from Callback to receive a callback when a predetermined checkpoint is reached in the training process. Others include on_epoch_begin, on_train_begin, on_train_end, on_train_batch_begin, and on_train_batch_end. You'll find a complete list, along with examples, in "Writing Your Own Callbacks" (*https://oreil.ly/X7lf4*) in the Keras documentation.

In addition to providing a base `Callback` class from which you can create your own callback classes, Keras provides several callback classes of its own. One of them is the `EarlyStopping` class (*https://oreil.ly/5DMHj*), which lets you stop training based on a specified criterion such as decreasing validation accuracy or increasing training loss without writing a lot of code. In the following example, training stops early if the validation accuracy fails to improve for five consecutive epochs (`patience=5`). When training is halted, the network's weights and biases are automatically restored to what they were when validation accuracy peaked in the final five epochs (`restore_best_weights=True`):

```
from tensorflow.keras.callbacks import EarlyStopping

callback = EarlyStopping(monitor='val_accuracy', patience=5,
                         restore_best_weights=True)
model.fit(x, y, validation_split=0.2, epochs=100, batch_size=20,
          callbacks=[callback])
```

Stopping the training process based on rising training loss rather than decreasing validation accuracy at the end of each epoch requires a minor code change:

```
from tensorflow.keras.callbacks import EarlyStopping

callback = EarlyStopping(monitor='loss', patience=5, restore_best_weights=True)
model.fit(x, y, validation_split=0.2, epochs=100, batch_size=20,
          callbacks=[callback])
```

The callback that is used perhaps more than any other is `ModelCheckpoint` (*https://oreil.ly/cLMwN*), which saves a model at specified intervals during training or, if you set `save_best_only` to `True`, saves the most accurate model. The next example trains a model for 100 epochs and saves the one that exhibits the highest validation accuracy in *best_model.h5*:

```
from tensorflow.keras.callbacks import ModelCheckpoint

callback = ModelCheckpoint(filepath='best_model.h5', monitor='val_accuracy',
                           save_best_only=True)
model.fit(x, y, validation_split=0.2, epochs=100, batch_size=20,
          callbacks=[callback])
```

Another frequently used callback is `TensorBoard` (*https://oreil.ly/h3sHp*), which logs a variety of information to a specified location in the filesystem as a model is trained. The following example logs to the *logs* subdirectory of the current directory:

```
from tensorflow.keras.callbacks import TensorBoard

callback = TensorBoard(log_dir='logs', histogram_freq=1)
```

```
model.fit(x_train, y_train, validation_split=0.2, epochs=100, batch_size=20,
          callbacks=[callback])
```

You can use a tool called TensorBoard to monitor accuracy and loss, changes in the model's weights and biases, and more while training takes place or after it has completed. You can launch TensorBoard from a Jupyter notebook and point it to the *logs* subdirectory with a command like this one:

```
%tensorboard --logdir logs
```

Or you can launch it from a command prompt by executing the same command without the percent sign. Then point your browser to http://localhost:6006 to open the TensorBoard console (Figure 9-5). "Get Started with TensorBoard" (*https://oreil.ly/1Q0G4*) in the TensorFlow documentation contains a helpful tutorial on the basics of TensorBoard. It's an indispensable tool in the hands of professionals, especially when training complex models that require hours, days, or even weeks to fully train.

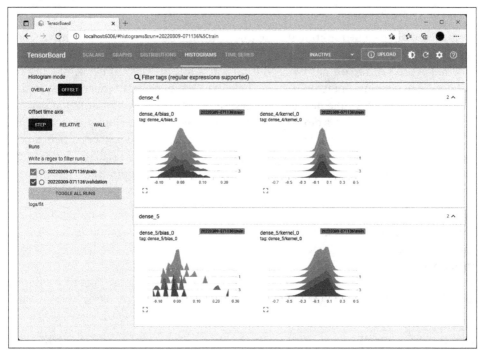

Figure 9-5. TensorBoard showing the results of training a neural network

Other Keras callback classes include LearningRateScheduler (*https://oreil.ly/ OMEMh*) for adjusting the learning rate at the beginning of each epoch and CSV Logger (*https://oreil.ly/BjUny*) for capturing the results of each training epoch in a CSV file. Refer to the callbacks API documentation (*https://oreil.ly/Mm3fY*) for a complete list. In addition, observe that the fit method's callbacks parameter is a Python list, which means you can specify multiple callbacks when you train a model. You could use one callback to stop training if certain conditions are met, for example, and another callback to log training metrics in a CSV file.

Summary

Keras and TensorFlow are widely used open source frameworks that facilitate building, training, saving, loading, and consuming (making predictions with) neural networks. You can build neural networks in a variety of programming languages using native TensorFlow APIs, but Keras abstracts those APIs and makes deep learning much more approachable. Keras apps are written in Python.

Neural networks, like traditional machine learning models, can be used to solve regression problems and classification problems. A network that performs regression has one neuron in the output layer with no activation function; the output from the network is a floating-point number. A network that performs binary classification also has one neuron in the output layer, but the sigmoid activation function ensures that the output is a value from 0.0 to 1.0 representing the probability that the input represents the positive class. For multiclass classification, the number of neurons in the output layer equals the number of classes the network can predict. The softmax activation function transforms the raw values assigned to the output neurons into an array of probabilities for each class.

When you find that a neural network is fitting too tightly to the training data, one way to combat overfitting and increase the network's ability to generalize is to reduce the complexity of the network: reduce the number of layers, the number of neurons in individual layers, or both. Another approach is to add dropout to the network. Dropout purposely impedes a network's ability to learn the training data by randomly ignoring a subset of the connections between layers when updating weights and biases.

Keras's callbacks API lets you customize the training process. By processing the callbacks that occur at the end of each training epoch, for example, you can check the model's accuracy and halt training if it has reached an acceptable level. Keras also includes a simple and easy-to-use API for saving and loading trained models. This is essential for operationalizing the models that you train—deploying them to production and using the predictive powers developed during training.

The facial recognition model in this chapter exceeded 80% in validation accuracy, but modern deep-learning models often achieve 99% accuracy on the same dataset. It won't surprise you to learn that there is more to deep learning than multilayer perceptrons. We'll take a deep dive into facial recognition in Chapter 11, but first we'll explore a different type of neural network—one that's particularly adept at solving computer-vision problems. It's called the *convolutional neural network*, and it is the subject of Chapter 10.

Image Classification with Convolutional Neural Networks

Computer vision (*https://oreil.ly/SKGgL*) is a branch of deep learning in which computers discern information from images. Real-world uses include identifying objects in photos, removing inappropriate images from social media sites, counting the cars in line at a tollbooth, and recognizing faces in photos. Computer-vision models can even be combined with natural language processing (NLP) models to caption photos. I snapped a photo while on vacation and asked Azure's Computer Vision service (*https://oreil.ly/lRCW5*) to caption it. The result is shown in Figure 10-1. It's somewhat remarkable given that no human intervention was required.

Figure 10-1. "A body of water with a dock and a building in the background"—Azure AI

The field of computer vision has advanced rapidly in recent years, mostly due to *convolutional neural networks*, also known as CNNs or *ConvNets*. In 2012, an eight-layer CNN called AlexNet (*https://oreil.ly/JCNBM*) outperformed traditional machine learning models entered in the annual ImageNet Large Scale Visual Recognition Challenge (ILSVRC) (*https://oreil.ly/8OK4o*) by achieving an error rate of 15.3% when identifying objects in photos. In 2015, ResNet-152 (*https://oreil.ly/4rBgb*) featuring a whopping 152 layers won the challenge with an error rate of just 3.5%, which exceeds a human's ability to classify images featured in the competition.

CNNs are magical because they treat images as *images* rather than just arrays of pixel values. They use a decades-old technology called convolution kernels (*https://oreil.ly/upd6y*) to extract "features" from images, allowing them to recognize the shape of a cat's head or the outline of a dog's tail. Moreover, they are easy to build with Keras and TensorFlow.

State-of-the-art CNNs such as ResNet-152 are trained at great expense with millions of images on GPUs, but there's a lot you can do with an ordinary CPU. In this chapter, you'll learn what CNNs are and how they work, and you'll build and train a few CNNs of your own. You'll also learn how to leverage advanced CNNs published for public consumption by companies such as Google and Microsoft, and how to use a technique called *transfer learning* to repurpose those CNNs to solve domain-specific problems.

Understanding CNNs

Figure 10-2 shows the topology of a basic CNN. It begins with one or more sets of *convolution layers* and *pooling layers*. Convolution layers extract features from images, generating transformed images that are commonly referred to as *feature maps* because they highlight distinguishing features such as shapes and contours. Pooling layers reduce the feature maps' size by half so that features can be extracted at various resolutions and are less sensitive to small changes in position. Output from the final pooling layer is flattened to one dimension and input to one or more dense layers for classification. The convolution and pooling layers are called *bottleneck layers* since they reduce the dimensionality of images input to them. They also account for the bulk of the computation time during training.

Convolution layers (*https://oreil.ly/dc5b7*) extract features from images by passing convolution kernels over them—the same technique used by image editing tools to blur, sharpen, and emboss images. A kernel is simply a matrix of values. It usually measures 3 × 3, but it can be larger. To process an image, you place the kernel in the upper-left corner of the image, multiply the kernel values by the pixel values underneath, and compute a new value for the center pixel by summing the products, as shown in Figure 10-3. Then you move the kernel one pixel to the right and repeat the process, continuing row by row and column by column until the entire image has been processed.

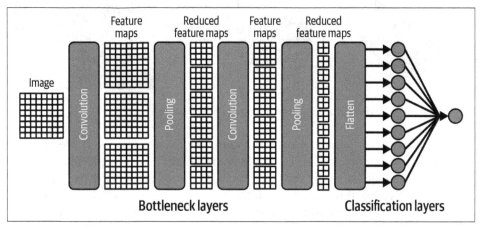

Figure 10-2. Convolutional neural network

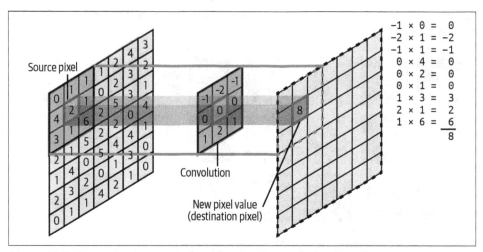

Figure 10-3. Processing image pixels with a 3 × 3 convolution kernel

Figure 10-4 shows what happens when you apply a 3 × 3 kernel to a hot dog image. This particular kernel is called a *bottom Sobel kernel*, and it's designed to do edge detection by highlighting edges as if a light were shined from the bottom. The convolution layers of a CNN use kernels like this one to extract features that help distinguish one class from another.

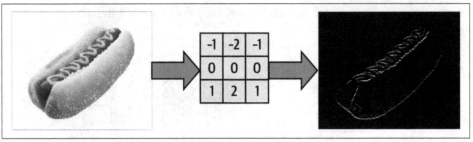

Figure 10-4. Processing an image with a bottom Sobel kernel

A convolution layer doesn't use just one kernel to process images. It uses many—sometimes 100 or more. The kernel values aren't determined ahead of time. They are initialized with random values and then learned (adjusted) as the CNN is trained, just as the weights connecting neurons in dense layers are learned. Each kernel also has a bias associated with it, just like a neuron in a dense layer. The images in Figure 10-5 were generated by the first convolution layer in a trained CNN. You can see how the various convolution kernels allow the network to view the same hot dog image in different ways, and how certain features such as the shape of the bun and the ribbon of mustard on top are highlighted.

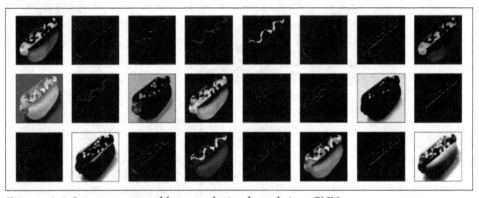

Figure 10-5. Images generated by convolution kernels in a CNN

Pooling layers (*https://oreil.ly/KW5rW*) downsample images to reduce their size. The most common resizing technique is *max pooling*, which divides images into 2×2 blocks of pixels and selects the highest of the four values in each block. An alternative is *average pooling*, which averages the values in each block.

Figure 10-6 shows how an image contracts as it passes through successive pooling layers. The first row came from the first pooling layer, the second row came from the second pooling layer, and so on.

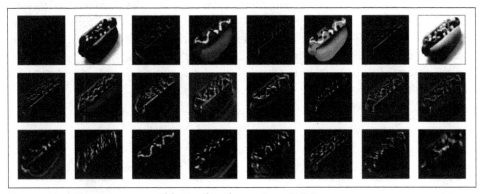

Figure 10-6. Images generated by pooling layers in a CNN

Pooling isn't the only way to downsize an image. While less common, reduction can be accomplished without pooling layers by setting a convolution layer's stride to 2. *Stride* is the number of pixels a convolution kernel moves as it passes over an image. It defaults to 1, but setting it to 2 halves the image size by ignoring every other row and every other column of pixels.

The dense layers at the end of the network classify features extracted from the bottleneck layers and are referred to as the CNN's *classification layers*. They are no different than the multilayer perceptrons featured in Chapter 9. For binary classification, the output layer contains one neuron and uses the `sigmoid` activation function. For multiclass classification, the output layer contains one neuron per class and uses the `soft max` activation function.

There's no law that says bottleneck layers *have* to be paired with classification layers. You could take the feature maps output from the bottleneck layers and classify them with a support vector machine rather than a multilayer perceptron. It's not as far-fetched as it sounds. In Chapter 12, I'll introduce one well-known model that does just that.

Using Keras and TensorFlow to Build CNNs

To simplify building CNNs that classify images, Keras offers the `Conv2D` class (*https://oreil.ly/zXy3K*), which models convolution layers, and the `MaxPooling2D` class (*https://oreil.ly/FreyC*), which implements max pooling layers. The following statements create a CNN with two pairs of convolution and pooling layers, a flatten layer to reshape the output into a 1D array for input to a dense layer, a dense layer to classify the features extracted from the bottleneck layers, and a softmax output layer for classification:

```
from tensorflow.keras.models import Sequential
from tensorflow.keras.layers import Conv2D, MaxPooling2D, Flatten, Dense

model = Sequential()
model.add(Conv2D(32, (3, 3), activation='relu', input_shape=(28, 28, 1)))
model.add(MaxPooling2D(2, 2))
model.add(Conv2D(64, (3, 3), activation='relu'))
model.add(MaxPooling2D(2, 2))
model.add(Flatten())
model.add(Dense(128, activation='relu'))
model.add(Dense(10, activation='softmax'))
```

The first parameter passed to the `Conv2D` function is the number of convolution kernels to include in the layer. More kernels means more fitting power, similar to the number of neurons in a dense layer. The second parameter is the dimensions of each kernel. You sometimes get greater accuracy from 5 × 5 kernels, but a kernel that size increases training time by requiring 25 multiplication operations for each pixel as opposed to nine for a 3 × 3 kernel. The `input_shape` parameter in the first layer specifies the size of the images input to the CNN: in this case, one-channel (grayscale) 28 × 28 images. All the images used to train a CNN must be the same size.

> `Conv2D` processes images, which are two-dimensional. Keras also offers the `Conv1D` class (*https://oreil.ly/TgzPk*) for processing 1D data and `Conv3D` (*https://oreil.ly/WshPD*) for 3D data. The former finds use processing text and time-series data (*https://oreil.ly/6gw8k*). Canonical use cases for `Conv3D` include analyzing video and 3D medical images.

Given a set of images with a relatively high degree of separation between classes, it's perfectly feasible to train a CNN to classify those images on a typical laptop or PC. A great example is the MNIST dataset (*https://oreil.ly/WAePv*), which contains 60,000 training images of scanned, handwritten digits, each measuring 28 × 28 pixels, plus 10,000 test images. Figure 10-7 shows the first 50 scans in the training set.

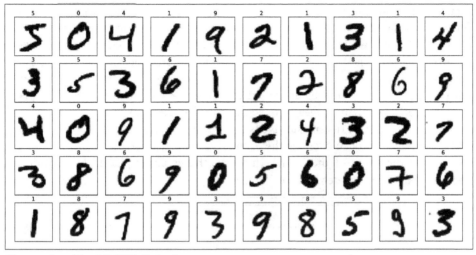

Figure 10-7. The MNIST digits dataset

Let's train a CNN to recognize digits in the MNIST dataset, which conveniently is one of several sample datasets (*https://oreil.ly/84c5G*) built into Keras. Begin by creating a new Jupyter notebook and using the following statements to load the dataset, reshape the 28 × 28 images into 28 × 28 × 1 arrays (28 × 28 images containing a single color channel), and divide the pixel values by 255 as a simple form of normalization:

```
from tensorflow.keras.datasets import mnist

(train_images, y_train), (test_images, y_test) = mnist.load_data()
x_train = train_images.reshape(60000, 28, 28, 1) / 255
x_test = test_images.reshape(10000, 28, 28, 1) / 255
```

Next, define a CNN that accepts 28 × 28 × 1 arrays of pixel values as input, contains two pairs of convolution and pooling layers, and has a softmax output layer with 10 neurons since the dataset contains scans of 10 different digits:

```
from tensorflow.keras.layers import Conv2D, MaxPooling2D, Dense, Flatten
from tensorflow.keras.models import Sequential

model = Sequential()
model.add(Conv2D(32, (3, 3), activation='relu', input_shape=(28, 28, 1)))
model.add(MaxPooling2D(2, 2))
model.add(Conv2D(64, (3, 3), activation='relu'))
model.add(MaxPooling2D(2, 2))
model.add(Flatten())
model.add(Dense(128, activation='relu'))
model.add(Dense(10, activation='softmax'))
model.compile(optimizer='adam', loss='sparse_categorical_crossentropy',
              metrics=['accuracy'])
model.summary(line_length=80)
```

Figure 10-8 shows the output from the call to summary on the final line. The summary reveals a lot about how this CNN processes images. Each pooling layer reduces the image size by half, while each convolution layer reduces the image's height and width by two pixels. Why is that? By default, a convolution kernel doesn't start with its center cell over the pixel in the upper-left corner of the image; rather, its upper-left corner is aligned with the image's upper-left corner. For a 3 × 3 kernel, there's a 1-pixel-wide border around the edges that doesn't survive the convolution. (For a 5 × 5 kernel, the border that doesn't survive is 2 pixels wide.) The term for this is *padding*, and if you'd like, you can override the default behavior to push the kernel's center cell right up to the edges of the image. In Keras, this is accomplished by including a padding='same' parameter in the call to Conv2D.

```
Layer (type)                    Output Shape              Param #
=================================================================
conv2d (Conv2D)                 (None, 26, 26, 32)        320

max_pooling2d (MaxPooling2D)    (None, 13, 13, 32)        0

conv2d_1 (Conv2D)               (None, 11, 11, 64)        18496

max_pooling2d_1 (MaxPooling2D)  (None, 5, 5, 64)          0

flatten (Flatten)               (None, 1600)              0

dense (Dense)                   (None, 128)               204928

dense_1 (Dense)                 (None, 10)                1290

=================================================================
Total params: 225,034
Trainable params: 225,034
Non-trainable params: 0
```

Figure 10-8. Output from the summary method

Another takeaway is that each 28 × 28 image exits the first convolution layer as a 3D array or *tensor* measuring 26 × 26 × 32: one 26 × 26 feature map for each of the 32 kernels. After max pooling, the tensor is reduced to 13 × 13 × 32 and input to the second convolution layer, where 64 more kernels filter features from the thirty-two 13 × 13 feature maps and combine them to produce 64 new feature maps (a tensor measuring 11 × 11 × 64). A final pooling layer reduces that to 5 × 5 × 64. These values are flattened into a 1D tensor containing 1,600 values and fed into a dense layer for classification.

The big picture here is that the CNN transforms each 28 × 28 image comprising 784 pixel values into an array of 1,600 floating-point numbers that (hopefully) distinguishes the contents of the image more clearly than ordinary pixel values do. That's what bottleneck layers do: they transform *matrices of integer pixel values* into *tensors of floating-point numbers* that better characterize the images input to them. As you'll see in Chapter 13, NLP networks use *word embeddings* to create dense vector representations of the words in a document. *Dense vector representation* is a term you'll encounter a lot in deep learning. It's nothing more than arrays of floating-point numbers that do more to characterize the input than the input data itself.

The output from summary would look exactly the same if the images input to the network were three-channel color images rather than one-channel grayscale images. Applying a convolution layer with *n* kernels to an image produces *n* feature maps regardless of image depth, just as applying a convolution layer featuring *n* kernels to the feature maps output by preceding layers produces *n* new feature maps regardless of input depth. Internally, CNNs use *tensor dot products* to produce 2D feature maps from 3D feature maps. Python's NumPy library (*https://oreil.ly/Hek8J*) includes a function named tensordot (*https://oreil.ly/6NyCk*) for computing tensor dot products quickly.

Now train the network and plot the training and validation accuracy:

```
%matplotlib inline
import matplotlib.pyplot as plt
import seaborn as sns
sns.set()

hist = model.fit(x_train, y_train,
                 validation_data=(x_test, y_test),
                 epochs=10, batch_size=50)
acc = hist.history['accuracy']
val_acc = hist.history['val_accuracy']
epochs = range(1, len(acc) + 1)

plt.plot(epochs, acc, '-', label='Training Accuracy')
plt.plot(epochs, val_acc, ':', label='Validation Accuracy')
plt.title('Training and Validation Accuracy')
plt.xlabel('Epoch')
plt.ylabel('Accuracy')
plt.legend(loc='lower right')
```

Once trained, this simple CNN can achieve 99% accuracy classifying handwritten digits:

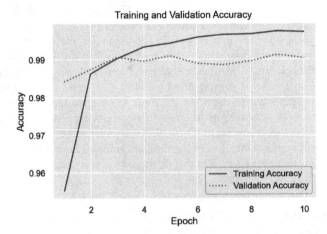

One reason it can attain such accuracy is the number of training samples—roughly 6,000 per class. (As a test, I trained the network with just 100 samples of each class and got 92% accuracy.) Another factor is that a 2 looks very different from, say, an 8. If a person can rather easily distinguish between the two, then a CNN can too.

Training a CNN to Recognize Arctic Wildlife

A basic CNN can easily achieve 99% accuracy on the MNIST dataset. But it isn't as easy when the problem is more perceptual—for example, when the goal is to determine whether a photo contains a dog or a cat. One reason is that most 8s look a lot alike, while dogs and cats come in many varieties. Another factor is that each digit in the MNIST dataset is carefully cropped to precisely fill the frame, whereas dogs and cats can appear anywhere in the frame and can be photographed in different poses and from an infinite number of angles.

To demonstrate, let's train a CNN to distinguish between Arctic foxes, polar bears, and walruses. For context, imagine you've been tasked with creating a system that uses AI to examine pictures snapped by motion-activated cameras deployed in the Arctic to document polar bear activity.

Start by downloading a ZIP file (*https://oreil.ly/p7Y0k*) containing images for training and testing the CNN. Unpack the ZIP file and place its contents in a subdirectory named *Wildlife* where your Jupyter notebooks are hosted. The ZIP file contains folders named *train*, *test*, and *samples*. Each folder contains subfolders named *arctic_fox*, *polar_bear*, and *walrus*. The training folders contain 100 images each, while the test folders contain 40 images each. Figure 10-9 shows some of the polar bear training images. These are public images that were downloaded from the internet and cropped and resized to 224 × 224 pixels.

Figure 10-9. Polar bear images

Now create a Jupyter notebook and use the following code to define a pair of helper functions—one to load a batch of images from a specified location in the filesystem and assign them labels, and another to show the first eight images in a batch of images:

```python
import os
from tensorflow.keras.preprocessing import image
import matplotlib.pyplot as plt
%matplotlib inline

def load_images_from_path(path, label):
    images, labels = [], []

    for file in os.listdir(path):
        img = image.load_img(os.path.join(path, file), target_size=(224, 224, 3))
        images.append(image.img_to_array(img))
        labels.append((label))

    return images, labels

def show_images(images):
    fig, axes = plt.subplots(1, 8, figsize=(20, 20),
                             subplot_kw={'xticks': [], 'yticks': []})

    for i, ax in enumerate(axes.flat):
        ax.imshow(images[i] / 255)

x_train, y_train, x_test, y_test = [], [], [], []
```

Use the following statements to load 100 Arctic fox training images and plot a subset of them:

```python
images, labels = load_images_from_path('Wildlife/train/arctic_fox', 0)
show_images(images)
```

```
x_train += images
y_train += labels
```

Do the same to load and label the polar bear training images:

```
images, labels = load_images_from_path('Wildlife/train/polar_bear', 1)
show_images(images)

x_train += images
y_train += labels
```

And then the walrus training images:

```
images, labels = load_images_from_path('Wildlife/train/walrus', 2)
show_images(images)

x_train += images
y_train += labels
```

You also need to load the images used to validate the CNN. Start with 40 Arctic fox test images:

```
images, labels = load_images_from_path('Wildlife/test/arctic_fox', 0)
show_images(images)

x_test += images
y_test += labels
```

Then the polar bear test images:

```
images, labels = load_images_from_path('Wildlife/test/polar_bear', 1)
show_images(images)

x_test += images
y_test += labels
```

And finally the walrus test images:

```
images, labels = load_images_from_path('Wildlife/test/walrus', 2)
show_images(images)

x_test += images
y_test += labels
```

The next step is to normalize the training and testing images by dividing their pixel values by 255:

```
import numpy as np

x_train = np.array(x_train) / 255
x_test = np.array(x_test) / 255

y_train = np.array(y_train)
y_test = np.array(y_test)
```

Now it's time to build a CNN. Since the images measure 224 × 224 and we want the final feature maps to compress as much information as possible into a small space, we'll use five pairs of convolution and pooling layers to extract features from the training images at five resolutions: 224 × 224, 111 × 111, 54 × 54, 26 × 26, and 12 × 12. We'll follow those with a dense layer and a softmax output layer containing three neurons—one for each of the three classes:

```
from tensorflow.keras.models import Sequential
from tensorflow.keras.layers import Conv2D, MaxPooling2D, Flatten, Dense

model = Sequential()
model.add(Conv2D(32, (3, 3), activation='relu', input_shape=(224, 224, 3)))
model.add(MaxPooling2D(2, 2))
model.add(Conv2D(64, (3, 3), activation='relu'))
model.add(MaxPooling2D(2, 2))
model.add(Conv2D(64, (3, 3), activation='relu'))
model.add(MaxPooling2D(2, 2))
model.add(Conv2D(128, (3, 3), activation='relu'))
model.add(MaxPooling2D(2, 2))
model.add(Conv2D(128, (3, 3), activation='relu'))
model.add(MaxPooling2D(2, 2))
model.add(Flatten())
model.add(Dense(1024, activation='relu'))
model.add(Dense(3, activation='softmax'))
model.compile(optimizer='adam', loss='sparse_categorical_crossentropy',
              metrics=['accuracy'])
model.summary(line_length=80)
```

Call `fit` to train the model:

```
hist = model.fit(x_train, y_train,
                 validation_data=(x_test, y_test),
                 batch_size=10, epochs=20)
```

If you train the model on a CPU, training will probably require from 10 to 20 seconds per epoch. (Think of all those pixel calculations taking place on all those images with all those convolution kernels.) When training is complete, use the following statements to plot the training and validation accuracy:

```
import seaborn as sns
sns.set()

acc = hist.history['accuracy']
val_acc = hist.history['val_accuracy']
epochs = range(1, len(acc) + 1)

plt.plot(epochs, acc, '-', label='Training Accuracy')
plt.plot(epochs, val_acc, ':', label='Validation Accuracy')
plt.title('Training and Validation Accuracy')
plt.xlabel('Epoch')
plt.ylabel('Accuracy')
```

```
plt.legend(loc='lower right')
plt.plot()
```

Here is the output:

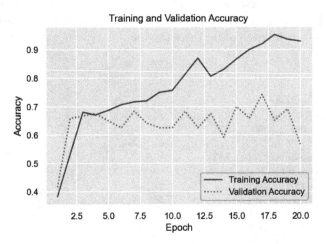

Were the results what you expected? The validation accuracy is decent, but it's not state of the art. It probably landed between 60% and 70%. Modern CNNs often do 95% or better classifying images such as these. You might be able to squeeze more out of this model by stacking convolution layers or increasing the number of kernels, and you might get it to generalize slightly better by introducing a dropout layer. But you won't reach 95% with this network and this dataset.

One of the reasons modern CNNs can do image classification so accurately is that they're trained with millions of images. You don't need millions of samples of each class, but you probably need at *least* an order of magnitude more—if not *two* orders of magnitude more—than the 300 you trained with here. You could scour the internet for more images, but more images means more training time. If the goal is to achieve an accuracy of 95% or more, you'll quickly get to the point where the CNN takes too long to train—or find yourself shopping for an NVIDIA GPU (*https://oreil.ly/fkWUQ*).

That doesn't mean CNNs aren't practical for solving business problems. It just means that there's more to learn. The next section is the first step in understanding how to attain high levels of accuracy without training a CNN from scratch.

Pretrained CNNs

Microsoft, Google, and other tech companies use a subset of the ImageNet dataset (*https://oreil.ly/UZuxt*) containing more than 1 million images to train state-of-the-art CNNs to recognize hundreds of objects, including Arctic foxes and polar bears.

Then they make them available for public consumption. Called *pretrained CNNs*, they are more sophisticated than anything you're likely to train yourself. And if that's not awesome enough, Keras reduces the process of loading a pretrained CNN to one line of code.

Keras provides classes that wrap more than two dozen popular pretrained CNNs. The full list is documented on the Keras website (*https://oreil.ly/Jpl0Q*). Most of these CNNs are documented in scholarly papers such as "Deep Residual Learning for Image Recognition" (*https://oreil.ly/7np6D*) and "EfficientNet: Rethinking Model Scaling for Convolutional Neural Networks" (*https://oreil.ly/GCdfo*). Some have won prestigious competitions such as the ImageNet Large Scale Visual Recognition Challenge and the COCO Detection Challenge (*https://oreil.ly/RWOSu*). Among the most notable are the ResNet family of networks from Microsoft and the Inception networks from Google. Also noteworthy is MobileNet (*https://oreil.ly/m1UFj*), which trades size for accuracy and is ideal for mobile devices due to its small memory footprint. You can learn more about it in the Google AI blog (*https://oreil.ly/eoofX*).

Pretrained CNN Architectures

Pretrained CNNs achieve impressive levels of accuracy not just because they're trained with millions of images but also because they are deeper and architecturally more advanced than the CNNs presented thus far. For example, they use consecutive convolution layers to further refine the features extracted from each image, and many use *batch normalization* (*https://oreil.ly/jHAkA*) to normalize (standardize to unit variance) values propagated between layers during training.

But pretrained CNNs are more sophisticated for other reasons as well. ResNets, for example, pioneered a concept called *residual layers*, which add their input to their output. This simple innovation helped mitigate the vanishing-gradient problem that causes updates to become increasingly smaller as the optimizer goes backward through the network updating weights and biases, and it made much deeper networks possible.

And then there is the Inception family of CNNs, whose chief innovation was the use of 1 × 1 convolution kernels to minimize the computational overhead of larger convolution kernels by reducing the depth of the feature maps input to them. Recent versions of Inception have incorporated features of ResNets as well, and a derivative architecture known as Xception improved on Inception by introducing *depthwise separable convolutions* (*https://oreil.ly/8wfRD*). If you'd like to learn more, check out "A Simple Guide to the Versions of the Inception Network" (*https://oreil.ly/iwv4L*) and "Xception: Deep Learning with Depthwise Separable Convolutions" (*https://oreil.ly/U2OCm*).

The following statement instantiates Keras's `MobileNetV2` class (*https://oreil.ly/Oq0yb*) and initializes it with the weights, biases, and kernel values arrived at when the network was trained on the ImageNet dataset:

```
from tensorflow.keras.applications import MobileNetV2

model = MobileNetV2(weights='imagenet')
```

The `weights='imagenet'` parameter tells Keras what parameters to load to re-create the network in its trained state. You can also pass a path to a file containing custom weights, but `imagenet` is the only set of predefined weights that are currently supported.

Before an image is submitted to a pretrained CNN for classification, it must be sized to the dimensions the CNN expects—typically 224 × 224—and preprocessed. Different CNNs expect images to be preprocessed in different ways, so Keras provides a `preprocess_input` function for each pretrained CNN. It also includes utility functions for loading and resizing images. The following statements load an image from the filesystem and preprocess it for input to the MobileNetV2 network:

```
import numpy as np
from tensorflow.keras.applications.mobilenet import preprocess_input
from tensorflow.keras.preprocessing import image

x = image.load_img('arctic_fox.jpg', target_size=(224, 224))
x = image.img_to_array(x)
x = np.expand_dims(x, axis=0)
x = preprocess_input(x)
```

In most cases, `preprocess_input` does all the work that's needed, which often involves applying unit variance to pixel values and converting RGB images to BGR format. In some cases, however, you still need to divide the pixel values by 255. ResNet50V2 (*https://oreil.ly/IanRd*) is one example:

```
import numpy as np
from tensorflow.keras.applications.resnet50 import preprocess_input
from tensorflow.keras.preprocessing import image

x = image.load_img('arctic_fox.jpg', target_size=(224, 224))
x = image.img_to_array(x)
x = np.expand_dims(x, axis=0)
x = preprocess_input(x) / 255
```

Once an image is preprocessed, making a prediction is as simple as calling the network's `predict` method:

```
y = model.predict(x)
```

To help you interpret the output, Keras also provides a network-specific decode
_predictions method. Figure 10-10 shows what that method returned for a photo
submitted to ResNet50V2.

Figure 10-10. Output from decode_predictions

ResNet50V2 is 89% sure the photo contains an Arctic fox—which, it so happens, it
does. MobileNetV2 predicted with 92% certainty that the photo contains an Arctic
fox. Both networks were trained on the same dataset, but different pretrained CNNs
classify images slightly differently.

Using ResNet50V2 to Classify Images

Let's use Keras to load a pretrained CNN and classify a pair of images. Fire up a notebook and use the following statements to load ResNet50V2:

```
from tensorflow.keras.applications import ResNet50V2

model = ResNet50V2(weights='imagenet')
model.summary()
```

Next, load an Arctic fox image and show it in the notebook:

```
%matplotlib inline
import matplotlib.pyplot as plt
from tensorflow.keras.preprocessing import image

x = image.load_img('Wildlife/samples/arctic_fox/arctic_fox_140.jpeg',
                   target_size=(224, 224))
plt.xticks([])
plt.yticks([])
plt.imshow(x)
```

Now preprocess the image (remember that for ResNet50V2, you also have to divide
all the pixel values by 255 after calling Keras's preprocess_input method) and pass it
to the CNN for classification:

```
import numpy as np
from tensorflow.keras.applications.resnet50 import preprocess_input
```

```
from tensorflow.keras.applications.resnet50 import decode_predictions

x = image.img_to_array(x)
x = np.expand_dims(x, axis=0)
x = preprocess_input(x) / 255

y = model.predict(x)
decode_predictions(y)
```

The output should look like this:

```
[[('n02120079', 'Arctic_fox', 0.9999944),
  ('n02114548', 'white_wolf', 4.760021e-06),
  ('n02119789', 'kit_fox', 2.3306782e-07),
  ('n02442845', 'mink', 1.2460312e-07),
  ('n02111889', 'Samoyed', 1.1914468e-07)]]
```

ResNet50V2 is virtually certain that the image contains an Arctic fox. But now load a walrus image:

```
x = image.load_img('Wildlife/samples/walrus/walrus_143.png',
                   target_size=(224, 224))
plt.xticks([])
plt.yticks([])
plt.imshow(x)
```

Ask ResNet50V2 to classify it:

```
x = image.img_to_array(x)
x = np.expand_dims(x, axis=0)
x = preprocess_input(x) / 255

y = model.predict(x)
decode_predictions(y)
```

Here's the output:

```
[[('n02454379', 'armadillo', 0.63758147),
  ('n01704323', 'triceratops', 0.16057032),
  ('n02113978', 'Mexican_hairless', 0.07795086),
  ('n02398521', 'hippopotamus', 0.022284042),
  ('n01817953', 'African_grey', 0.016944142)]]
```

ResNet50V2 thinks the image is most likely an armadillo, but it's not even very sure about that. Can you guess why?

ResNet50V2 was trained with almost 1.3 million images. None of them, however, contained a walrus. The ImageNet 1000 Class List (*https://oreil.ly/mj9Zs*) shows a complete list of classes it *was* trained to recognize. A pretrained CNN is great when you need it to classify images using the classes it was trained with, but it is powerless to handle domain-specific tasks that it wasn't trained for.

But all is not lost. A technique called *transfer learning* enables pretrained CNNs to be repurposed to solve domain-specific problems. The repurposing can be done on an ordinary CPU; no GPU required. Transfer learning sometimes achieves 95% accuracy with just a few hundred training images. Once you learn about it, you'll have a completely different perspective on the efficacy of using CNNs to solve business problems.

Transfer Learning

Earlier, you used a dataset with photos of Arctic foxes, polar bears, and walruses to train a CNN to recognize Artic wildlife. Trained with 300 images—100 for each of the three classes—the CNN achieved an accuracy of around 60%. That's not sufficient for most purposes.

One solution is to train the CNN with tens of thousands of photos. A better solution—one that can deliver world-class accuracy with the 300 photos you have and doesn't require expensive hardware—is transfer learning. In the hands of software developers and engineers, transfer learning makes CNNs a practical solution for a variety of computer-vision problems. And it requires orders of magnitude less time and compute power than CNNs trained from scratch. Let's take a moment to understand what transfer learning is and how it works—and then put it to work identifying Arctic wildlife.

Pretrained CNNs trained on the ImageNet dataset can identify Arctic foxes and polar bears, but they can't identify walruses because they weren't trained with walrus images. Transfer learning lets you repurpose pretrained CNNs to identify objects they weren't originally trained to identify. It leverages the intelligence baked into pretrained CNNs, but it repurposes that intelligence to solve new problems.

Recall that a CNN has two groups of layers: bottleneck layers containing the convolution and pooling layers that extract features from images at various resolutions, and classification layers, which classify features output from the bottleneck layers as belonging to an Arctic fox, a polar bear, or something else. Convolution layers use convolution kernels to extract features, and the values in the convolution kernels are learned during training. This learning accounts for the bulk of the training time. When sophisticated CNNs are trained with millions of images, the convolution kernels become very efficient at extracting features. But that efficiency comes at a cost.

The premise behind transfer learning is shown in Figure 10-11. You load the bottleneck layers of a pretrained CNN, but you *don't* load the classification layers. Instead, you provide your own, which train orders of magnitude more quickly than an entire CNN. Then you pass the training images through the bottleneck layers for feature extraction and train the classification layers on the resulting features. The pretrained CNN might have been trained to extract features from pictures of apples and oranges, but those same layers are probably pretty good at extracting features from photos of dogs and cats too. By using the pretrained bottleneck layers to extract features and then using those features to train your own classification layers, you can teach the model that a certain feature extracted from an image might be indicative of a dog rather than an apple.

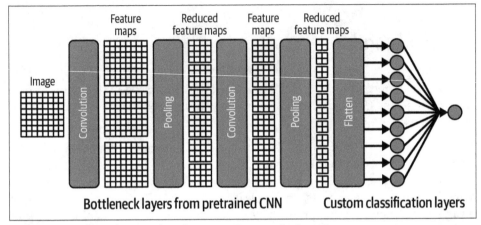

Figure 10-11. Neural network architecture for transfer learning

Transfer learning is relatively simple to implement with Keras and TensorFlow. Recall that the following statement loads ResNet50V2 and initializes it with the weights (including kernel values) and biases that were arrived at when the network was trained on a subset of the ImageNet dataset:

```
base_model = ResNet50V2(weights='imagenet')
```

To load ResNet50V2 (or any other pretrained CNN that Keras supports) without the classification layers, you simply add an `include_top=False` attribute:

```
base_model = ResNet50V2(weights='imagenet', include_top=False)
```

From that point, there are two ways to go about transfer learning. The first involves appending classification layers to the base model's bottleneck layers and setting each base layer's `trainable` attribute to `False` so that the weights, biases, and convolution kernels won't be updated when the network is trained:

```
for layer in base_model.layers:
    layer.trainable = False

model = Sequential()
model.add(base_model)
model.add(Flatten())
model.add(Dense(1024, activation='relu'))
model.add(Dense(3, activation='softmax'))
model.compile(optimizer='adam', loss='sparse_categorical_crossentropy',
              metrics=['accuracy'])

model.fit(x, y, validation_split=0.2, epochs=10, batch_size=10)
```

The second technique is to run all the training images through the base model for feature extraction, and then run the features through a separate network containing your classification layers:

```
features = base_model.predict(x)

model = Sequential()
model.add(Flatten())
model.add(Dense(128, activation='relu'))
model.add(Dense(3, activation='softmax'))
model.compile(optimizer='adam', loss='sparse_categorical_crossentropy',
              metrics=['accuracy'])

model.fit(features, y, validation_split=0.2, epochs=10, batch_size=10)
```

Which technique is better? The second is faster because the training images go through the bottleneck layers for feature extraction just one time rather than once per epoch. It's the technique you should use in the absence of a compelling reason to do otherwise. The first technique is slightly slower, but it lends itself to *fine-tuning* (*https://oreil.ly/pIQrN*), in which you unfreeze one or more bottleneck layers after training is complete and train for a few more epochs with a very low learning rate. It also facilitates *data augmentation*, which I'll introduce in the next section.

 Fine-tuning is frequently applied to transfer-learning models after training is complete in an effort to squeeze out an extra percentage point or two of accuracy. We will use fine-tuning in Chapter 13 to increase the accuracy of an NLP model that utilizes a pretrained neural network.

If you use the first technique to implement transfer learning, you make predictions by preprocessing the images and passing them to the model's `predict` method. For the second technique, making predictions is a two-step process. After preprocessing the images, you pass them to the base model's `predict` method, and then you pass the output from that method to your model's `predict` method:

```
x = image.img_to_array(x)
x = np.expand_dims(x, axis=0)
x = preprocess_input(x) / 255

features = base_model.predict(x)
predictions = model.predict(features)
```

And with that, transfer learning is complete. All that remains is to put it in practice.

Using Transfer Learning to Identify Arctic Wildlife

Let's use transfer learning to solve the same problem that we attempted to solve earlier with a scratch-built CNN: building a model that determines whether a photo contains an Arctic fox, a polar bear, or a walrus.

Create a Jupyter notebook and use the same code you used earlier to load the training and test images and assign labels to them: 0 for Arctic foxes, 1 for polar bears, and 2 for walruses. Once that's done, the next step is to preprocess the images. We'll use ResNet50V2 as our pretrained CNN, so use the ResNet version of `preprocess_input` to preprocess the pixels. Then divide the pixel values by 255:

```python
import numpy as np
from tensorflow.keras.applications.resnet50 import preprocess_input

x_train = preprocess_input(np.array(x_train)) / 255
x_test = preprocess_input(np.array(x_test)) / 255

y_train = np.array(y_train)
y_test = np.array(y_test)
```

The next step is to load ResNet50V2, being careful to load the bottleneck layers but *not* the classification layers, and use it to extract features from the training and test images:

```python
from tensorflow.keras.applications import ResNet50V2

base_model = ResNet50V2(weights='imagenet', include_top=False)

x_train = base_model.predict(x_train)
x_test = base_model.predict(x_test)
```

Now train a neural network to classify features extracted from the training images:

```python
from tensorflow.keras.models import Sequential
from tensorflow.keras.layers import Flatten, Dense

model = Sequential()
model.add(Flatten())
model.add(Dense(1024, activation='relu'))
model.add(Dense(3, activation='softmax'))
model.compile(optimizer='adam', loss='sparse_categorical_crossentropy',
              metrics=['accuracy'])

hist = model.fit(x_train, y_train,
                 validation_data=(x_test, y_test),
                 batch_size=10, epochs=10)
```

How well did the network train? Plot the training accuracy and validation accuracy for each epoch:

```
import seaborn as sns
sns.set()

acc = hist.history['accuracy']
val_acc = hist.history['val_accuracy']
epochs = range(1, len(acc) + 1)

plt.plot(epochs, acc, '-', label='Training Accuracy')
plt.plot(epochs, val_acc, ':', label='Validation Accuracy')
plt.title('Training and Validation Accuracy')
plt.xlabel('Epoch')
plt.ylabel('Accuracy')
plt.legend(loc='lower right')
plt.plot()
```

Your results will differ from mine, but I got about 97% accuracy. If you didn't quite get there, try training the network again:

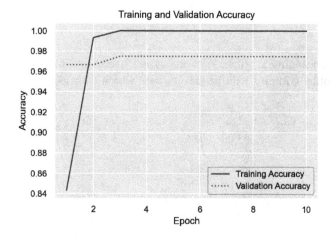

Finally, use a confusion matrix to visualize how well the network distinguishes between classes:

```
from sklearn.metrics import ConfusionMatrixDisplay as cmd

sns.reset_orig()
fig, ax = plt.subplots(figsize=(4, 4))
ax.grid(False)

y_pred = model.predict(x_test)
class_labels = ['arctic fox', 'polar bear', 'walrus']

cmd.from_predictions(y_test, y_pred.argmax(axis=1),
                     display_labels=class_labels, colorbar=False,
                     cmap='Blues', xticks_rotation='vertical', ax=ax)
```

Here's how it turned out for me:

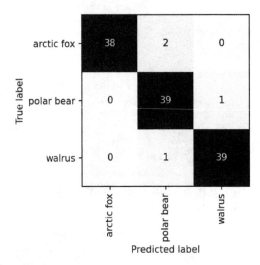

To see transfer learning at work, load one of the Arctic fox images from the *samples* folder. That folder contains wildlife images with which the model was neither trained nor validated:

```
x = image.load_img('Wildlife/samples/arctic_fox/arctic_fox_140.jpeg',
                   target_size=(224, 224))
plt.xticks([])
plt.yticks([])
plt.imshow(x)
```

Now preprocess the image, run it through ResNet50V2's feature extraction layers, and run the output through the newly trained classification layers:

```
x = image.img_to_array(x)
x = np.expand_dims(x, axis=0)
x = preprocess_input(x) / 255

y = base_model.predict(x)
predictions = model.predict(y)

for i, label in enumerate(class_labels):
    print(f'{label}: {predictions[0][i]}')
```

For me, the network predicted with almost 100% confidence that the image contains an Arctic fox:

```
arctic fox: 1.0
polar bear: 0.0
walrus: 0.0
```

Perhaps that's not surprising, since ResNet50V2 was trained with Arctic fox images. But now let's load a walrus image, which, you'll recall, ResNet50V2 was unable to classify:

```
x = image.load_img('Wildlife/samples/walrus/walrus_143.png',
                   target_size=(224, 224))
plt.xticks([])
plt.yticks([])
plt.imshow(x)
```

Preprocess the image and make a prediction:

```
x = image.img_to_array(x)
x = np.expand_dims(x, axis=0)
x = preprocess_input(x) / 255

y = base_model.predict(x)
predictions = model.predict(y)

for i, label in enumerate(class_labels):
    print(f'{label}: {predictions[0][i]}')
```

Here's how it turned out this time:

```
arctic fox: 0.0
polar bear: 0.0
walrus: 1.0
```

ResNet50V2 wasn't trained to recognize walruses, but your network was. That's transfer learning in a nutshell. It's the deep-learning equivalent of having your cake and eating it too. And it's the secret sauce that makes CNNs a viable tool for anyone with a laptop and a few hundred training images.

That's not to say that transfer learning will *always* get you 97% accuracy with 100 images per class. It won't. If a dataset lacks the information to achieve that level of separation, neither scratch-built CNNs nor transfer learning will magically make it happen. That's always true in machine learning and AI. You can't get water from a rock. And you can't build an accurate model from data that doesn't support it.

Data Augmentation

The previous example demonstrated how to use transfer learning to build a model that, with just 300 training images, can classify photos of three different types of Arctic wildlife with 97% accuracy. One of the benefits of transfer learning is that it can do more with fewer images. This feature is also a bug, however. With just 100 or so samples of each class, there is little diversity among images. A model might be able to recognize a polar bear if the bear's head is perfectly aligned in the center of the photo. But if the training images don't include photos with the bear's head aligned differently or tilted at different angles, the model might have difficulty classifying the photo.

One solution is *data augmentation*. Rather than scare up more training images, you can rotate, translate, and scale the images you have. It doesn't *always* increase a CNN's accuracy, but it frequently does, especially with small datasets. Keras makes it easy to randomly transform training images provided to a network. Images are transformed differently in each epoch, so if you train for 10 epochs, the network sees 10 different variations of each training image. This can increase a model's ability to generalize with little impact on training time. Figure 10-12 shows the effect of applying random transforms to a hot dog image. You can see why presenting the same image to a model in different ways might make the model more adept at recognizing hot dogs, regardless of how the hot dog is framed.

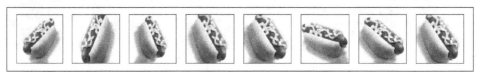

Figure 10-12. Hot dog image with random transforms applied

Keras has built-in support for data augmentation with images. Let's look at a couple of ways to put image augmentation to work, and then apply it to the Arctic wildlife model.

Image Augmentation with ImageDataGenerator

One way to apply image augmentation when training a model is to use Keras's Image DataGenerator class (*https://oreil.ly/C78UD*). ImageDataGenerator generates batches of training images on the fly, either from images you've loaded (for example, with Keras's load_img function) or from a specified location in the filesystem. The latter is especially useful when training CNNs with millions of images because it loads images into memory in batches rather than all at once. Regardless of where the images come from, however, ImageDataGenerator is happy to apply transforms as it serves them up.

Here's a simple example that you can try yourself. Use the following code to load an image from your filesystem, wrap an ImageDataGenerator around it, and generate 24 versions of the image:

```
import numpy as np
from tensorflow.keras.preprocessing import image
from tensorflow.keras.preprocessing.image import ImageDataGenerator
import matplotlib.pyplot as plt
%matplotlib inline

# Load an image
x = image.load_img('Wildlife/train/polar_bear/polar_bear_010.jpeg')
x = image.img_to_array(x)
x = np.expand_dims(x, axis=0)
```

```
# Wrap an ImageDataGenerator around it
idg = ImageDataGenerator(rescale=1./255,
                         horizontal_flip=True,
                         rotation_range=30,
                         width_shift_range=0.2,
                         height_shift_range=0.2,
                         zoom_range=0.2)
idg.fit(x)

# Generate 24 versions of the image
generator = idg.flow(x, [0], batch_size=1, seed=0)
fig, axes = plt.subplots(3, 8, figsize=(16, 6),
                         subplot_kw={'xticks': [], 'yticks': []})

for i, ax in enumerate(axes.flat):
    img, label = generator.next()
    ax.imshow(img[0])
```

Here's the result:

The parameters passed to `ImageDataGenerator` tell it how to transform each image it delivers:

`rescale=1./255`
    Divides each pixel value by 255

`horizontal_flip=True`
    Randomly flips the image horizontally (around the vertical axis)

`rotation_range=30`
    Randomly rotates the image by –30 to 30 degrees

`width_shift_range=0.2` *and* `height_shift_range=0.2`
    Randomly translates the image by –20% to 20%

```
zoom_range=0.2
```
Randomly scales the image by –20% to 20%

There are other parameters that you can use, such as `vertical_flip`, `shear_range`, and `brightness_range`, but you get the picture. The `flow` method (*https://oreil.ly/djx5s*) used in this example generates images from the images you pass to `fit`. The related `flow_from_directory` method (*https://oreil.ly/RBleY*) loads images from the filesystem and optionally labels them based on the subdirectories they're in.

The generator returned by `flow` can be passed directly to a model's `fit` method to provide randomly transformed images to the model as it is trained. Assume that `x_train` and `y_train` hold a collection of training images and labels. The following code wraps an `ImageDataGenerator` around them and uses them to train a model:

```
idg = ImageDataGenerator(rescale=1./255,
                         horizontal_flip=True,
                         rotation_range=30,
                         width_shift_range=0.2,
                         height_shift_range=0.2,
                         zoom_range=0.2)

idg.fit(x_train)
image_batch_size = 10
generator = idg.flow(x_train, y_train, batch_size=image_batch_size, seed=0)

model.fit(generator,
          steps_per_epoch=len(x_train) // image_batch_size,
          validation_data=(x_test, y_test),
          batch_size=20,
          epochs=10)
```

The `steps_per_epoch` parameter is key because an `ImageDataGenerator` can provide an infinite number of versions of each image. In this example, the `batch_size` parameter passed to `flow` tells the generator to create 10 images in each batch. Dividing the number of images by the image batch size to calculate `steps_per_epoch` ensures that in each training epoch, the model is provided with one transformed version of each image in the dataset.

 Versions of Keras prior to 2.1 didn't allow a generator to be passed to the `fit` method. Instead, they provided a separate method named `fit_generator`. That method is deprecated and will be removed in a future release.

Observe that the call to `fit` includes a `validation_data` parameter identifying a separate set of images and labels for validating the network during training. You

---

generally *don't* want to augment validation images, so you should avoid using `validation_split` when passing a generator to `fit`.

Image Augmentation with Augmentation Layers

You can use `ImageDataGenerator` to provide transformed images to a model, but recent versions of Keras provide an alternative in the form of image preprocessing layers (*https://oreil.ly/scoED*) and image augmentation layers (*https://oreil.ly/ikQMF*). Rather than transform training images separately, you can integrate the transforms directly into the model. Here's an example:

```python
from tensorflow.keras.models import Sequential
from tensorflow.keras.layers import Conv2D, MaxPooling2D, Flatten, Dense
from tensorflow.keras.layers import Rescaling, RandomFlip, RandomRotation
from tensorflow.keras.layers import RandomTranslation, RandomZoom

model = Sequential()
model.add(Rescaling(1./255))
model.add(RandomFlip(mode='horizontal'))
model.add(RandomTranslation(0.2, 0.2))
model.add(RandomRotation(0.2))
model.add(RandomZoom(0.2))
model.add(Conv2D(32, (3, 3), activation='relu', input_shape=(224, 224, 3)))
model.add(MaxPooling2D(2, 2))
model.add(Conv2D(128, (3, 3), activation='relu'))
model.add(MaxPooling2D(2, 2))
model.add(Flatten())
model.add(Dense(128, activation='relu'))
model.add(Dense(3, activation='softmax'))
```

Each image used to train the CNN has its pixel values divided by 255 and is then randomly flipped, translated, rotated, and scaled. Significantly, the `RandomFlip`, `RandomTranslation`, `RandomRotation`, and `RandomZoom` layers operate only on training images. They are inactive when the network is validated or asked to make predictions. Consequently, it's fine to use `validation_split` when training a model that contains image augmentation layers. The `Rescaling` layer is active at all times, meaning you no longer have to remember to divide pixel values by 255 before training the model or submitting an image for classification.

Applying Image Augmentation to Arctic Wildlife

Would image augmentation make transfer learning even better? There's one way to find out.

Create a Jupyter notebook and copy the code that loads the training and test images from the transfer learning example. Then use the following statements to prepare the data. Note that there is no need to divide by 255 this time because a `Rescaling` layer will take care of that:

```
import numpy as np
from tensorflow.keras.applications.resnet50 import preprocess_input

x_train = preprocess_input(np.array(x_train))
x_test = preprocess_input(np.array(x_test))

y_train = np.array(y_train)
y_test = np.array(y_test)
```

Now load ResNet50V2 without the classification layers and initialize it with the ImageNet weights. A key element here is preventing the bottleneck layers from training when the network is trained by setting their trainable attributes to False, effectively freezing those layers. Rather than setting each individual layer's trainable attribute to False, we'll set trainable to False on the model itself and allow that setting to be "inherited" by the individual layers:

```
from tensorflow.keras.applications import ResNet50V2

base_model = ResNet50V2(weights='imagenet', include_top=False)
base_model.trainable = False
```

Define a network that incorporates rescaling and augmentation layers, ResNet50V2's bottleneck layers, and dense layers for classification. Then train the network:

```
from tensorflow.keras.models import Sequential
from tensorflow.keras.layers import Flatten, Dense, Rescaling, RandomFlip
from tensorflow.keras.layers import RandomRotation, RandomTranslation, RandomZoom

model = Sequential()
model.add(Rescaling(1./255))
model.add(RandomFlip(mode='horizontal'))
model.add(RandomTranslation(0.2, 0.2))
model.add(RandomRotation(0.2))
model.add(RandomZoom(0.2))
model.add(base_model)
model.add(Flatten())
model.add(Dense(1024, activation='relu'))
model.add(Dense(3, activation='softmax'))
model.compile(optimizer='adam', loss='sparse_categorical_crossentropy',
              metrics=['accuracy'])

hist = model.fit(x_train, y_train,
                 validation_data=(x_test, y_test),
                 batch_size=10, epochs=10)
```

How well did the network train? Plot the training accuracy and validation accuracy for each epoch:

```
import seaborn as sns
sns.set()

acc = hist.history['accuracy']
```

```
val_acc = hist.history['val_accuracy']
epochs = range(1, len(acc) + 1)

plt.plot(epochs, acc, '-', label='Training Accuracy')
plt.plot(epochs, val_acc, ':', label='Validation Accuracy')
plt.title('Training and Validation Accuracy')
plt.xlabel('Epoch')
plt.ylabel('Accuracy')
plt.legend(loc='lower right')
plt.plot()
```

With a little luck, the accuracy slightly exceeded that of the model trained without data augmentation:

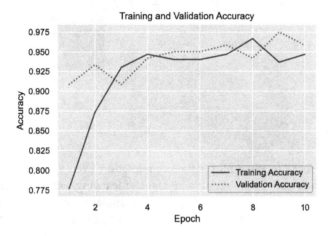

 You may find that the version of the model that uses data augmentation is *less* accurate than the version that doesn't. To be sure, I trained each version 10 times and averaged the results. I found that the data augmentation version delivered, on average, about 0.5% more accuracy than the version that lacks augmentation. That's not a lot, but data scientists frequently go to great lengths to improve accuracy by just a fraction of a percentage point.

Use a confusion matrix to visualize how well the network performed during testing:

```
from sklearn.metrics import ConfusionMatrixDisplay as cmd

sns.reset_orig()
fig, ax = plt.subplots(figsize=(4, 4))
ax.grid(False)

y_pred = model.predict(x_test)
class_labels = ['arctic fox', 'polar bear', 'walrus']
```

```
cmd.from_predictions(y_test, y_pred.argmax(axis=1),
                     display_labels=class_labels, colorbar=False,
                     cmap='Blues', xticks_rotation='vertical', ax=ax)
```

Here's how it turned out for me:

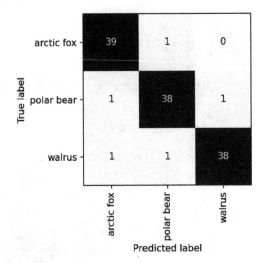

Data scientists sometimes employ data augmentation even when they're training a CNN from scratch rather than employing transfer learning, especially when the dataset is relatively small. It's a useful tool to know about, and one that could make a difference when you're trying to squeeze every last ounce of accuracy out of a deep-learning model.

Global Pooling

The purpose of including a `Flatten` layer in a CNN is to reshape the 3D tensors containing the final feature maps into 1D tensors suitable for input to a `Dense` layer. But `Flatten` isn't the only way to do it. Flattening sometimes leads to overfitting by providing too much information to the classification layers.

One way to combat overfitting is to introduce a `Dropout` layer. Another strategy is to reduce the width of the `Dense` layer. A third option is to replace the `Flatten` layer with a `GlobalMaxPooling2D` layer (*https://oreil.ly/hjPM3*) or a `GlobalAverage Pooling2D` layer (*https://oreil.ly/8hOPF*). They, too, output 1D tensors, but they generate them in a different way. And that way is less prone to overfitting.

To demonstrate, modify the MNIST dataset example earlier in this chapter to use a `GlobalMaxPooling2D` layer rather than a `Flatten` layer:

```
from tensorflow.keras.models import Sequential
from tensorflow.keras.layers import Conv2D, MaxPooling2D, \
```

```
GlobalMaxPooling2D, Dense

model = Sequential()
model.add(Conv2D(32, (3, 3), activation='relu', input_shape=(28, 28, 1)))
model.add(MaxPooling2D(2, 2))
model.add(Conv2D(64, (3, 3), activation='relu'))
model.add(MaxPooling2D(2, 2))
model.add(GlobalMaxPooling2D()) # In lieu of Flatten()
model.add(Dense(128, activation='relu'))
model.add(Dense(10, activation='softmax'))
model.compile(optimizer='adam', loss='sparse_categorical_crossentropy',
              metrics=['accuracy'])
model.summary(line_length=120)
```

The summary (Figure 10-13) shows that the output from the GlobalMaxPooling2D layer is a tensor containing 64 values—one per feature map emitted by the final Max Pooling2D layer—rather than $5 \times 5 \times 64$, or 1,600, values, as it was for the Flatten layer. Each value is the maximum of the 25 values in each 5×5 feature map. Had you used GlobalAveragePooling2D instead, each value would have been the average of the 25 values in each feature map.

Layer (type)	Output Shape	Param #
conv2d (Conv2D)	(None, 26, 26, 32)	320
max_pooling2d (MaxPooling2D)	(None, 13, 13, 32)	0
conv2d_1 (Conv2D)	(None, 11, 11, 64)	18496
max_pooling2d_1 (MaxPooling2D)	(None, 5, 5, 64)	0
global_max_pooling2d (GlobalMaxPooling2D)	(None, 64)	0
dense (Dense)	(None, 128)	8320
dense_1 (Dense)	(None, 10)	1290

```
Total params: 28,426
Trainable params: 28,426
Non-trainable params: 0
```

Figure 10-13. Output from the summary method

Global pooling sometimes increases a CNN's ability to generalize and sometimes does not. For the MNIST dataset, it slightly diminishes accuracy. As is so often the case in machine learning, the only way to know is to try. And due to the randomness inherent in training neural networks, it's always advisable to train the network several times in each configuration and average the results before drawing conclusions.

Audio Classification with CNNs

Imagine that you're the leader of a group of climate scientists concerned about the planet's dwindling rainforests. The world loses up to 10 million acres of old-growth rainforests each year, much of it due to illegal logging. Your team plans to convert thousands of discarded smartphones into solar-powered listening devices and position them throughout the Amazon to transmit alerts in response to the sounds of chainsaws and truck engines. You need software that uses AI to identify such sounds in real time. And you need it fast, because climate change won't wait.

An effective way to perform audio classification is to convert audio streams into spectrogram images (*https://oreil.ly/gpCij*), which provide visual representations of spectrums of frequencies as they vary over time, and use CNNs to classify the spectrograms. The spectrograms in Figure 10-14 were generated from WAV files containing chainsaw sounds. Let's use transfer learning to create a model that can identify the telltale sounds of logging operations and distinguish them from ambient sounds such as wildlife and thunderstorms.

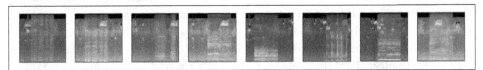

Figure 10-14. Spectrograms generated from audio files containing chainsaw sounds

The tutorial in this section was inspired by the Rainforest Connection (*https://oreil.ly/SkoZO*), which uses recycled Android phones to monitor rainforests for sounds of illegal activity. A TensorFlow CNN hosted in the cloud analyzes audio from the phones and may one day run on the phones themselves (*https://oreil.ly/cN3Cv*) with an assist from TensorFlow Lite (*https://oreil.ly/2q9qJ*), a smaller version of TensorFlow designed for mobile, embedded, and edge devices. For more information, see "The Fight Against Illegal Deforestation with TensorFlow" (*https://oreil.ly/aL4Ga*) in the Google AI blog. It's just one example of how AI is making the world a better place.

Begin by downloading a ZIP file (*https://oreil.ly/dkSeJ*) containing a dataset of rainforest sounds. (Warning: it's a 666 MB download.) Create a subdirectory named *Sounds* in the directory where your notebooks are hosted, and copy the contents of the ZIP file into the subdirectory. *Sounds* now contains subdirectories named *background*, *chainsaw*, *engine*, and *storm*. Each subdirectory contains 100 WAV files. The WAV files in the *background* directory contain rainforest background noises only, while the files in the other subdirectories include the sounds of chainsaws, engines, and thunderstorms overlaid on the background noises. I generated these files by using a soundscape synthesis package named Scaper (*https://oreil.ly/fS6os*) to combine sounds in the public UrbanSound8K dataset (*https://oreil.ly/mNQt2*) with rainforest sounds. Play a few of the WAV files on your computer to get a feel for the sounds they contain.

Now create a Jupyter notebook and paste the following code into the first cell:

```
import numpy as np
import librosa.display, os
import matplotlib.pyplot as plt
%matplotlib inline

def create_spectrogram(audio_file, image_file):
    fig = plt.figure()
    ax = fig.add_subplot(1, 1, 1)
    fig.subplots_adjust(left=0, right=1, bottom=0, top=1)

    y, sr = librosa.load(audio_file)
    ms = librosa.feature.melspectrogram(y=y, sr=sr)
    log_ms = librosa.power_to_db(ms, ref=np.max)
    librosa.display.specshow(log_ms, sr=sr)

    fig.savefig(image_file)
    plt.close(fig)

def create_pngs_from_wavs(input_path, output_path):
    if not os.path.exists(output_path):
        os.makedirs(output_path)

    dir = os.listdir(input_path)

    for i, file in enumerate(dir):
        input_file = os.path.join(input_path, file)
        output_file = os.path.join(output_path, file.replace('.wav', '.png'))
        create_spectrogram(input_file, output_file)
```

This code defines a pair of functions to help convert WAV files into spectrogram images. `create_spectrogram` uses a Python package named Librosa (*https://oreil.ly/vbAc8*) to create a spectrogram image from a WAV file. `create_pngs_from_wavs` converts all the WAV files in a specified directory into spectrogram images. You will need to install Librosa if it isn't installed already.

Use the following statements to create PNG files containing spectrograms from all the WAV files in the *Sounds* directory's subdirectories:

```
create_pngs_from_wavs('Sounds/background', 'Spectrograms/background')
create_pngs_from_wavs('Sounds/chainsaw', 'Spectrograms/chainsaw')
create_pngs_from_wavs('Sounds/engine', 'Spectrograms/engine')
create_pngs_from_wavs('Sounds/storm', 'Spectrograms/storm')
```

Check the *Spectrograms* directory for subdirectories containing spectrograms and confirm that each subdirectory contains 100 PNG files. Then use the following code to define two new helper functions for loading and displaying spectrograms, and declare two Python lists—one to store spectrogram images and another to store class labels:

```
from tensorflow.keras.preprocessing import image

def load_images_from_path(path, label):
    images, labels = [], []

    for file in os.listdir(path):
        images.append(image.img_to_array(image.load_img(os.path.join(path, file),
                     target_size=(224, 224, 3))))
        labels.append((label))

    return images, labels

def show_images(images):
    fig, axes = plt.subplots(1, 8, figsize=(20, 20),
                             subplot_kw={'xticks': [], 'yticks': []})

    for i, ax in enumerate(axes.flat):
        ax.imshow(images[i] / 255)

x, y = [], []
```

Use the following statements to load the background spectrogram images, add them to the list named x, and label them with 0s:

```
images, labels = load_images_from_path('Spectrograms/background', 0)
show_images(images)

x += images
y += labels
```

Repeat this process to load chainsaw spectrograms from the *Spectrograms/chainsaw* directory, engine spectrograms from the *Spectrograms/engine* directory, and thunderstorm spectrograms from the *Spectrograms/storm* directory. Label chainsaw spectrograms with 1s, engine spectrograms with 2s, and thunderstorm spectrograms with 3s. Here are the labels for the four classes of images:

Spectrogram type	Label
Background	0
Chainsaw	1
Engine	2
Storm	3

Since this model may one day run on mobile phones, we'll use MobileNetV2 as the base network. Use the following code to preprocess the pixels and split the images and labels into two datasets—one for training and one for testing:

```
from sklearn.model_selection import train_test_split
from tensorflow.keras.applications.mobilenet import preprocess_input

x = preprocess_input(np.array(x))
```

```
y = np.array(y)

x_train, x_test, y_train, y_test = train_test_split(x, y, stratify=y,
                                                    test_size=0.3,
                                                    random_state=0)
```

Call Keras's `MobileNetV2` function (*https://oreil.ly/UyRAc*) to instantiate MobileNetV2 without the classification layers. Then run the training data and test data through MobileNetV2 to extract features from the spectrogram images:

```
from tensorflow.keras.applications import MobileNetV2

base_model = MobileNetV2(weights='imagenet', include_top=False,
                         input_shape=(224, 224, 3))

train_features = base_model.predict(x_train)
test_features = base_model.predict(x_test)
```

Define a neural network to classify features extracted by MobileNetV2:

```
from tensorflow.keras.models import Sequential
from tensorflow.keras.layers import Dense, Flatten

model = Sequential()
model.add(Flatten())
model.add(Dense(512, activation='relu'))
model.add(Dense(4, activation='softmax'))
model.compile(optimizer='adam', loss='sparse_categorical_crossentropy',
              metrics=['accuracy'])
```

 As an experiment, I replaced the `Flatten` layer with a `Global AveragePooling2D` layer. Validation accuracy improved slightly, but the model didn't generalize as well when tested with audio extracted from a documentary video. This underscores an important point from Chapter 9: you can have full trust and confidence in a model only when it's tested with data it has never seen before—preferably data that comes from a different source.

Train the network with the features:

```
hist = model.fit(train_features, y_train,
                 validation_data=(test_features, y_test),
                 batch_size=10, epochs=10)
```

Plot the training and validation accuracy:

```
import seaborn as sns
sns.set()

acc = hist.history['accuracy']
val_acc = hist.history['val_accuracy']
epochs = range(1, len(acc) + 1)
```

```
plt.plot(epochs, acc, '-', label='Training Accuracy')
plt.plot(epochs, val_acc, ':', label='Validation Accuracy')
plt.title('Training and Validation Accuracy')
plt.xlabel('Epoch')
plt.ylabel('Accuracy')
plt.legend(loc='lower right')
plt.plot()
```

The validation accuracy should reach 95% or higher:

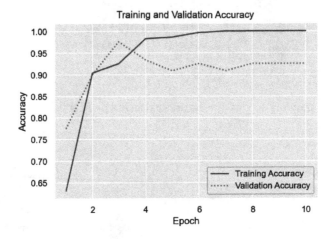

Run the test images through the network and use a confusion matrix to assess the results:

```
from sklearn.metrics import ConfusionMatrixDisplay as cmd

sns.reset_orig()
fig, ax = plt.subplots(figsize=(4, 4))
ax.grid(False)

y_pred = model.predict(test_features)
class_labels = ['background', 'chainsaw', 'engine', 'storm']

cmd.from_predictions(y_test, y_pred.argmax(axis=1),
                     display_labels=class_labels, colorbar=False,
                     cmap='Blues', xticks_rotation='vertical', ax=ax)
```

The network is reasonably adept at identifying clips that don't contain the sounds of chainsaws or engines. It sometimes confuses chainsaw sounds and engine sounds. That's OK, because the presence of either might indicate illicit activity in a rainforest:

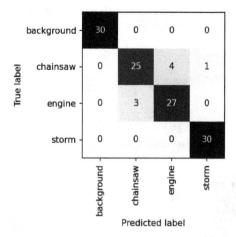

The *Sounds* directory has a subdirectory named *samples* containing WAV files with which the CNN was neither trained nor validated. The WAV files bear no relation to the samples used for training and testing; they come from a YouTube video documenting Brazil's efforts to curb illegal logging. Let's use the model you just trained to analyze these files for sounds of logging activity.

Start by creating a spectrogram from the first sample WAV file, which contains audio of loggers cutting down trees in the Amazon:

```
create_spectrogram('Sounds/samples/sample1.wav', 'Spectrograms/sample1.png')

x = image.load_img('Spectrograms/sample1.png', target_size=(224, 224))
plt.xticks([])
plt.yticks([])
plt.imshow(x)
```

Preprocess the spectrogram image, pass it to MobileNetV2 for feature extraction, and classify the features:

```
x = image.img_to_array(x)
x = np.expand_dims(x, axis=0)
x = preprocess_input(x)

y = base_model.predict(x)
predictions = model.predict(y)

for i, label in enumerate(class_labels):
    print(f'{label}: {predictions[0][i]}')
```

Now create a spectrogram from a WAV file that features the sound of a logging truck rumbling through the rainforest:

```
create_spectrogram('Sounds/samples/sample2.wav', 'Spectrograms/sample2.png')

x = image.load_img('Spectrograms/sample2.png', target_size=(224, 224))
plt.xticks([])
plt.yticks([])
plt.imshow(x)
```

Preprocess the image, pass it to MobileNetV2 for feature extraction, and classify the features:

```
x = image.img_to_array(x)
x = np.expand_dims(x, axis=0)
x = preprocess_input(x)

y = base_model.predict(x)
predictions = model.predict(y)

for i, label in enumerate(class_labels):
    print(f'{label}: {predictions[0][i]}')
```

If the network got either of the samples wrong, try training it again. Remember that a neural network will train differently every time, in part because Keras initializes the weights with small random values. In the real world, data scientists often train a neural network several times and average the results to quantify its accuracy.

Summary

Convolutional neural networks excel at image classification because they use convolution kernels to extract features from images at different resolutions—features intended to accentuate differences between classes. Convolution layers use convolution kernels to extract features, and pooling layers reduce the size of the feature maps output from the convolution layers. Output from these layers is input to fully connected layers for classification. Keras provides implementations of convolution and pooling layers in classes such as Conv2D and MaxPooling2D.

Training a CNN from scratch when there is a relatively high degree of separation between classes—for example, the MNIST dataset—is feasible on an ordinary laptop or PC. Training a CNN to solve a more perceptual problem requires more training images and commensurately more compute power. Transfer learning is a practical alternative to training CNNs from scratch. It uses the intelligence already present in the bottleneck layers of pretrained CNNs to extract features from images, and then uses its own classification layers to interpret the results.

Data augmentation can increase the accuracy of a CNN trained with a relatively small number of images and is especially useful with transfer learning. Augmentation involves applying random transforms such as translations and rotations to the training images. You can transform images before inputting them to the network with Keras's `ImageDataGenerator` class, or you can build the transforms into the network with layers such as `RandomRotation` and `RandomTranslation`. Layers that transform images are active at training time but inactive when the network makes predictions.

CNNs are applicable to a wide variety of computer-vision problems and are almost single-handedly responsible for the rapid advancements made in that field in the past decade. They play an important role in modern facial recognition systems too. Want to know more? Detecting and identifying faces in photographs is the subject of the next chapter.

## CHAPTER 11

Face Detection and Recognition

Not long ago, I boarded a flight to Europe and was surprised that I didn't have to show my passport. I passed in front of a camera and was promptly welcomed aboard the flight. It was part of an early pilot for Delta Air Lines' effort to push forward with facial recognition (*https://oreil.ly/tRoSq*) and offer a touchless curb-to-gate travel experience.

Facial recognition is everywhere. It's one of the most common, and sometimes controversial, applications for AI. Meta, formerly known as Facebook, uses it to tag friends in photos—at least it did until it killed the feature due to privacy concerns (*https://oreil.ly/ffn8n*). Apple uses it to allow users to unlock their iPhones (*https:// oreil.ly/X9j5J*), while Microsoft uses it to unlock Windows PCs. Uber uses it to confirm the identity of its drivers (*https://oreil.ly/ZkRAw*). Used properly, facial recognition has vast potential to make the world a better, safer, and more secure place.

Suppose you want to build a system that identifies people in photos or video frames. Perhaps it's part of a security system that restricts access to college dorms (*https:// oreil.ly/NXt8q*) to students and staff who are authorized to enter. Or perhaps you're writing an app that searches your hard disk for photos of people you know. ("Show me all the photos of me and my daughter.") Building systems such as these requires algorithms or models capable of:

- Finding faces in photos or video frames, a process known as *face detection*
- Identifying the faces detected, a process known as *facial recognition* or *face identification*

Numerous well-known algorithms exist for finding and identifying faces in photos. Some rely on deep learning—in particular, convolutional neural networks—while some do not. Facial recognition, after all, predated the explosion of deep learning by decades. But deep learning has supercharged the science of facial recognition and made it more practical than ever before.

This chapter begins by introducing two popular face detection methods. Then it moves on to facial recognition and introduces transfer learning as a means for recognizing faces. It concludes with a tutorial in which you put the pieces together and build a facial recognition system of your own. Sound like fun? Then let's get started.

Face Detection

The sections that follow introduce two widely used algorithms for face detection—one that relies on machine learning and another that uses deep learning—as well as libraries that implement them. The goal is to be able to find all the faces in a photo or video frame like the one in Figure 11-1. Afterward, I'll present an easy-to-use function that you can call to extract all the facial images from a photo and save them to disk or submit them to a facial recognition model.

Figure 11-1. Face detection

Face Detection with Viola-Jones

One of the fastest and most popular algorithms for detecting faces in photos stems from a paper published in 2001 titled "Rapid Object Detection Using a Boosted Cascade of Simple Features" (*https://oreil.ly/uLz4E*). Sometimes known as Viola-Jones (the authors of the paper), the algorithm keys on the relative intensities of adjacent blocks of pixels. For example, the average pixel intensity in a rectangle around the eyes is typically darker than the average pixel intensity in a rectangle immediately below that area. Similarly, the bridge of the nose is usually lighter than the region around the eyes, so two dark rectangles with a bright rectangle in the middle might represent two eyes and a nose. The presence of many such Haar-like features (*https:// oreil.ly/zrFxR*) in a frame at the right locations is an indicator that the frame contains a face (Figure 11-2).

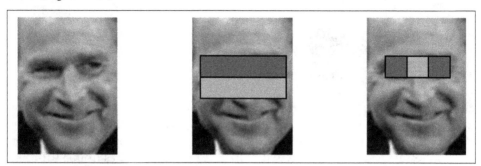

Figure 11-2. Face detection using Haar-like features

Viola-Jones works by sliding windows of various sizes over an image looking for frames with Haar-like features in the right places. At each stop, the pixels in the window are scaled to a specified size (typically 24 × 24), and features are extracted and fed into a binary classifier that returns positive indicating the frame contains a face or negative indicating it does not. Then the window slides to the next location and the detection regimen begins again.

The key to Viola-Jones's performance is the binary classifier. A frame that is 24 pixels wide and 24 pixels high contains more than 160,000 combinations of rectangles representing potential Haar-like features. Rather than compute values for every combination, Viola-Jones computes only those that the classifier requires. Furthermore, how many features the classifier requires depends on the content of the frame. The classifier is actually several binary classifiers arranged in stages. The first stage might require just one feature. The second stage might require 10, the third might require 20, and so on. Features are extracted and passed to stage *n* only if stage *n* − 1 returns positive, giving rise to the term *cascade classifier*.

Figure 11-3 depicts a three-stage cascade classifier. Each stage is carefully tuned to achieve a 100% detection rate using a limited number of features even if the false-positive rate is high. In the first stage, one feature determines whether the frame contains a face. A positive response means the frame *might* contain a face; a negative response means that it most certainly doesn't, in which case no further checks are performed. If stage 1 returns positive, however, 10 other features are extracted and passed to stage 2. A frame is judged to contain a face only if *all* stages return positive, yielding a cumulative false-positive rate near zero. In machine learning, this is a design pattern known as *high recall then precision* because while individual stages are tuned for high recall, the cumulative effect is one of high precision.

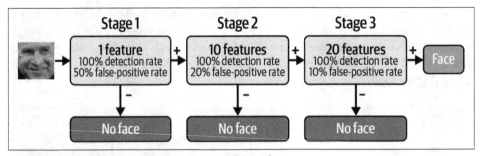

Figure 11-3. Face detection using a cascade classifier

One benefit of this architecture is that frames lacking faces tend to fall out fast because they evoke a negative response early in the cascade. Because most frames *don't* contain faces, the algorithm runs very quickly until it encounters a frame that does. In testing with a 38-stage classifier trained on 6,061 features from 4,916 facial images, Viola and Jones found that, on average, just 10 features were extracted from each frame.

The efficacy of Viola-Jones depends on the cascade classifier, which is essentially a machine learning model trained with facial and nonfacial images. Training is slow, but predictions are fast. In some respects, Viola-Jones acts like a CNN handcrafted to extract the minimum number of features needed to determine whether a frame contains a face. To speed feature extraction, Viola-Jones uses a clever mathematical trick called *integral images* (*https://oreil.ly/7Z79V*) to rapidly compute the difference in intensity between two blocks of pixels. The result is a system that can identify bounding boxes surrounding faces in an image with a relatively high degree of accuracy, and it can do so quickly enough to detect faces in live video frames.

Using the OpenCV Implementation of Viola-Jones

OpenCV (*https://oreil.ly/IWvGS*) is a popular open source computer-vision library that's free for commercial use. It provides an implementation of Viola-Jones in its `CascadeClassifier` class (*https://oreil.ly/NJAIz*), along with an XML file containing a cascade classifier trained to detect faces. The following statements use `Cascade Classifier` in a Jupyter notebook to detect faces in an image and draw rectangles around the faces. You can use an image of your own or download the one featured in my example (*https://oreil.ly/PM24X*) from GitHub:

```
import cv2
from cv2 import CascadeClassifier
from matplotlib.patches import Rectangle
import matplotlib.pyplot as plt
%matplotlib inline
```

```
image = plt.imread('Data/Amsterdam.jpg')
fig, ax = plt.subplots(figsize=(12, 8), subplot_kw={'xticks': [], 'yticks': []})
ax.imshow(image)

model = CascadeClassifier(cv2.data.haarcascades +
                          'haarcascade_frontalface_default.xml')
faces = model.detectMultiScale(image)

for face in faces:
    x, y, w, h = face
    rect = Rectangle((x, y), w, h, color='red', fill=False, lw=2)
    ax.add_patch(rect)
```

Here's the output with a photo of a mother and her daughter taken in Amsterdam a few years ago:

CascadeClassifier detected the two faces in the photo, but it also suffered a number of false positives. One way to mitigate that is to use the minNeighbors parameter. It defaults to 3, but higher values make CascadeClassifier more selective. With min Neighbors=20, detectMultiScale finds just the faces of the two people:

```
faces = model.detectMultiScale(image, minNeighbors=20)
```

Here is the output:

As detectMultiScale analyzes an image, it typically detects a face multiple times, each defined by a bounding box that's aligned slightly differently. The minNeighbors parameter specifies the minimum number of times a face must be detected to be reported as a face. Higher values deliver higher precision (fewer false positives), but at the cost of lower recall, which means some faces might not be detected.

CascadeClassifier frequently requires tuning in this manner to strike the right balance between finding too many faces and finding too few. With that in mind, it is among the fastest face detection algorithms in existence. It can also be used to detect objects other than faces by loading XML files containing other pretrained classifiers. In OpenCV's GitHub repo (*https://oreil.ly/Aj0At*), you'll find XML files for detecting silverware and other objects using Haar-like features, and XML files that detect objects using a different type of discriminator called local binary patterns (*https://oreil.ly/a9p27*).

Face Detection with Convolutional Neural Networks

While more computationally expensive, deep-learning methods often do a better job of detecting faces in images than Viola-Jones. In particular, *multitask cascaded convolutional neural networks*, or MTCNNs, have proven adept at face detection in a variety of benchmarks. They also identify facial landmarks such as the eyes, the nose, and the mouth.

Figure 11-4 is adapted from a diagram in the 2016 paper titled "Joint Face Detection and Alignment Using Multitask Cascaded Convolutional Networks" (*https://oreil.ly/QxjCe*) that proposed MTCNNs. An MTCNN uses three CNNs arranged in a series to detect faces. The first one, called the *Proposal Network*, or P-Net, is a shallow CNN that searches the image at various resolutions looking for features indicative of faces. Rectangles identified by P-Net are combined to form candidate face rectangles and are input to the Refine Network, or R-Net, which is a deeper CNN that examines each rectangle more closely and rejects those that lack faces. Finally, output from R-Net is input to the Output Network (O-Net), which further filters candidate rectangles and identifies facial landmarks. MTCNNs are multitask CNNs because they produce three outputs each—a classification output indicating the confidence level that the rectangle contains a face, and two regression outputs locating the face and facial landmarks—rather than just one. And they're cascaded like Viola-Jones classifiers to quickly rule out frames that don't contain faces.

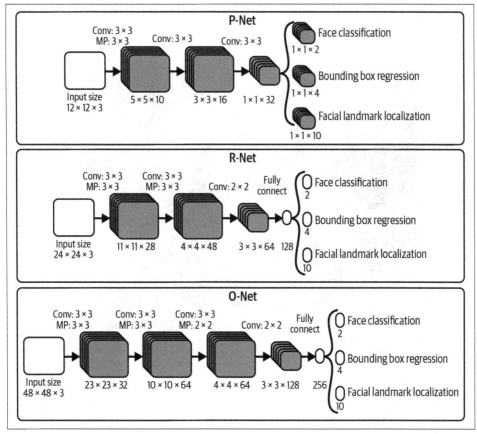

Figure 11-4. Multitask cascaded convolutional neural network

A handy MTCNN implementation is available in the Python package named MTCNN. The following statements use it to detect faces in the same photo featured in the previous example:

```
import matplotlib.pyplot as plt
from matplotlib.patches import Rectangle
from mtcnn.mtcnn import MTCNN
%matplotlib inline

image = plt.imread('Data/Amsterdam.jpg')
fig, ax = plt.subplots(figsize=(12, 8), subplot_kw={'xticks': [], 'yticks': []})
ax.imshow(image)

detector = MTCNN()
faces = detector.detect_faces(image)

for face in faces:
    x, y, w, h = face['box']
    rect = Rectangle((x, y), w, h, color='red', fill=False, lw=2)
    ax.add_patch(rect)
```

Here's the result:

The MTCNN detected not only the faces of the two people but also the face of a statue reflected in the door behind them. Here's what detect_faces actually returned—a list containing three dictionaries, each corresponding to one of the faces in the photo:

```
[
  {
    'box': [723, 248, 204, 258],
    'confidence': 0.9997798800468445,
    'keypoints': {
      'left_eye': (765, 341),
      'right_eye': (858, 343),
      'nose': (800, 408),
      'mouth_left': (770, 432),
      'mouth_right': (864, 433)
    }
  },
  {
    'box': [538, 258, 183, 232],
    'confidence': 0.9997591376304626,
    'keypoints': {
      'left_eye': (601, 353),
      'right_eye': (685, 344),
      'nose': (662, 394),
      'mouth_left': (614, 433),
      'mouth_right': (689, 424)
    }
  },
  {
    'box': [1099, 84, 40, 41],
    'confidence': 0.8863282203674316,
    'keypoints': {
      'left_eye': (1108, 101),
      'right_eye': (1123, 96),
      'nose': (1116, 102),
      'mouth_left': (1114, 115),
      'mouth_right': (1127, 111)
    }
  }
]
```

You can eliminate the face in the reflection in either of two ways: by ignoring faces with a confidence level below a certain threshold, or by passing a min_face_size parameter to the MTCNN function so that detect_faces ignores faces smaller than a specified size. Here's a modified for loop that does the former:

```
for face in faces:
    if face['confidence'] > 0.9:
        x, y, w, h = face['box']
        rect = Rectangle((x, y), w, h, color='red', fill=False, lw=2)
        ax.add_patch(rect)
```

And here's the result:

The facial rectangles in Figure 11-1 were generated using MTCNN's default settings—that is, without any filtering based on confidence levels or face sizes. Generally speaking, it does a better job out of the box than `CascadeClassifier` at detecting faces.

Extracting Faces from Photos

Once you know how to find faces in photos, it's a simple matter to extract facial images in order to train a model or submit them to a trained model for identification. Example 11-1 presents a Python function that accepts a path to an image file and returns a list of facial images. By default, it crops facial images so that they're square (perfect for passing them to a CNN), but you can disable cropping by passing the function a `crop=False` parameter. You can also specify a minimum confidence level with a `min_confidence` parameter, which defaults to 0.9.

Example 11-1. Function for extracting facial images from a photo

```
import numpy as np
from PIL import Image, ImageOps
from mtcnn.mtcnn import MTCNN

def extract_faces(input_file, min_confidence=0.9, crop=True):
    # Load the image and orient it correctly
    pil_image = Image.open(input_file)
    exif = pil_image.getexif()
```

```
for k in exif.keys():
    if k != 0x0112:
        exif[k] = None
        del exif[k]

pil_image.info["exif"] = exif.tobytes()
pil_image = ImageOps.exif_transpose(pil_image)
image = np.array(pil_image)

# Find the faces in the image
detector = MTCNN()
faces = detector.detect_faces(image)
faces = [face for face in faces if face['confidence'] >= min_confidence]
results = []

for face in faces:
    x1, y1, w, h = face['box']

    if (crop):
        # Compute crop coordinates
        if w > h:
            x1 = x1 + ((w - h) // 2)
            w = h
        elif h > w:
            y1 = y1 + ((h - w) // 2)
            h = w

    # Extract the facial image and add it to the list
    x2 = x1 + w
    y2 = y1 + h
    results.append(Image.fromarray(image[y1:y2, x1:x2]))

# Return all the facial images
return results
```

I passed the photo (*https://oreil.ly/gI4r2*) in Figure 11-5 to the function, and it returned the faces underneath. The items returned from `extract_faces` are Python Imaging Library (PIL) (*https://oreil.ly/VsHPz*) images, so you can resize them or save them to disk with a single line of code. Here's a code snippet that extracts all the faces from a photo, resizes them to 224 × 224 pixels, and saves the resized images:

```
faces = extract_faces('PATH_TO_IMAGE_FILE')

for i, face in enumerate(faces):
    face.resize((224, 224)).save(f'face{i}.jpg')
```

Figure 11-5. Facial images extracted from a photo

With `extract_faces` to lend a hand, it's a relatively simple matter to generate a set of facial images for training a CNN from a batch of photos on your hard disk, or to extract faces from a photo and submit them to a CNN for identification.

Facial Recognition

Now that you know how to detect faces in photos, the next step is to learn how to identify them. Several algorithms for recognizing faces in photos (*https://oreil.ly/E7D32*) have been developed over the years. Some rely on biometrics, such as the distance between the eyes or the texture of the skin, while others take a more holistic approach by treating facial identification as a pattern recognition problem. State-of-the-art models today frequently rely on deep convolutional neural networks. One of the primary benchmarks for facial recognition models is the Labeled Faces in the Wild (LFW) dataset (*https://oreil.ly/0hylZ*) pictured in Figure 11-6, which contains more than 13,000 facial images of more than 5,000 people collected from the web. Deep-learning models such as MobiFace (*https://oreil.ly/9TwRP*) and FaceNet (*https://oreil.ly/cXlF6*) routinely achieve greater than 99% accuracy on the dataset.

This equals or exceeds a human's ability to identify faces (*https://oreil.ly/WEMnj*) in LFW photos.

Figure 11-6. The Labeled Faces in the Wild dataset

Chapter 5 presented a support vector machine (SVM) that achieved 85% accuracy using a subset of 500 images—100 each of five famous people—from the dataset. Chapter 9 tackled the same problem with a neural network, with similar results. These models merely scratch the surface of what modern facial recognition can accomplish. Let's apply CNNs and transfer learning to the same LFW subset and see if they can do better at recognizing faces in photos. Along the way, you'll learn a valuable lesson about pretrained CNNs and the specificity of the weights that are generated when those CNNs are trained.

Applying Transfer Learning to Facial Recognition

The first step in exploring CNN-based facial recognition is to load the LFW dataset. This time, we'll load full-size color images and crop them to 128 × 128 pixels. Here's the code:

```
import pandas as pd
from sklearn.datasets import fetch_lfw_people

faces = fetch_lfw_people(min_faces_per_person=100, slice_=None, resize=1.0,
                         color=True)
faces.images = faces.images[:, 60:188, 60:188]
faces.data = faces.images.reshape(faces.images.shape[0], faces.images.shape[1] *
                                  faces.images.shape[2], faces.images.shape[3])
class_count = len(faces.target_names)
```

```
print(faces.target_names)
print(faces.images.shape)
```

Because we set `min_faces_per_person` to 100, a total of 1,140 facial images corresponding to five people were loaded. Use the following statements to show the first several images and the labels that go with them:

```
import matplotlib.pyplot as plt
%matplotlib inline

fig, ax = plt.subplots(3, 6, figsize=(18, 10))

for i, axi in enumerate(ax.flat):
    axi.imshow(faces.images[i])
    axi.set(xticks=[], yticks=[], xlabel=faces.target_names[faces.target[i]])
```

The dataset is imbalanced, containing almost as many photos of George W. Bush as of everyone else combined. Use the following code to reduce the dataset to 100 images of each person, for a total of 500 facial images:

```
import numpy as np

mask = np.zeros(faces.target.shape, dtype=bool)

for target in np.unique(faces.target):
    mask[np.where(faces.target == target)[0][:100]] = 1

x_faces = faces.data[mask]
y_faces = faces.target[mask]
x_faces = np.reshape(x_faces, (x_faces.shape[0], faces.images.shape[1],
                     faces.images.shape[2], faces.images.shape[3]))

x_faces.shape
```

Now preprocess the pixel values for input to a pretrained ResNet50 CNN and use Scikit-Learn's `train_test_split` function to split the dataset, yielding 400 training samples and 100 test samples:

```
from sklearn.model_selection import train_test_split
from tensorflow.keras.applications.resnet50 import preprocess_input

face_images = preprocess_input(np.array(x_faces * 255))

x_train, x_test, y_train, y_test = train_test_split(
    face_images, y_faces, train_size=0.8, stratify=y_faces,
    random_state=0)
```

If you wanted, you could divide the preprocessed pixel values by 255 and train a CNN from scratch right now with this data. Here's how you'd go about it (in case you care to give it a try):

```
from tensorflow.keras.models import Sequential
from tensorflow.keras.layers import Dense, Flatten, Conv2D, MaxPooling2D
```

```
model = Sequential()
model.add(Conv2D(32, (3, 3), activation='relu', input_shape=(x_train.shape[1:])))
model.add(MaxPooling2D(2, 2))
model.add(Conv2D(64, (3, 3), activation='relu'))
model.add(MaxPooling2D(2, 2))
model.add(Conv2D(64, (3, 3), activation='relu'))
model.add(MaxPooling2D(2, 2))
model.add(Flatten())
model.add(Dense(1024, activation='relu'))
model.add(Dense(class_count, activation='softmax'))
model.compile(optimizer='adam', loss='sparse_categorical_crossentropy',
              metrics=['accuracy'])

model.fit(x_train / 255, y_train, validation_data=(x_test / 255, y_test),
          epochs=20, batch_size=10)
```

I did it and then plotted the training and validation accuracy:

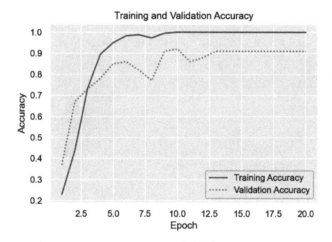

The validation accuracy is better than that of an SVM or a conventional neural network, but it's nowhere near what modern CNNs achieve on the LFW dataset. So clearly there is a better way.

That better way, of course, is transfer learning, which we covered in Chapter 10. ResNet50 was trained with more than 1 million images from the ImageNet dataset, so it should be adept at extracting features from photos—more so than a handcrafted CNN trained with 400 images. Let's see if that's the case. Use the following statements to load ResNet50's feature extraction layers, initialize them with the ImageNet weights, and freeze them so that the weights aren't adjusted during training:

```
from tensorflow.keras.applications import ResNet50

base_model = ResNet50(weights='imagenet', include_top=False)
base_model.trainable = False
```

Now add classification layers to the base model and include a `Resizing` layer to resize images input to the network to the size that ResNet50 expects:

```python
from tensorflow.keras.models import Sequential
from tensorflow.keras.layers import Flatten, Dense, Resizing

model = Sequential()
model.add(Resizing(224, 224))
model.add(base_model)
model.add(Flatten())
model.add(Dense(1024, activation='relu'))
model.add(Dense(class_count, activation='softmax'))
model.compile(optimizer='adam', loss='sparse_categorical_crossentropy',
              metrics=['accuracy'])
```

Train the model and plot the training and validation accuracy:

```python
import seaborn as sns
sns.set()

hist = model.fit(x_train, y_train, validation_data=(x_test, y_test),
                 batch_size=10, epochs=10)

acc = hist.history['accuracy']
val_acc = hist.history['val_accuracy']
epochs = range(1, len(acc) + 1)

plt.plot(epochs, acc, '-', label='Training Accuracy')
plt.plot(epochs, val_acc, ':', label='Validation Accuracy')
plt.title('Training and Validation Accuracy')
plt.xlabel('Epoch')
plt.ylabel('Accuracy')
plt.legend(loc='lower right')
plt.plot()
```

Results will vary, but my run produced a validation accuracy around 94%:

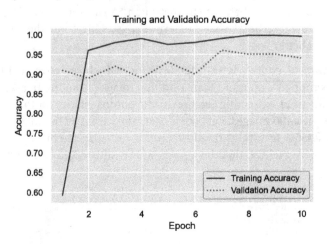

This is an improvement over a CNN trained from scratch, and it's an indication that ResNet50 does a better job of extracting features from facial images. But it's still not state of the art. Is it possible to do even better?

Boosting Transfer Learning with Task-Specific Weights

Initialized with ImageNet weights, ResNet50 does a credible job of feature extraction. Those weights were arrived at when ResNet50 was trained on more than 1 million photos of objects ranging from basketballs to butterflies. It was not, however, trained with facial images. Would it be better at extracting features from facial images if it were *trained* with facial images?

In 2017, a group of researchers at the University of Oxford's Visual Geometry Group published a paper titled "VGGFace2: A Dataset for Recognising Faces Across Pose and Age" (*https://oreil.ly/ssMNy*). After assembling a dataset comprising several million facial images, they trained two variations of ResNet50 with it and published the results. They also published the weights, which are wrapped in a handy Python library named Keras-vggface (*https://oreil.ly/46cYu*). That library includes a class named VGGFace that encapsulates ResNet50 with TensorFlow-compatible weights. Out of the box, the VGGFace model is capable of recognizing the faces of thousands of celebrities ranging from Brie Larson to Jennifer Aniston. But its real value lies in using transfer learning to repurpose it to recognize faces it wasn't trained to recognize before.

To simplify matters, I installed Keras-vggface, created an instance of VGGFace without the classification layers, initialized the weights, and saved the model to an H5 file named *vggface.h5*. Download that file (*https://oreil.ly/DV3Wc*) and drop it into your notebooks' *Data* subdirectory. Then use the following code to create an instance of VGGFace built on top of ResNet50, and add custom classification layers:

```
from tensorflow.keras.models import load_model

base_model = load_model('Data/vggface.h5')
base_model.trainable = False

model = Sequential()
model.add(Resizing(224, 224))
model.add(base_model)
model.add(Flatten())
model.add(Dense(1024, activation='relu'))
model.add(Dense(class_count, activation='softmax'))
model.compile(optimizer='adam', loss='sparse_categorical_crossentropy',
              metrics=['accuracy'])
```

Next, train the model and plot the training and validation accuracy:

```
hist = model.fit(x_train, y_train, validation_data=(x_test, y_test),
                 batch_size=10, epochs=10)
```

```
acc = hist.history['accuracy']
val_acc = hist.history['val_accuracy']
epochs = range(1, len(acc) + 1)

plt.plot(epochs, acc, '-', label='Training Accuracy')
plt.plot(epochs, val_acc, ':', label='Validation Accuracy')
plt.title('Training and Validation Accuracy')
plt.xlabel('Epoch')
plt.ylabel('Accuracy')
plt.legend(loc='lower right')
plt.plot()
```

The results are spectacular:

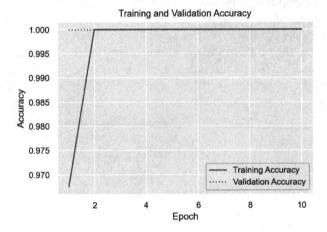

To be sure, run the test data through the network and use a confusion matrix to assess the results:

```
from sklearn.metrics import ConfusionMatrixDisplay as cmd

sns.reset_orig()
y_pred = model.predict(x_test)
fig, ax = plt.subplots(figsize=(5, 5))
ax.grid(False)

cmd.from_predictions(y_test, y_pred.argmax(axis=1),
                     display_labels=faces.target_names, colorbar=False,
                     cmap='Blues', xticks_rotation='vertical', ax=ax)
```

Because VGGFace was tuned to extract features from facial images, it achieves a perfect score on the 100 test images. That's not to say that it will never fail to recognize a face. It does indicate that, on the dataset you trained it with, it is remarkably adept at extracting features from facial images:

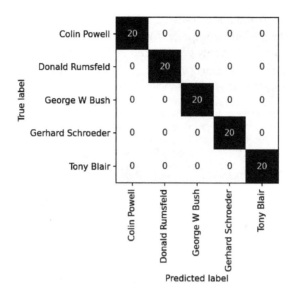

And therein lies an important lesson. CNNs that are trained in task-specific ways frequently provide a better base for transfer learning than CNNs trained in a more generic fashion. If the goal is to perform facial recognition, you'll almost always do better with a CNN trained with facial images than a CNN trained with photos of thousands of dissimilar objects. For a neural network, it's all about the weights.

ArcFace

VGGFace isn't the only pretrained CNN that excels at extracting features from facial images. Another is ArcFace, which was introduced in a 2019 paper titled "ArcFace: Additive Angular Margin Loss for Deep Face Recognition" (*https://oreil.ly/fWe7Y*). A handy implementation is available in a Python package named Arcface (*https://oreil.ly/KySCA*).

Each facial image submitted to ArcFace is transformed into a dense vector of 512 values known as a *face embedding*. The code for creating an embedding is simple:

```
from arcface import ArcFace

af = ArcFace.ArcFace()
embedding = af.calc_emb(image)
```

Embeddings can be used to train machine learning models, and they can be used to make predictions with those models. Thanks to the loss function named in the title of the paper, embeddings created by ArcFace often do a better job of capturing the uniqueness of a face than embeddings generated by conventional CNNs.

Another use for the embeddings created by ArcFace is *face verification*, which compares two facial images and computes the probability that they represent the same person. The following statements generate embeddings for two facial images and use the `cosine_similarity` function (*https://oreil.ly/iGDKx*) introduced in Chapter 4 to quantify the similarity between the two:

```
af = ArcFace.ArcFace()
face_emb1 = af.calc_emb(image1)
face_emb2 = af.calc_emb(image2)
sim = cosine_similarity([face_emb1, face_emb2])[0][1]
```

The result is a value from 0.0 to 1.0, with higher values reflecting greater similarity between the faces.

Putting It All Together: Detecting and Recognizing Faces in Photos

Any time a model scores perfectly in testing, you should be skeptical. No model is perfect, and even if it achieves 100% accuracy against a test dataset, it won't duplicate that in the wild. Given that VGGFace was trained with images of some of the same famous people found in the LFW dataset, is it possible that it's biased toward those people? That transfer learning with VGGFace wouldn't do as well if trained with images of ordinary people? And how would it perform with just a handful of training images?

To answer these questions, let's build a notebook that trains a facial recognition model based on VGGFace, uses an MTCNN to detect faces in photos, and uses the model to identify the faces it detects. The dataset you'll use contains eight pictures each of three ordinary people in slightly different poses, at ages up to 20 years apart, with and without glasses (Figure 11-7). These images were extracted from photos using the `extract_faces` function in Example 11-1 and resized to 224 × 224.

Figure 11-7. Photos for training a facial recognition model

Begin by downloading a ZIP file (*https://oreil.ly/AA6Gd*) containing the facial images and copying the contents of the ZIP file into a subdirectory named *Faces* where your notebooks are hosted. The ZIP file contains four folders: one named *Jeff,* one named *Lori,* one named *Abby,* and one named *Samples* that contains uncropped photos for testing.

Now create a new notebook and run the following code in the first cell to define helper functions for loading and displaying facial images from the subdirectories you copied them to, and declare a pair of Python lists to hold the images and labels:

```python
import os
from tensorflow.keras.preprocessing import image
import matplotlib.pyplot as plt
%matplotlib inline

def load_images_from_path(path, label):
    images, labels = [], []

    for file in os.listdir(path):
        images.append(image.img_to_array(image.load_img(os.path.join(path, file),
                    target_size=(224, 224, 3))))
        labels.append((label))

    return images, labels

def show_images(images):
    fig, axes = plt.subplots(1, 8, figsize=(20, 20),
                        subplot_kw={'xticks': [], 'yticks': []})

    for i, ax in enumerate(axes.flat):
        ax.imshow(images[i] / 255)

x, y = [], []
```

Next, load the images of Jeff and label them with 0s:

```python
images, labels = load_images_from_path('Faces/Jeff', 0)
show_images(images)

x += images
y += labels
```

Load the images of Lori and label them with 1s:

```python
images, labels = load_images_from_path('Faces/Lori', 1)
show_images(images)

x += images
y += labels
```

Load the images of Abby and label them with 2s:

```
images, labels = load_images_from_path('Faces/Abby', 2)
show_images(images)

x += images
y += labels
```

Finally, preprocess the pixels for the ResNet50 version of VGGFace and split the data fifty-fifty so that the network will be trained with four randomly selected images of each person and validated with the same number of images:

```
import numpy as np
from sklearn.model_selection import train_test_split
from tensorflow.keras.applications.resnet50 import preprocess_input

faces = preprocess_input(np.array(x))
labels = np.array(y)

x_train, x_test, y_train, y_test = train_test_split(
    faces, labels, train_size=0.5, stratify=labels,
    random_state=0)
```

The next step is to load the saved VGGFace model and freeze the bottleneck layers. If you didn't download *vggface.h5* earlier, download it now (*https://oreil.ly/noZoo*) and drop it into your notebooks' *Data* subdirectory. Then execute the following code:

```
from tensorflow.keras.models import load_model

base_model = load_model('Data/vggface.h5')
base_model.trainable = False
```

Now define a network that uses transfer learning with VGGFace to identify faces. The Resizing layer ensures that each image measures exactly 224 × 224 pixels. The Dense layer contains just eight neurons because the training dataset is small and we don't want the network to fit too tightly to it:

```
from tensorflow.keras.models import Sequential
from tensorflow.keras.layers import Flatten, Dense, Resizing

model = Sequential()
model.add(Resizing(224, 224))
model.add(base_model)
model.add(Flatten())
model.add(Dense(8, activation='relu'))
model.add(Dense(3, activation='softmax'))
model.compile(optimizer='adam', loss='sparse_categorical_crossentropy',
              metrics=['accuracy'])
```

Train the model:

```
hist = model.fit(x_train, y_train, validation_data=(x_test, y_test),
                 batch_size=2, epochs=10)
```

Plot the training and validation accuracy:

```
import seaborn as sns
sns.set()

acc = hist.history['accuracy']
val_acc = hist.history['val_accuracy']
epochs = range(1, len(acc) + 1)

plt.plot(epochs, acc, '-', label='Training Accuracy')
plt.plot(epochs, val_acc, ':', label='Validation Accuracy')
plt.title('Training and Validation Accuracy')
plt.xlabel('Epoch')
plt.ylabel('Accuracy')
plt.legend(loc='lower right')
plt.plot()
```

Hopefully you got something like this:

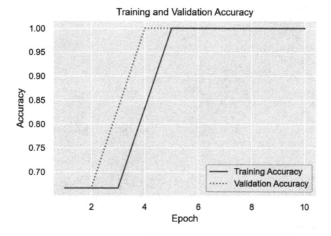

Now comes the fun part: using an MTCNN to detect the faces in a photo and the trained model to identify those faces. First make sure the MTCNN package is installed in your environment. Then define a pair of helper functions—one that retrieves a face from a specified location in an image (get_face), and another that loads a photo and annotates faces in the photo with names and confidence levels (label_faces):

```
from mtcnn.mtcnn import MTCNN
from PIL import Image, ImageOps
from tensorflow.keras.preprocessing import image
from matplotlib.patches import Rectangle

def get_face(image, face):
    x1, y1, w, h = face['box']
```

```
        if w > h:
            x1 = x1 + ((w - h) // 2)
            w = h
        elif h > w:
            y1 = y1 + ((h - w) // 2)
            h = w

        x2 = x1 + h
        y2 = y1 + w

        return image[y1:y2, x1:x2]

def label_faces(path, model, names, face_threshold=0.9, prediction_threshold=0.9,
                show_outline=True, size=(12, 8)):
    # Load the image and orient it correctly
    pil_image = Image.open(path)
    exif = pil_image.getexif()

    for k in exif.keys():
        if k != 0x0112:
            exif[k] = None
            del exif[k]

    pil_image.info["exif"] = exif.tobytes()
    pil_image = ImageOps.exif_transpose(pil_image)
    np_image = np.array(pil_image)

    fig, ax = plt.subplots(figsize=size, subplot_kw={'xticks': [], 'yticks': []})
    ax.imshow(np_image)

    detector = MTCNN()
    faces = detector.detect_faces(np_image)
    faces = [face for face in faces if face['confidence'] > face_threshold]

    for face in faces:
        x, y, w, h = face['box']

        # Use the model to identify the face
        face_image = get_face(np_image, face)
        face_image = image.array_to_img(face_image)
        face_image = preprocess_input(np.array(face_image))
        predictions = model.predict(np.expand_dims(face_image, axis=0))
        confidence = np.max(predictions)

        if (confidence > prediction_threshold):
            # Optionally draw a box around the face
            if show_outline:
                rect = Rectangle((x, y), w, h, color='red', fill=False, lw=2)
                ax.add_patch(rect)

            # Label the face
            index = int(np.argmax(predictions))
```

```
    text = f'{names[index]} ({confidence:.1%})'
ax.text(x + (w / 2), y, text, color='white', backgroundcolor='red',
        ha='center', va='bottom', fontweight='bold',
        bbox=dict(color='red'))
```

Now pass the first sample image in the *Samples* folder to `label_faces`:

```
labels = ['Jeff', 'Lori', 'Abby']
label_faces('Faces/Samples/Sample-1.jpg', model, labels)
```

The output should look like this, although your percentages might be different:

Try it again, but this time with a different photo:

```
label_faces('Faces/Samples/Sample-2.jpg', model, labels)
```

Here's the output:

Finally, submit a photo containing all three individuals that the model was trained with:

```
label_faces('Faces/Samples/Sample-3.jpg', model, labels)
```

Trained with just 12 facial images—four of each person—the model does a credible job of identifying faces in photos. Of course, you could generate a dataset of your own by passing photos of friends and family members through the function in Example 11-1 and training the model with the resulting images.

Handling Unknown Faces: Closed-Set Versus Open-Set Classification

Now for some bad news. A VGGFace facial recognition model is adept at identifying faces it was trained with, but it doesn't know what to do when it encounters a face it *wasn't* trained with. Try it: pass in a photo of yourself. The model will probably identify you as Jeff, Lori, or Abby, and it might do so with a high level of confidence. It literally doesn't know what it doesn't know. This is especially true when the dataset is small and the network is given room to overfit.

The reason why has nothing to do with VGGFace. It has everything to do with the fact that a neural network with a softmax output layer is a *closed-set classifier*, meaning it classifies any sample presented to it for predictions as one of the classes it was trained with. (Remember that softmax ensures that the sum of the probabilities for all classes is 1.0.) The alternative is an *open-set classifier* (Figure 11-8), which has the ability to say "this sample doesn't belong to any of the classes I was trained with."

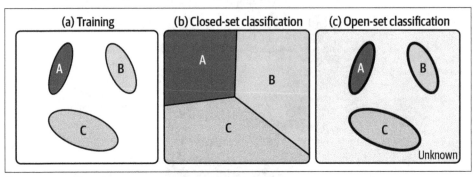

Figure 11-8. Closed-set versus open-set classification

There is not a one-size-fits-all solution for building open-set classifiers in the deep-learning community today. A 2016 paper titled "Towards Open Set Deep Networks" (*https://oreil.ly/IFmIF*) proposed one solution in the form of *openmax* output layers, which replace softmax output layers and "estimate the probability of an input being from an unknown class." Essentially, if the network is trained with 10 classes, the openmax output layer adds an 11th output representing the unknown class. It works by taking the activations from the final classification layer and adjusting them using a Weibull distribution (*https://oreil.ly/plWN2*) rather than normalizing the probabilities as softmax does.

Another potential solution was put forth in a 2018 paper titled "Reducing Network Agnostophobia" (*https://oreil.ly/hL9jk*) from researchers at the University of Colorado. It proposed replacing cross-entropy loss with a new loss function called *entropic open-set loss* that drives softmax scores for unknown classes toward a uniform probability distribution. Using this technique, you could more reliably detect samples belonging to an unknown class using probability thresholds paired with conventional softmax output layers. For a great summary of the problems posed by open-set classification in deep learning and an overview of openmax and entropic open-set loss, see "Does a Neural Network Know When It Doesn't Know?" (*https://oreil.ly/dHb7m*) by Tivadar Danka.

Yet another solution is to use ArcFace to verify each face the model identifies by comparing an embedding generated from that face to a reference embedding for the same person. You could reject the model's conclusion if cosine similarity falls below a predetermined threshold.

A more naive approach is to prevent the network from learning the training data too well in hopes that unknown classes will yield lower softmax probabilities. That's why I included just eight neurons in the classification layer in the previous example. (You could go even further by introducing a dropout layer.) It works up to a point, but it isn't perfect. The `label_faces` function has a default prediction threshold of 0.9, meaning it labels a face only if the model classifies it with at least 90% confidence. You could set `prediction_threshold` to 0.99 to rule out more unknown faces, but at the cost of failing to identify more known faces. Tuning in this manner to strike the right balance between recognizing known faces while ignoring unknowns is an inevitable part of readying a facial recognition model for production.

Summary

Building an end-to-end facial recognition system requires a means for detecting faces in photos as well as a means for classifying (identifying) those faces. One way to detect faces is the Viola-Jones algorithm, for which the OpenCV library provides a convenient implementation. An alternative that relies on deep learning is a multitask cascaded convolutional neural network, or MTCNN. The Python package named MTCNN contains a ready-to-use MTCNN implementation. Viola-Jones is faster and more suitable for real-time applications (for example, identifying faces in a live webcam feed), but MTCNNs are generally more accurate and incur fewer false positives.

Deep learning can be applied to the task of facial recognition by employing convolutional neural networks. Transfer learning with a pretrained CNN such as ResNet50 can identify faces with a relatively high degree of accuracy, but transfer learning with a CNN that was trained with millions of facial images delivers unparalleled accuracy. The primary reason is that the CNN's bottleneck layers are optimized for extracting features from facial images.

With a highly optimized set of weights, a neural network can do almost anything. Generic weights will suffice when nothing better is available, but given a task-specific set of weights to start with, facial recognition via transfer learning can achieve human-like accuracy.

## CHAPTER 12

Object Detection

The previous chapter introduced two popular algorithms for detecting faces in photographs: Viola-Jones, which relies on machine learning, and MTCNNs, which rely on deep learning. Face detection is a special case of *object detection* (*https://oreil.ly/ 4H8yw*), in which computers detect and identify objects in images. Identifying an object is an image classification problem, something at which CNNs excel. But finding objects to identify poses a different challenge.

Object detection is challenging because objects aren't assumed to be perfectly cropped and aligned as they are for image classification tasks. Nor are they limited to one per image. Figure 12-1 shows what a self-driving car might see as it scans video frames from a forward-pointing camera. A CNN trained to do conventional image classification using carefully prepared training images is powerless to help. It might be able to classify the image as one of a city street, but it can't determine that the image contains cars, people, and traffic lights, much less pinpoint their locations.

Object detection has grown in speed and accuracy in recent years, and state-of-the-art methods rely on deep learning. In particular, they employ CNNs, which were introduced in Chapter 10. Let's discuss how CNNs do object detection and identification and try our hand at it using cutting-edge object detection models.

Figure 12-1. Detecting and identifying objects in a scene

R-CNNs

One way to apply deep learning to the task of object detection is to use *region-based CNNs* (*https://oreil.ly/ozIoy*), also known as *region CNNs* or simply *R-CNNs*. The first R-CNN was introduced in a 2014 paper titled "Rich Feature Hierarchies for Accurate Object Detection and Semantic Segmentation" (*https://oreil.ly/TFYN6*). The model described in the paper comprises three stages. The first stage scans the image and identifies up to 2,000 bounding boxes representing regions of interest—regions that *might* contain objects. The second stage is a deep CNN that extracts features from regions of interest. The third is a support vector machine that classifies the features. The output is a collection of bounding boxes with class labels and confidence scores. An algorithm called *non-maximum suppression* (NMS) (*https://oreil.ly/ku8SU*) filters the output and selects the best bounding box for each object.

NMS is a crucial element of virtually all modern object detection systems. A detector invariably emits several bounding boxes for each object. If a photo contains one instance of a given class—for example, one zebra—NMS selects the bounding box with the highest confidence score. If the photo contains two zebras (Figure 12-2), NMS divides the bounding boxes into two groups and selects the box with the highest confidence score in each group. It groups boxes based on the amount of overlap between them. Overlap is computed by dividing the area of intersection between two boxes by the area formed by the union of the boxes. If the resulting *intersection-over-union* (IoU) score is greater than a predetermined threshold (typically 0.5), NMS assigns the boxes to the same group. Otherwise, it assigns them to separate groups.

When two objects of the same class have little or no overlap, NMS easily separates the two. When two instances of the same class overlap significantly (picture one zebra standing behind the other), the IoU threshold might have to be adjusted to achieve separation. IoU threshold is a hyperparameter that can be tuned to strike the right balance between being overly aggressive at separating overlapping objects and not aggressive enough.

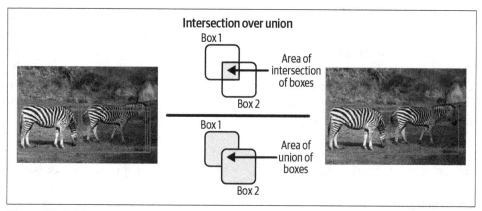

Figure 12-2. Non-maximum suppression

The first stage of most R-CNN implementations uses an algorithm called *selective search* (*https://oreil.ly/3S1K0*) to identify regions of interest by keying on similarities in color, texture, shape, and size. Figure 12-3 shows the first 500 bounding boxes generated when OpenCV's implementation of selective search (*https://oreil.ly/7XDD8*) examines an image. Submitting a targeted list of regions to the CNN is faster than a brute-force sliding-window approach that inputs the contents of the window to the CNN at every stop.

Figure 12-3. Bounding boxes generated by selective search

Even with selective search narrowing the list of candidate regions input to stage 2, an R-CNN can't do object detection in real time. Why? Because the CNN individually processes the 2,000 or so regions of interest identified in stage 1. These regions invariably overlap, so the CNN processes the same pixels multiple times.

A 2015 paper titled "Fast R-CNN" (*https://oreil.ly/ih0x2*) addressed this by proposing a modified architecture in which the entire image passes through the CNN one time (Figure 12-4). Selective search or a similar algorithm identifies regions of interest in the image, and those regions are projected onto the feature map generated by the CNN. A *region of interest (ROI) pooling layer* then uses a form of max pooling to reduce the features in each region of interest to a fixed-length vector, independent of the region's size and shape. (By contrast, R-CNN scales each region to an image of predetermined size before submitting it to the CNN, which is substantially more expensive than ROI pooling.) Classification of the feature vectors is performed by fully connected layers rather than SVMs, and the output is split to include both a softmax classifier and a bounding-box regressor. NMS filters the bounding boxes down to the ones that matter. The result is a system that trains an order of magnitude faster than R-CNN, makes predictions two orders of magnitude faster, and is slightly more accurate than R-CNN.

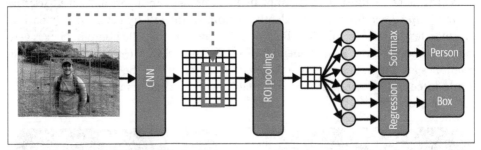

Figure 12-4. Fast R-CNN architecture

 ROI pooling reduces any region of interest to a feature vector of a specified size, regardless of the region's height and width. Imagine that you have an 8 × 16 region of a feature map and you want to reduce it to 4 × 4. You can divide the 8 × 16 region into a 4 × 4 grid, with each cell in the grid measuring 2 × 4. You can then take the maximum of the eight values in each cell and plug them into the 4 × 4 grid. That's ROI pooling. It's simple, fast, and effective. And it works with regions of any size and aspect ratio.

A 2016 paper titled "Faster R-CNN: Towards Real-Time Object Detection with Region Proposal Networks" (*https://oreil.ly/wWMOy*) further boosted performance by replacing selective search with a *region proposal network*, or RPN. The RPN is a shallow CNN that shares layers with the main CNN (Figure 12-5). To generate region proposals, it slides a window over the feature map generated by the last shared layer. At each stop in the window's travel, the RPN evaluates *n* candidate regions called *anchors* or *anchor boxes* and computes an *objectness score* for each based on IoU scores with ground-truth boxes. Objectness scores and anchor boxes are input to fully connected layers for classification (does the anchor contain an object, and if so, what type?) and regression. The output from these layers ultimately determines the regions of interest projected onto the feature map generated by the main CNN and forwarded to the ROI pooling layer.

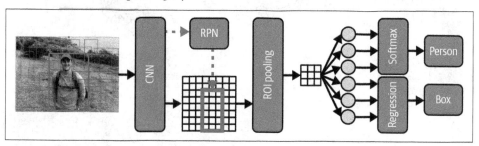

Figure 12-5. Faster R-CNN architecture

Faster R-CNN can perform 10 times faster than Fast R-CNN and do object detection in near real time. It's typically more accurate than Fast R-CNNs too, thanks to the RPN's superior ability to identify candidate regions. Selective search is static, but an RPN gets smarter as the network is trained. RPNs also execute in parallel to the main CNN, which means region proposals have little impact on performance.

Mask R-CNN

The next chapter in the R-CNN story came in 2017 in a paper titled "Mask R-CNN" (*https://oreil.ly/33ktB*). Mask R-CNNs extend Faster R-CNNs by adding *instance segmentation*, which identifies the shapes of objects detected in an image using *segmentation masks* like the ones in Figure 12-6. Performance impact is minimal because instance segmentation is performed in parallel with region evaluation. The benefit of Mask R-CNNs is that they provide more detail about the objects they detect. For example, you can tell whether a person's arms are extended—something you can't discern from a simple bounding box. They are also slightly more accurate than Faster R-CNNs because they replace ROI pooling with *ROI alignment*, which discards less information when generating feature vectors whose boundaries don't perfectly align with the boundaries of the regions they represent.

Figure 12-6. Bounding boxes and segmentation masks generated by a Mask R-CNN

Zoom uses instance segmentation to display custom backgrounds behind you in video feeds. Instance segmentation also powers AI-based image editing tools that crop figures from the foreground and place them on different backgrounds. Let's demonstrate using a pretrained Mask R-CNN.

Start by downloading a ZIP file (*https://oreil.ly/NKGlg*) containing the files needed for this exercise. The files are:

adam.jpg
   A photo of a young man wearing a backpack

maui.jpg
   A photo taken poolside in Hawaii

mask.py
   Contains helper functions for preprocessing images before they're input to a Mask R-CNN and postprocessing functions for interpreting the results

MaskRCNN-12-int8.onnx
   Contains a pretrained Mask R-CNN saved in ONNX format

Place *mask.py* in the directory where your Jupyter notebooks are hosted, and the remaining files in your notebooks' *Data* subdirectory.

The critical file in this ensemble is *MaskRCNN-12-int8.onnx*. Chapter 7 introduced Open Neural Network Exchange (ONNX) (*https://oreil.ly/UwqD3*) as a means for loading Scikit-Learn models written in Python and consuming them in other languages. But ONNX does more than bridge the language gap to Scikit. It's a neutral format that enables neural networks written with one deep-learning framework to be exported to others. For example, you can save a PyTorch model to a *.onnx* file, and then convert the *.onnx* file to a TensorFlow model and use it as if it had been written with TensorFlow from the outset. Or you can install an ONNX runtime and load and call the model without converting it to TensorFlow. Chapter 7 demonstrated how to save a Scikit model in ONNX format and use the Python ONNX runtime (*https://oreil.ly/U79N2*) to load the model and call it from Python. It also used the .NET ONNX runtime (*https://oreil.ly/u5yYL*) to load the model and call it from C#. As noted in Chapter 7, ONNX runtimes are available for a variety of platforms and programming languages.

MaskRCNN-12-int8.onnx contains a sophisticated Mask R-CNN (*https://oreil.ly/bz31O*) implementation from Facebook Research. It uses a ResNet50 backbone to extract features from the images input to it, and it includes a branch that computes instance segmentation masks in parallel with bounding box computations. It was trained with more than 200,000 images from the COCO dataset (*https://oreil.ly/C22xv*). Those images contain more than 1.5 million objects that fall into 80 categories (*https://oreil.ly/D7Bzd*) ranging from cats and dogs to people, bicycles, and cars. And because the trained model was published in the ONNX model zoo (*https://oreil.ly/ytRcn*) on GitHub, you can load it and pass images to it with just a few lines of code.

The first step in using the model in Python is to make sure the Python ONNX runtime is installed in your environment. For this exercise, make sure OpenCV is installed too. Then create a Jupyter notebook and run the following code in the first cell to load and display *adam.jpg*:

```python
import matplotlib.pyplot as plt
from PIL import Image
%matplotlib inline

image = Image.open('Data/adam.jpg')
fig, ax = plt.subplots(figsize=(12, 8), subplot_kw={'xticks': [], 'yticks': []})
ax.imshow(image)
```

Here's the output:

The goal is to run this image through Mask R-CNN, detect the objects in it, and annotate the image accordingly. To that end, use the following code to preprocess the image pixels the way the model expects, load Mask R-CNN from the ONNX file, and submit the preprocessed image to it:

```
from mask import *
import onnxruntime as rt

image_data = preprocess(image)
session = rt.InferenceSession('Data/MaskRCNN-12-int8.onnx')
input_name = session.get_inputs()[0].name
result = session.run(None, { input_name: image_data })
```

The value returned by the run method is an array of four arrays. The first array contains bounding boxes for the objects detected in the image. In this example, it detected a total of 13 objects:

```
array([[ 377.50537,  174.36874,  801.177  ,  787.7125 ],
       [ 428.15994,  344.70105,  699.99097,  599.8499 ],
       [ 757.1859 ,  529.4938 ,  814.2776 ,  658.8041 ],
       [ 432.57123,  351.229  ,  672.2242 ,  484.63702],
       [ 435.53653,  357.34235,  494.73557,  516.4939 ],
       [ 672.56714,  516.2991 ,  822.82153,  663.68726],
       [ 608.1904 ,  361.0429 ,  686.1207 ,  552.70105],
       [ 629.84894,  355.3375 ,  799.4963 ,  718.9458 ],
       [ 437.99078,  331.19318,  709.4969 ,  748.2785 ],
       [ 438.44028,  348.14404,  579.8056 ,  546.0588 ],
       [1152.0518 ,  203.89978, 1163.7583 ,  223.2485 ],
       [1151.8087 ,  132.45256, 1164.0013 ,  150.6876 ],
       [1151.8087 ,  404.0875 , 1164.0013 ,  423.058  ],
```

```
[ 416.03802,   396.8123 ,   694.8975 ,   511.93896],
[ 683.35236,   765.0078 ,   701.3403 ,   784.47546]], dtype=float32)
```

The second array is an array of integer class identifiers corresponding to classes in the COCO dataset: 1 for a person, 2 for a bicycle, and so on. These identify the objects in the image:

```
array([ 1, 25, 27, 25, 25, 27, 25, 27, 27, 25,  1,  1,  1, 25, 75],
      dtype=int64)
```

The third array contains confidence scores for the 13 objects. They range from a high of 99.9% for the person in the image to a low of 5.5% for the logo on his cap, which the model thinks might be a clock. Observe that only two objects scored a confidence level of 70% or higher:

```
array([0.99911565, 0.8009716 , 0.4624603 , 0.19400069, 0.17042953,
       0.11972303, 0.11626374, 0.10352837, 0.08243025, 0.08089489,
       0.07266886, 0.07266886, 0.07266886, 0.06278796, 0.0549276 ],
      dtype=float32)
```

Finally, the fourth array contains segmentation masks for each object. Each mask is a 28 × 28 array of floating-point values sometimes referred to as a *soft mask* due to the soft edges. Figure 12-7 shows the mask for the person detected in the image. When mapped back to the image's original size and aspect ratio, it does a reasonable job of defining which pixels correspond to that person.

Figure 12-7. A 28 × 28 soft mask corresponding to the person in the preceding photo

The next step is to grab the bounding boxes, predicted class labels, confidence scores, and segmentation masks and pass them to the annotate_image function to visualize the results:

```
boxes = result[0]   # Bounding boxes
labels = result[1]  # Class labels
scores = result[2]  # Confidence scores
masks = result[3]   # Segmentation masks

annotate_image(image, boxes, labels, scores, masks)
```

Here's the result:

By default, `annotate_image` ignores objects identified with less than 70% confidence. You can override that by including a `min_confidence` parameter in the call. The model detected two objects in the image with a confidence that equaled or exceeded 70%: a person and a backpack. `annotate_image` draws the bounding boxes and annotates them with class names and confidence levels. It also shades the objects by overlaying them with partially transparent pixels. These are the segmentation masks returned by `run`.

`annotate_image` is one of several helper functions found in *mask.py*. Another is `preprocess`, which preps each image in the way Mask R-CNN expects by resizing the image, making sure the width and height are multiples of 32, converting the image to BGR format, and normalizing the pixel values. I brought both of these functions over from the model's GitHub page (*https://oreil.ly/06Nmo*) with some minor modifications. Since Mask R-CNNs are often used to crop objects from images, I also added a `change_background` function that extracts all the objects detected in an image with a specified confidence level (default = 0.7) and composites them onto a new background. To see for yourself, run the following statements in the notebook's next cell:

```
fg_image = Image.open('Data/adam.jpg')
bg_image = Image.open('Data/maui.jpg')

change_background(session, fg_image, bg_image)
```

The output should look like this:

`change_background` submits the foreground image (the one passed in the function's second parameter) to the model and copies all the pixels inside the segmentation masks to the background image. It works best when the foreground and background images are the same size, but it will resize the background image if needed to match the foreground image. The background will be distorted if its aspect ratio differs from that of the foreground image.

The fact that segmentation masks produced by Mask R-CNN measure just 28 × 28 explains the "tearing" around the pixels copied from the foreground image: the resolution of the mask is lower than that of the person in the photo. You could modify Mask R-CNN to use higher-resolution masks, but you'd have to retrain the model from scratch. That's a time-consuming endeavor with a model as complex as this one and a dataset as large as COCO, even if you do the training on GPUs.

YOLO

While the R-CNN family of object detection systems delivers unparalleled accuracy, it leaves something to be desired when it comes to real-time object detection of the type required by, say, self-driving cars. A paper titled "You Only Look Once: Unified, Real-Time Object Detection" (*https://oreil.ly/qtiJj*) published in 2015 proposed an alternative to R-CNNs known as YOLO (You Only Look Once) that revolutionized the way engineers think about object detection. From the paper's introduction:

Humans glance at an image and instantly know what objects are in the image, where they are, and how they interact. The human visual system is fast and accurate, allowing us to perform complex tasks like driving with little conscious thought. Fast, accurate algorithms for object detection would allow computers to drive cars without specialized sensors, enable assistive devices to convey real-time scene information to human users, and unlock the potential for general purpose, responsive robotic systems.

Current detection systems repurpose classifiers to perform detection. To detect an object, these systems take a classifier for that object and evaluate it at various locations and scales in a test image. Systems like deformable parts models (DPM) use a sliding window approach where the classifier is run at evenly spaced locations over the entire image.

More recent approaches like R-CNN use region proposal methods to first generate potential bounding boxes in an image and then run a classifier on these proposed boxes. After classification, post-processing is used to refine the bounding boxes, eliminate duplicate detections, and rescore the boxes based on other objects in the scene. These complex pipelines are slow and hard to optimize because each individual component must be trained separately.

We reframe object detection as a single regression problem, straight from image pixels to bounding box coordinates and class probabilities. Using our system, you only look once (YOLO) at an image to predict what objects are present and where they are.

YOLO systems are characterized by their performance:

Our base network runs at 45 frames per second with no batch processing on a Titan X GPU and a fast version runs at more than 150 fps. This means we can process streaming video in real-time with less than 25 milliseconds of latency. Furthermore, YOLO achieves more than twice the mean average precision of other real-time systems.

At a high level, YOLO works by dividing feature maps into grids of cells and evaluating each cell for the presence of an object, as shown in Figure 12-8. Using the anchor-box concept borrowed from Faster R-CNN, YOLO analyzes bounding boxes of various shapes and sizes around each cell and assigns classes and probabilities to the boxes. One CNN handles everything, including feature extraction, classification, and regression designating the sizes and locations of the bounding boxes, and an image goes through the CNN just once. At the end, NMS reduces the number of bounding boxes to one per object, and each bounding box is attributed with a class as well as a confidence level—the probability that the box actually contains an object of the specified class.

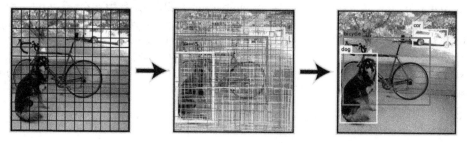

Figure 12-8. YOLO's approach to analyzing an image (https://oreil.ly/4dfXD)

There are currently seven versions of YOLO referred to as YOLOv1 through YOLOv7. Each new version improves on the previous version in terms of accuracy and performance. There are also variations such as PP-YOLO (*https://oreil.ly/AMxGm*) and YOLO9000 (*https://oreil.ly/GOBMc*). YOLOv3 (*https://oreil.ly/jBHYy*) was the last version that YOLO creator Joseph Redmon contributed to (*https://oreil.ly/MnoZw*) and is considered a reference implementation of sorts. By extracting feature maps from certain layers of the CNN, YOLOv3 analyzes the image using a 13 × 13 grid, a 26 × 26 grid, and a 52 × 52 grid in an effort to detect objects of various sizes. It uses anchors to predict nine bounding boxes per cell. YOLO's primary weakness is that it has difficulty detecting very small objects that are close together, although YOLOv3 improved on YOLOv1 and YOLOv2 in this regard. More information about YOLO can be found on its creator's website (*https://oreil.ly/x3EzO*). A separate article titled "Digging Deep into YOLO V3" (*https://oreil.ly/eRCUE*) offers a deep dive into the YOLOv3 architecture.

YOLOv3 and Keras

YOLO was originally written using a deep-learning framework called Darknet (*https://oreil.ly/YfeXf*), but it can be implemented with other frameworks as well. The keras-yolo3 project (*https://oreil.ly/yZZMt*) on GitHub contains a Keras implementation of YOLOv3 that can be trained from scratch or initialized with predefined weights and used for inference—that is, to make predictions. This version accepts 416 × 416 images. To simplify usage, I created a file named *yolov3.py* containing helper classes and helper functions. It is a modified version of a file (*https://oreil.ly/pTlE6*) that's available in the keras-yolo3 project. I removed elements that weren't needed for making predictions, rewrote some of the code for improved utility and performance, and added a little code of my own, but most of the credit goes to Huynh Ngoc Anh, whose GitHub ID is experiencor (*https://oreil.ly/UxTFz*). With *yolov3.py* and a set of weights to lend a hand, you can load YOLOv3 and make predictions with just a few lines of code.

To see YOLOv3 in action, begin by downloading a ZIP file (*https://oreil.ly/mQPZS*) containing the files you need. Open the ZIP file and place *yolov3.py* in the directory with your notebooks. Place the other files in your notebooks' *Data* subdirectory. Next, download *yolov3.weights* (*https://oreil.ly/ukHHF*), which contains the weights arrived at when YOLOv3 was trained on the COCO dataset, and place it in the *Data* subdirectory as well.

Now create a new Jupyter notebook and paste the following code into the first cell:

```
from yolov3 import *

model = make_yolov3_model()
weight_reader = WeightReader('Data/yolov3.weights')
weight_reader.load_weights(model)
model.summary()
```

Run the code to import the helper classes and functions in *yolov3.py*, create the model, and initialize it with the COCO weights. Then use the following statements to load a photo of a couple biking the city wall surrounding Xian, China:

```
import matplotlib.pyplot as plt
%matplotlib inline

image = plt.imread('Data/xian.jpg')
width, height = image.shape[1], image.shape[0]
fig, ax = plt.subplots(figsize=(12, 8), subplot_kw={'xticks': [], 'yticks': []})
ax.imshow(image)
```

Now let's see if YOLOv3 can detect objects in the image. Use the following code to load the image again, resize it to 416 × 416, preprocess the pixels, and submit the resulting image to the model for prediction:

```
import numpy as np
from tensorflow.keras.preprocessing.image import load_img, img_to_array

x = load_img('Data/xian.jpg', target_size=(YOLO3.width, YOLO3.height))
x = img_to_array(x) / 255
x = np.expand_dims(x, axis=0)
y = model.predict(x)
```

predict returns arrays containing information about objects detected in the image at three resolutions (that is, with grid cells measuring 8, 16, and 32 pixels square), but the arrays need to be decoded into bounding boxes and the boxes filtered with NMS. To help, *yolov3.py* contains a function called decode_predictions (inspired by Keras's decode_predictions function (*https://oreil.ly/hQwXP*)) to do the post-processing. decode_predictions requires the width and height of the original image as input so that it can scale the bounding boxes to match the original image dimensions. The return value is a list of BoundingBox objects, each containing the pixel coordinates of a box surrounding an object detected in the scene, along with a label identifying the object and a confidence value from 0.0 to 1.0.

The next step, then, is to pass the predictions returned by `predict` to `decode_predictions` and list the bounding boxes:

```
boxes = decode_predictions(y, width, height)

for box in boxes:
    print(f'({box.xmin}, {box.ymin}), ({box.xmax}, {box.ymax}), ' +
          f'{box.label}, {box.score}')
```

Here's the output:

```
(692, 232), (1303, 1490), person, 0.9970048069953918
(1314, 327), (1920, 1496), person, 0.9957388639450073
(716, 786), (1277, 1634), bicycle, 0.9924144744873047
(1210, 845), (2397, 1600), bicycle, 0.9957170486450195
```

The model detected four objects in the image: two people and two bikes. The labels come from the COCO dataset. There are 80 in all, and they're built into the `YOLO3` class in *yolov3.py*. Use the following command to list them and see all the different types of objects the model can detect:

```
YOLO3.labels
```

yolov3.py also contains a helper function named `annotate_image` that loads an image from the filesystem and draws the bounding boxes returned by `decode_predictions` as well as labels and confidence values. Use the following statement to visualize what the model found in the image:

```
annotate_image('Data/xian.jpg', boxes)
```

The output should look like this:

Now let's try it with another image—this time, a photo of a young woman and the family dog. First show the image and save its width and height:

```
image = plt.imread('Data/abby-lady.jpg')
width, height = image.shape[1], image.shape[0]
fig, ax = plt.subplots(figsize=(12, 8), subplot_kw={'xticks': [], 'yticks': []})
ax.imshow(image)
```

Preprocess the image and pass it to the model's `predict` method:

```
x = load_img('Data/abby-lady.jpg', target_size=(YOLO3.width, YOLO3.height))
x = img_to_array(x) / 255
x = np.expand_dims(x, axis=0)
y = model.predict(x)
```

Show the image again, this time annotated with the objects detected in it:

```
boxes = decode_predictions(y, width, height)
annotate_image('Data/abby-lady.jpg', boxes)
```

Here is the output:

The model detected the girl and her laptop, but it didn't detect the dog even though the COCO training images included dogs. Under the hood, it *did* detect the dog, but with less than 90% confidence. The model's `predict` method typically returns information about thousands of bounding boxes, most of which can be ignored because the confidence levels are so low. By default, `decode_predictions` ignores bounding boxes with confidence scores less than 0.9, but you can override that by including a `min_score` parameter in the call. Use the following statements to decode the predictions and visualize them again, this time with a minimum confidence level of 55%:

```
boxes = decode_predictions(y, width, height, min_score=0.55)
annotate_image('Data/abby-lady.jpg', boxes)
```

With the confidence threshold lowered to 0.55, the model not only detected the dog, but also the sofa:

Because the model was trained on the COCO dataset, it can detect lots of other objects too, including traffic lights, stop signs, various types of food and animals, and even bottles and wine glasses. Try it with some images of your own to experience state-of-the-art object detection firsthand.

Custom Object Detection

An object detection model trained on the COCO dataset can readily detect and identify 80 different types of objects. But what if you need an object detection model that identifies other objects? What if, for example, you need a model that detects license plates on cars, packages left on your front porch, or cancer cells in tissue samples? All of these are ways in which object detection is employed in industry today. And none of them rely on models trained solely with COCO images.

Training an object detection model from scratch is a heavy lift even for researchers at Microsoft, Facebook, and Google. You can short-circuit the process a bit by starting with the COCO weights and incrementally training the network with new classes, but even that is compute and time intensive.

There is an easier way. The premise underlying this entire book is that you're a software developer or engineer, not a PhD data scientist. You want to use machine learning (or deep learning) to add value to your business. You want results, not equations on a blackboard. You may or may not own a GPU, but you have a job to do, and in this case, that job involves custom object detection.

The "easier way" is Azure Cognitive Services (*https://oreil.ly/98V37*)—specifically, a member of the Cognitive Services family called the Custom Vision service (*https://oreil.ly/cMAHe*). I will formally introduce Azure Cognitive Services in Chapter 14, but if your goal is to build a custom object detector, now's the right time for your first foray into the world of AI as a service. In my opinion, nothing makes crafting custom object detection models easier than the Custom Vision service. Little expertise is required, and at the end, you're left with a trained model that you can easily consume in Python or C# or other programming languages such as Swift.

Azure's Custom Vision service is one of the best-kept secrets in AI. Let's use it to train a custom object detection model—and then don our marine biologist cap and put the model to work.

Training a Custom Object Detection Model with the Custom Vision Service

The Custom Vision service is one of more than 20 services and APIs that make up Azure Cognitive Services. Cognitive Services enables you to build intelligence into apps without requiring deep expertise in machine learning and AI. The Custom Vision service lets you build image classification models and object detection models and train them in the cloud on GPUs. When training is complete, you can either deploy the models as web services and call them using REST APIs, or download them in various formats, including TensorFlow and Core ML (*https://oreil.ly/uSkBE*), and consume them locally. Included in the download are sample source code files demonstrating how to consume the model.

To use the Custom Vision service, you need a Microsoft account and an Azure subscription. If you don't have a Microsoft account, you can create one for free (*https://oreil.ly/n5T1n*). If you don't have an Azure subscription, you can create a free trial subscription (*https://oreil.ly/izl8G*). You have to provide a credit card or debit card, but you get $200 in free credits to use for 30 days and access to many free service tiers. Even if you have a paid subscription or let your free trial roll over into a paid subscription, you won't necessarily incur any charges for using the Custom Vision service—especially if you export the models you create.

If you'd prefer not to create an Azure subscription, you can skip the remainder of this section and pick up again with the next section, where we use the trained model to detect objects. All the files you need in order to utilize the model are present in the *Chapter 12* folder (*https://oreil.ly/IPwtZ*) of this book's GitHub repo.

Ready to build a custom object detection model? How about one that spots sea turtles in the wild? Start by downloading a dataset (*https://oreil.ly/5L2g4*) of sea turtle images. I used Bing image search to find these photos and limited the search to images that are free to be shared and used commercially. Copy the images from the ZIP file to a convenient location on your hard disk.

Next, navigate to the Custom Vision portal (*https://oreil.ly/a9aMy*) in your browser and log in with your Microsoft account. Click "+ NEW PROJECT," and in the "Create new project" dialog, click "create new" to create a new Azure resource for the project (Figure 12-9).

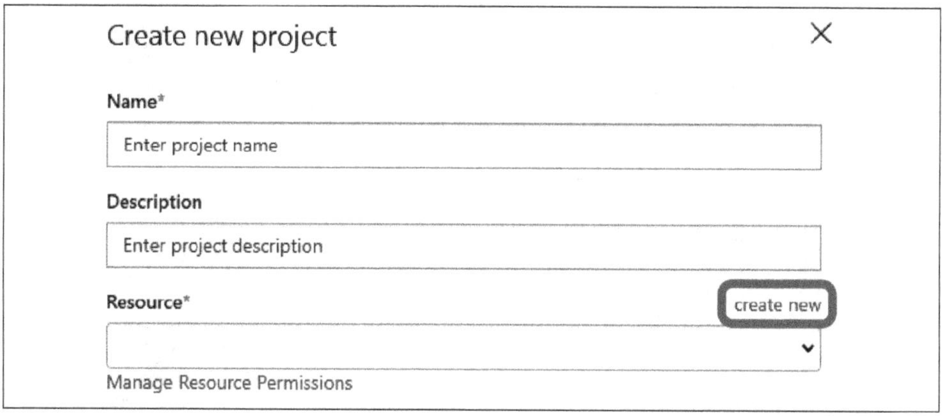

Figure 12-9. Creating a new Azure resource in a Custom Vision service project

Enter a name such as "turtle-detector" for the resource and select your Azure subscription, as shown in Figure 12-10. Select an existing Azure resource group for the new resource or click "create new" and create a new one. Make sure CognitiveServices is selected as the resource type and either accept the default Azure location or choose one that's closer to you. Finally, select F0 as the pricing tier if it's available, or S1 if it's not. (F0 is a free tier and is generally limited to one per Azure subscription.) Then click the "Create resource" button.

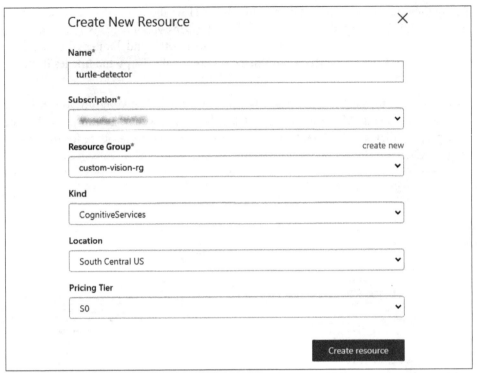

Figure 12-10. Creating a new Cognitive Services resource

 In Azure, a *resource group* is a collection of related resources that serve a common purpose and share a common lifetime. Deleting a resource group deletes all the resources in that group. In addition, you can easily get billing information for the group as a whole rather than add up the costs of the individual resources.

In the "Create new project" dialog, enter a name for the project and select the resource that you just created (Figure 12-11). Select Object Detection as the project type and "General (compact)" as the domain. Selecting one of the "compact" domains is essential if the goal is to train a model that can be exported and consumed locally. Make sure "Basic platforms" is selected under Export Capabilities, and then click "Create project."

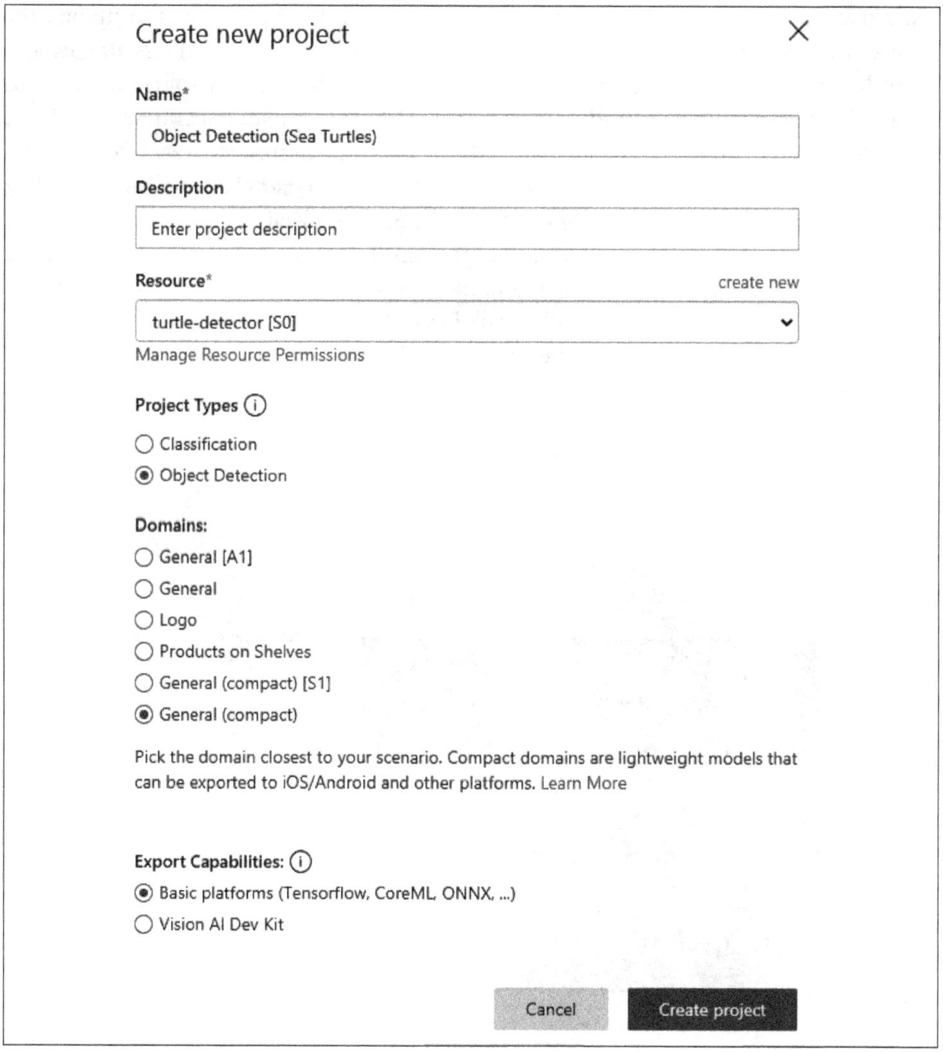

Figure 12-11. Creating a project containing an object detector

Now that the Custom Vision project has been created, the next step is to upload the images that you'll use to train the model and annotate them with class names and bounding boxes. To begin, click "Add images." Then navigate to the folder where your sea turtle images are stored and upload all 50. After the upload is complete, click the first image to open it in the image detail editor. Use your mouse to draw a box around the sea turtle, as shown in Figure 12-12. Then type **sea-turtle** into the tag editor that pops up, and click the plus sign to the right of it. Now click the right arrow at the far right to move to the next image, and repeat this process for the remaining training images. You don't have to enter the tag name again; you simply select it from a list of existing tags. In addition, you usually don't have to draw boxes around the sea turtles. Hover your cursor over a sea turtle and the editor will use a little AI of its own to suggest a bounding box.

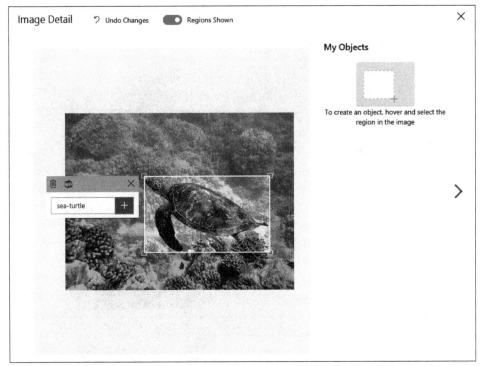

Figure 12-12. Identifying sea turtles in training images

The most tedious part of training a custom object detection model is identifying objects and bounding boxes in all the training images. The Custom Vision service helps by providing an intelligent editor to minimize the amount of clicking and dragging required. Still, imagine doing this for hundreds of thousands of images!

In cases in which a photo contains two or more sea turtles, be sure to tag them all, as shown in Figure 12-13. It's fine if the bounding boxes overlap. Objects frequently do overlap in the real world, after all. Data scientists refer to overlapping bounding boxes as *occlusions*.

Figure 12-13. Sea turtles with overlapping bounding boxes

After you've tagged all 50 training images, click the green Train button in the upper-right corner of the page. When asked to choose a training type (Figure 12-14), select Quick Training to conserve time and money, and then click the Train button in the lower-right corner of the dialog to commence training. The Advanced Training option usually produces a more accurate model, but it takes longer and charges you for up to the number of hours you budget. Advanced training might not be an option if you selected the free F0 tier when you created the model. F0 provides one hour of training per month at no charge.

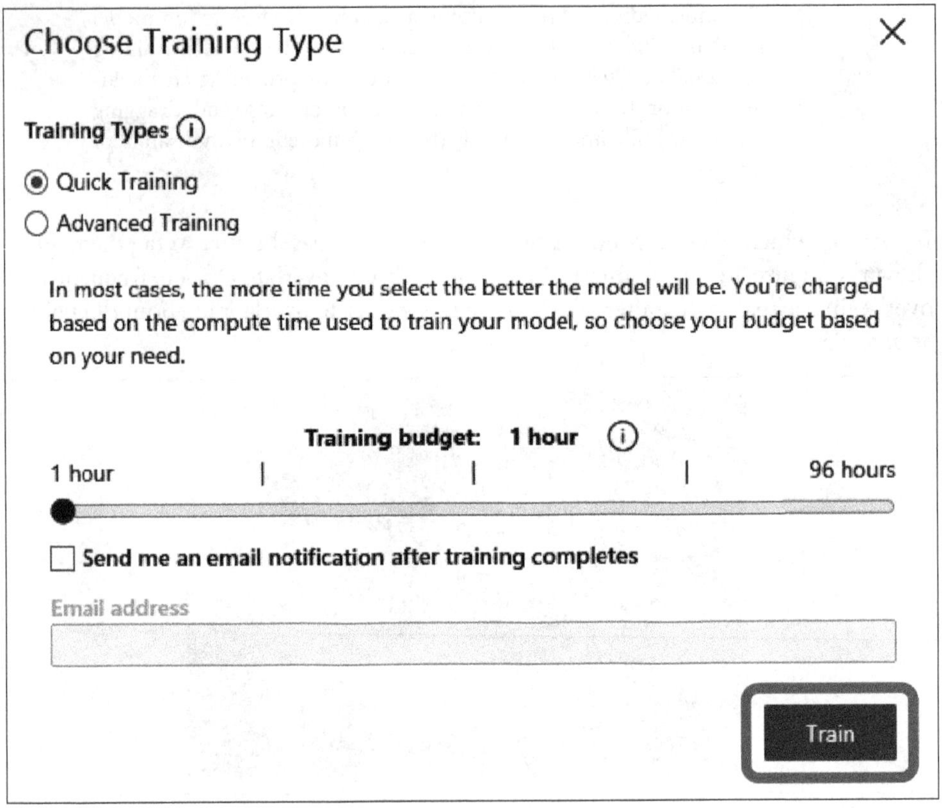

Choose Training Type ✕

Training Types ⓘ

◉ Quick Training

◯ Advanced Training

In most cases, the more time you select the better the model will be. You're charged based on the compute time used to train your model, so choose your budget based on your need.

Training budget: 1 hour ⓘ

1 hour | | | 96 hours

☐ **Send me an email notification after training completes**

Email address

[   ]

[Train]

Figure 12-14. Choosing a training option

Quick training usually takes 5 to 10 minutes. When training is complete, you'll see a summary like the one in Figure 12-15. You already know what precision and recall are. The third metric, *mean average precision* (*mAP*), is a common metric for gauging the accuracy of object detection models. It reflects the model's ability to classify objects *and* identify the objects' bounding boxes. The higher the percentage, the better. If the number isn't sufficiently high, then you have two choices: train with more images, or try advanced training rather than quick training (or both). State-of-the-art object detection models are usually trained with thousands of examples of each class.

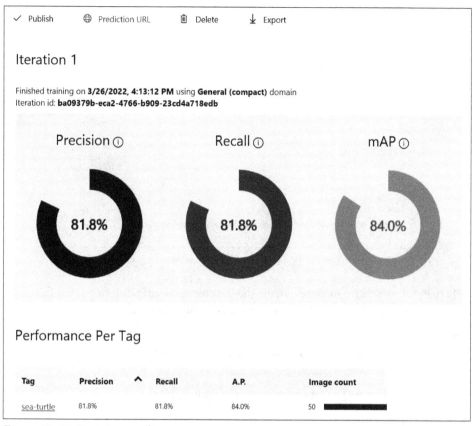

Figure 12-15. Training results

 As an experiment, I trained the same model using the advanced training option and budgeted one hour for training. Training took less than 30 minutes and mAP reached 97.2%. At the time of this writing, advanced training costs $10 per hour, so the cost to my Azure subscription was less than $5. The Custom Vision pricing page (*https://oreil.ly/DCAvD*) has more information on pricing.

You can test the model's accuracy by clicking the Quick Test button at the top of the page and selecting a photo or two that the model wasn't trained with. Or you can click Publish in the upper-left corner of the page and deploy the model as a web service. But our goal is to download the model and run it locally. To that end, click Export at the top of the page. Choose TensorFlow as the export type and then click the Export button. Wait for the export to be prepared, and then click the Download button (Figure 12-16).

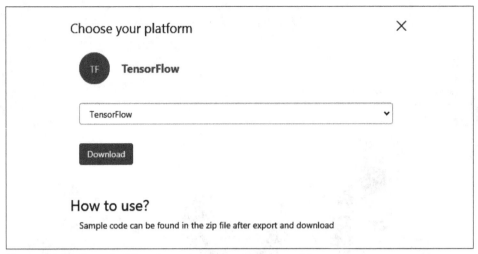

Figure 12-16. Downloading the trained TensorFlow model

When the download completes, open the downloaded ZIP file. Copy the following files from the ZIP file to the directory where your Jupyter notebooks are hosted:

labels.txt
Contains the label you created (sea-turtle) when you labeled the images.

model.pb
Contains the trained and serialized TensorFlow model.

object_detection.py
Found in the ZIP file's *python* folder, this file contains helper functions for utilizing the model in Python apps.

Once these files are downloaded, delete the project in the Custom Vision portal unless you want to go back and refine the model by training it with more images (or granting it more training time). To ensure that no additional charges to your Azure subscription will be incurred, delete the Azure resource group containing the model. You can delete resource groups through the Azure portal (*https://oreil.ly/wYcWH*).

Note that when preparing a model with the Custom Vision service, you aren't limited to one class of object. You can upload images with dozens of different classes and tag them accordingly in the image detail editor. In that case, *labels.txt* will contain all the labels rather than just one.

Using the Exported Model

Now you're ready to put the model to work. Begin by creating a new Jupyter notebook and pasting the following code into the first cell. These statements define a class named `SeaTurtleDetector` that represents a TensorFlow model and a function named `annotate_image` that annotates an image to show what the model detected. The `SeaTurtleDetector` class is a child of the `ObjectDetection` class implemented in *object_detection.py*. It has a `predict` method that you can call to detect objects in an image:

```python
import sys
import tensorflow as tf
import matplotlib.pyplot as plt
from matplotlib.patches import Rectangle
from object_detection import ObjectDetection
from PIL import Image
import numpy as np


class SeaTurtleDetector(ObjectDetection):
    def __init__(self, graph_def, labels):
        super(SeaTurtleDetector, self).__init__(labels)
        self.graph = tf.compat.v1.Graph()
        with self.graph.as_default():
            input_data = tf.compat.v1.placeholder(
                tf.float32, [1, None, None, 3], name='Placeholder')
            tf.import_graph_def(graph_def,
                                input_map={"Placeholder:0": input_data},
                                name="")

    def predict(self, preprocessed_image):
        inputs = np.array(preprocessed_image, dtype=float)[:, :, (2, 1, 0)]

        with tf.compat.v1.Session(graph=self.graph) as sess:
            output_tensor = sess.graph.get_tensor_by_name('model_outputs:0')
            outputs = sess.run(output_tensor,
                               {'Placeholder:0': inputs[np.newaxis, ...]})
            return outputs[0]

def annotate_image(image, predictions, min_score=0.7, figsize=(12, 8)):
    fig, ax = plt.subplots(figsize=figsize,
                           subplot_kw={'xticks': [], 'yticks': []})
    img_width, img_height = image.size
    ax.imshow(image)

    for p in predictions:
        score = p['probability']

        if score >= min_score:
            x1 = p['boundingBox']['left'] * img_width
            y1 = p['boundingBox']['top'] * img_height
            width = p['boundingBox']['width'] * img_width
            height = p['boundingBox']['height'] * img_height
            label = p['tagName']

            rect = Rectangle((x1, y1), width, height, fill=False,
```

```
                        color='red', lw=2)
        ax.add_patch(rect)
        label = f'{label} ({score:.0%})'
        ax.text(x1 + (width / 2), y1, label, color='white',
                backgroundcolor='red', ha='center', va='bottom',
                fontweight='bold', bbox=dict(color='red'))
```

Next, use the following statements to load the serialized model in *model.pb* and the label in *labels.txt* and initialize an object detection model:

```
# Load the serialized model graph
graph_def = tf.compat.v1.GraphDef()
with tf.io.gfile.GFile('model.pb', 'rb') as f:
    graph_def.ParseFromString(f.read())

# Load the labels
with open('labels.txt', 'r') as f:
    labels = [l.strip() for l in f.readlines()]

# Create the model
model = SeaTurtleDetector(graph_def, labels)
```

Finally, go out to the internet and download a sea turtle photo. Then load the photo into your notebook and see if the model detects the sea turtle in it:

```
%matplotlib inline

image = Image.open('PATH_TO_IMAGE_FILE')
predictions = model.predict_image(image)
annotate_image(image, predictions)
```

Did the model get it right? It did for me:

Keep in mind that state-of-the-art accuracy in object detection is a product of training with *lots* of labeled images. The 50 you trained with here are a good start, but to train a model to a sufficient level of accuracy for real-world use, you'll probably need a minimum of 10 times more. The good news is that once you've collected the images, the Custom Vision service makes the process of labeling the images and training the model relatively easy.

Summary

Object detection represents the tip of the spear in the world of computer vision. It's how self-driving cars "see" objects in front of them, and it's the technology underlying numerous real-world applications. State-of-the-art object detection is accomplished with deep learning using specially crafted CNNs that do more than mere image classification. You can use pretrained models such as YOLO and Mask R-CNN to detect objects in images and video frames. With Mask R-CNN, you can even generate segmentation masks revealing additional information about those objects.

Custom object detection requires you to train models like these with images of your own carefully labeled to denote objects and bounding boxes. Azure's Custom Vision service simplifies the process of training custom object detection models, and allows the models you train to be hosted in the cloud and called using REST APIs or downloaded and consumed locally. It's one of several services that comprise Azure Cognitive Services, and it's the right tool for the job when the job calls for detecting objects that pretrained models weren't trained to detect.

## CHAPTER 13

Natural Language Processing

It's not difficult to use Scikit-Learn to build machine learning models that analyze text for sentiment, identify spam, and classify text in other ways. But today, state-of-the-art text classification is most often performed with neural networks. You already know how to build neural networks that accept numbers and images as input. Let's build on that to learn how to construct deep-learning models that process text—a segment of deep learning known as *natural language processing*, or NLP for short.

NLP encompasses a variety of activities including text classification, named-entity recognition, keyword extraction, question answering, and language translation. The accuracy of NLP models has improved in recent years for a variety of reasons, not the least of which are newer and better ways of converting words and sentences into dense vector representations that incorporate meaning, and a relatively new neural network architecture called the *transformer* that can zero in on the most meaningful words and even differentiate between different meanings of the same word.

One element that virtually all neural networks that process text have in common is an *embedding layer*, which uses *word embeddings* to transform arrays, or *sequences*, of scalar values representing words into arrays of floating-point numbers called *word vectors*. These vectors encode information about the meanings of words and the relationships between them. Output from an embedding layer can be input to a classification layer, or it can be input to other types of neural network layers to tease more meaning from it before subjecting it to further processing.

Transformers? Embedding layers? Word vectors? There's a lot to unpack here, but once you wrap your head around a few basic concepts, neural networks that process language are pure magic. Let's dive in. We'll start by learning how to prepare text for processing by a deep-learning model and how to create word embeddings. Then we'll put that knowledge to work building neural networks that classify text, translate text from one language to another, and more—all classic applications of NLP.

Text Preparation

Chapter 4 introduced Scikit-Learn's `CountVectorizer` class (*https://oreil.ly/EcATK*), which converts rows of text into rows of word counts that a machine learning model can consume. `CountVectorizer` also converts characters to lowercase, removes numbers and punctuation symbols, and optionally removes stop words—common words such as *and* and *the* that are likely to have little influence on the outcome.

Text must be cleaned and vectorized before it's used to train a neural network too, but vectorization is typically performed differently. Rather than create a table of word counts, you create a table of sequences containing *tokens* representing individual words. Tokens are often indices into a dictionary, or *vocabulary*, built from the corpus of words in the dataset. To help, Keras provides the `Tokenizer` class (*https://oreil.ly/nfVsh*), which you can think of as the deep-learning equivalent of `CountVectorizer`. Here's an example that uses `Tokenizer` to create sequences from four lines of text:

```
from tensorflow.keras.preprocessing.text import Tokenizer

lines = [
    'The quick brown fox',
    'Jumps over $$$ the lazy brown dog',
    'Who jumps high into the blue sky after counting 123',
    'And quickly returns to earth'
]

tokenizer = Tokenizer()
tokenizer.fit_on_texts(lines)
sequences = tokenizer.texts_to_sequences(lines)
```

`fit_on_texts` (*https://oreil.ly/NbbbA*) creates a dictionary containing all the words in the input text. `texts_to_sequences` (*https://oreil.ly/JijdJ*) returns a list of sequences, which are simply arrays of indices into the dictionary:

```
[[1, 4, 2, 5],
 [3, 6, 1, 7, 2, 8],
 [9, 3, 10, 11, 1, 12, 13, 14, 15, 16],
 [17, 18, 19, 20, 21]]
```

The word *brown* appears in lines 1 and 2 and is represented by the index 2. Therefore, 2 appears in both sequences. Similarly, a 3 representing the word *jumps* appears in sequences 2 and 3. The index 0 isn't used to denote words; it's reserved to serve as padding. More on this in a moment.

You can use Tokenizer's `sequences_to_texts` method (*https://oreil.ly/cxSdq*) to reverse the process and convert the sequences back into text:

```
['the quick brown fox',
 'jumps over the lazy brown dog',
```

```
'who jumps high into the blue sky after counting 123',
'and quickly returns to earth']
```

One revelation that comes from this is that `Tokenizer` converts text to lowercase and removes symbols, but it doesn't remove stop words or numbers. If you want to remove stop words, you can use a separate library such as the Natural Language Toolkit (NLTK) (*https://oreil.ly/Ic2ON*). You can also remove words containing numbers while you're at it:

```python
from tensorflow.keras.preprocessing.text import Tokenizer
from nltk.tokenize import word_tokenize
from nltk.corpus import stopwords

lines = [
    'The quick brown fox',
    'Jumps over $$$ the lazy brown dog',
    'Who jumps high into the blue sky after counting 123',
    'And quickly returns to earth'
]

def remove_stop_words(text):
    text = word_tokenize(text.lower())
    stop_words = set(stopwords.words('english'))
    text = [word for word in text if word.isalpha() and not word in stop_words]
    return ' '.join(text)

lines = list(map(remove_stop_words, lines))

tokenizer = Tokenizer()
tokenizer.fit_on_texts(lines)
tokenizer.texts_to_sequences(lines)
```

The resulting sequences look like this:

```
[[3, 1, 4],
 [2, 5, 1, 6],
 [2, 7, 8, 9, 10],
 [11, 12, 13]]
```

which, converted back to text, are as follows:

```
['quick brown fox',
 'jumps lazy brown dog',
 'jumps high blue sky counting',
 'quickly returns earth']
```

The sequences range from three to five values in length, but a neural network expects all sequences to be the same length. Keras's `pad_sequences` function (*https://oreil.ly/nQoQZ*) performs this final step, truncating sequences longer than the specified length and padding sequences shorter than the specified length with 0s:

```
from tensorflow.keras.preprocessing.sequence import pad_sequences

padded_sequences = pad_sequences(sequences, maxlen=4)
```

The resulting padded sequences look like this:

```
array([[ 0,  3,  1,  4],
       [ 2,  5,  1,  6],
       [ 7,  8,  9, 10],
       [ 0, 11, 12, 13]])
```

Converting these sequences back to text yields this:

```
['quick brown fox',
 'jumps lazy brown dog',
 'high blue sky counting',
 'quickly returns earth']
```

By default, `pad_sequences` pads and truncates on the left, but you can include a `padding='post'` parameter if you prefer to pad and truncate on the right. Padding on the right is sometimes important when using neural networks to translate text.

> Removing stop words frequently has little or no effect on text classification tasks. If you simply want to remove numbers from text input to a neural network without removing stop words, create a `Tokenizer` this way:
>
> ```
> tokenizer = Tokenizer(
>     filters='!"#$%&()*+,-./:;<=>?@[\\]^_` ' \
>     '{|}~\t\n0123456789')
> ```
>
> The `filters` parameter tells `Tokenizer` what characters to remove. It defaults to `'!"#$%&()*+,-./:;<=>?@[\\]^_`{|}~\t\n'`. This code simply adds 0–9 to the list. Note that `Tokenizer` does *not* remove apostrophes by default, but you can remove them by adding an apostrophe to the `filters` list.

It's important that text input to a neural network for predictions be tokenized and padded in the same way as text input to the model for training. If you're thinking it sure would be nice not to have to do the tokenization and sequencing manually, there is a way around it. Rather than write a lot of code, you can include a `TextVectorization` layer (*https://oreil.ly/rydux*) in the model. I'll demonstrate how momentarily. But first, you need to learn about embedding layers.

Word Embeddings

Once text is tokenized and converted into padded sequences, it is ready for training a neural network. But you probably won't get very far training on the raw padded sequences.

One of the crucial elements of a neural network that processes text is an embedding layer (*https://oreil.ly/vd83A*) whose job is to convert padded sequences of word tokens into arrays of word vectors (*https://oreil.ly/bA39w*), which represent each word with an array (vector) of floating-point numbers rather than a single integer. Each word in the input text is represented by a vector in the embedding layer, and as the network is trained, vectors representing individual words are adjusted to reflect their relationship to one another. If you're building a sentiment analysis model and words such as *excellent* and *amazing* have similar connotations, then the vectors representing those words in the embedding space should be relatively close together so that phrases such as "excellent service" and "amazing service" score similarly.

Implementing an embedding layer by hand is a complex undertaking (especially the training aspect), so Keras offers the `Embedding` class (*https://oreil.ly/OQMzf*). With Keras, creating a trainable embedding layer requires just one line of code:

```
Embedding(input_dim=10000, output_dim=32, input_length=100)
```

In order, the three parameters passed to the `Embedding` function are:

- The vocabulary size, or the number of words in the vocabulary built by `Tokenizer`
- The number of dimensions m in the embedding space
- The length n of each padded sequence

You pick the number of dimensions m, and each word gets encoded in the embedding space as an m-dimensional vector. More dimensions provide more fitting power, but also increase training time. In practice, m is usually a number from 32 to 512.

The vectors that represent individual words in an embedding layer are learned during training, just as the weights connecting neurons in adjacent dense layers are learned. If the number of training samples is sufficiently high, training the network usually creates effective vector representations of all the words. However, if you have only a few hundred training samples, the embedding layer might not have enough information to properly vectorize the corpus of text.

In that case, you can elect to initialize the embedding layer with *pretrained word embeddings* rather than rely on it to learn the word embeddings on its own. Several popular pretrained word embeddings exist in the public domain, including the GloVe word vectors (*https://oreil.ly/9xBhl*) developed by Stanford and Google's own Word2Vec (*https://oreil.ly/NzrSo*). Pretrained embeddings tend to model semantic relationships between words, recognizing, for example, that *king* and *queen* are related terms while *stairs* and *zebra* are not. While that can be beneficial, a network trained to classify text usually performs better when word embeddings are learned from the training data because such embeddings are task specific. For an example showing how to use pretrained embeddings, see "Using Pre-trained Word Embed-dings" (*https://oreil.ly/0VC3Q*) by the author of Keras.

Text Classification

Figure 13-1 shows the baseline architecture for a neural network that classifies text. Tokenized text sequences are input to the embedding layer. The output from the embedding layer is a 2D matrix of floating-point values measuring *m* by *n*, where *m* is the number of dimensions in the embedding space and *n* is the sequence length. The Flatten layer (*https://oreil.ly/2fsvZ*) following the embedding layer "flattens" the 2D output into a 1D array suitable for input to a dense layer, and the dense layer classifies the values emitted from the flatten layer. You can experiment with different dimensions in the embedding layer and different widths of the dense layer to maximize accuracy. You can also add more dense layers if needed.

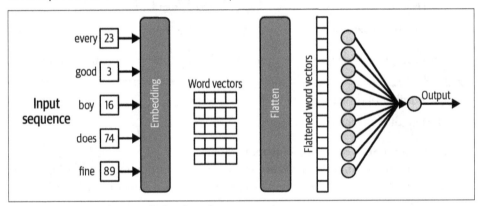

Figure 13-1. Neural network for classifying text

A neural network like the one in Figure 13-1 can be implemented this way:

```
from tensorflow.keras.models import Sequential
from tensorflow.keras.layers import Dense, Flatten, Embedding

model = Sequential()
model.add(Embedding(10000, 32, input_length=100))
model.add(Flatten())
model.add(Dense(128, activation='relu'))
model.add(Dense(1, activation='sigmoid'))
model.compile(optimizer='adam', loss='binary_crossentropy', metrics=['accuracy'])
```

One application for a neural network that classifies text is spam filtering. Chapter 4 demonstrated how to use Scikit to build a machine learning model that separates spam from legitimate emails. Let's build an equivalent deep-learning model with Keras and TensorFlow. We'll use the same dataset we used before: one containing

1,000 emails, half of which are spam (indicated by 1s in the label column) and half of which are not (indicated by 0s in the label column).

Begin by downloading the dataset (*https://oreil.ly/jWEyN*) and copying it into the *Data* subdirectory where your Jupyter notebooks are hosted. Then use the following statements to load the dataset and shuffle the rows to distribute positive and negative samples throughout. Shuffling is important because rather than use `train_test_split` to create a validation dataset, we'll use `fit`'s `validation_split` parameter. It doesn't shuffle the data as `train_test_split` does:

```
import pandas as pd

df = pd.read_csv('Data/ham-spam.csv')
df = df.sample(frac=1, random_state=0)
df.head()
```

Use the following statements to remove any duplicate rows from the dataset and check for balance:

```
df = df.drop_duplicates()
df.groupby('IsSpam').describe()
```

Next, extract the emails from the `DataFrame`'s `Text` column and labels from the `IsSpam` column. Then use Keras's `Tokenizer` class to tokenize the text and convert it into sequences, and `pad_sequences` to produce sequences of equal length. There's no need to remove stop words because doing so doesn't impact the outcome:

```
from tensorflow.keras.preprocessing.text import Tokenizer
from tensorflow.keras.preprocessing.sequence import pad_sequences

x = df['Text']
y = df['IsSpam']

max_words = 10000 # Limit the vocabulary to the 10,000 most common words
max_length = 500

tokenizer = Tokenizer(num_words=max_words)
tokenizer.fit_on_texts(x)
sequences = tokenizer.texts_to_sequences(x)
x = pad_sequences(sequences, maxlen=max_length)
```

Define a binary classification model that contains an embedding layer with 32 dimensions, a flatten layer to flatten output from the embedding layer, a dense layer for classification, and an output layer with a single neuron and sigmoid activation:

```
from tensorflow.keras.models import Sequential
from tensorflow.keras.layers import Dense, Flatten, Embedding

model = Sequential()
model.add(Embedding(max_words, 32, input_length=max_length))
model.add(Flatten())
```

```
model.add(Dense(128, activation='relu'))
model.add(Dense(1, activation='sigmoid'))
model.compile(loss='binary_crossentropy', optimizer='adam', metrics=['accuracy'])
model.summary()
```

Train the network and allow Keras to use 20% of the training samples for validation:

```
hist = model.fit(x, y, validation_split=0.2, epochs=5, batch_size=20)
```

Use the history object returned by `fit` to plot the training and validation accuracy in each epoch:

```
import seaborn as sns
import matplotlib.pyplot as plt
%matplotlib inline
sns.set()

acc = hist.history['accuracy']
val = hist.history['val_accuracy']
epochs = range(1, len(acc) + 1)

plt.plot(epochs, acc, '-', label='Training accuracy')
plt.plot(epochs, val, ':', label='Validation accuracy')
plt.title('Training and Validation Accuracy')
plt.xlabel('Epoch')
plt.ylabel('Accuracy')
plt.legend(loc='lower right')
plt.plot()
```

Hopefully, the network achieved a validation accuracy exceeding 95%. If it didn't, train it again. Here's how it turned out for me:

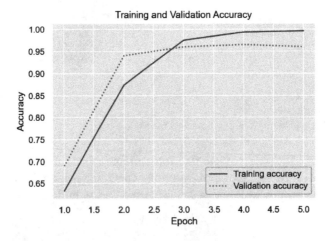

Once you're satisfied with the accuracy, use the following statements to compute the probability that an email regarding a code review is spam:

```
text = 'Can you attend a code review on Tuesday? ' \
       'Need to make sure the logic is rock solid.'

sequence = tokenizer.texts_to_sequences([text])
padded_sequence = pad_sequences(sequence, maxlen=max_length)
model.predict(padded_sequence)[0][0]
```

Then do the same for another email:

```
text = 'Why pay more for expensive meds when ' \
       'you can order them online and save $$$?'

sequence = tokenizer.texts_to_sequences([text])
padded_sequence = pad_sequences(sequence, maxlen=max_length)
model.predict(padded_sequence)[0][0]
```

What did the network predict for the first email? What about the second? Do you agree with the predictions? Remember that a number close to 0.0 indicates that the email is not spam, while a number close to 1.0 indicates that it is.

Automating Text Vectorization

Rather than run `Tokenizer` and `pad_sequences` manually, you can preface an embedding layer with a `TextVectorization` layer (*https://oreil.ly/1hDHa*). Here's an example:

```
from tensorflow.keras.models import Sequential
from tensorflow.keras.layers import Dense, Flatten, Embedding
from tensorflow.keras.layers import TextVectorization, InputLayer
import tensorflow as tf

model = Sequential()
model.add(InputLayer(input_shape=(1,), dtype=tf.string))
model.add(TextVectorization(max_tokens=max_words,
                            output_sequence_length=max_length))
model.add(Embedding(max_words, 32, input_length=max_length))
model.add(Flatten())
model.add(Dense(128, activation='relu'))
model.add(Dense(1, activation='sigmoid'))
model.compile(loss='binary_crossentropy', optimizer='adam',
              metrics=['accuracy'])
model.summary()
```

Note that the input layer (an instance of `InputLayer` (*https://oreil.ly/uZ5ic*)) is now explicitly defined, and it's configured to accept string input. In addition, before training the model, the `TextVectorization` layer must be fit to the input data by calling `adapt` (*https://oreil.ly/BxRAa*):

```
model.layers[0].adapt(x)
```

Now you no longer need to preprocess the training text, and you can pass raw text strings to `predict`:

```
text = 'Why pay more for expensive meds when ' \
       'you can order them online and save $$$?'
model.predict([text])[0][0]
```

TextVectorization doesn't remove stop words, so if you want them removed, you can do that separately or use the TextVectorization function's standardize parameter to identify a callback function that does it for you.

Using TextVectorization in a Sentiment Analysis Model

To demonstrate how TextVectorization layers simplify text processing, let's use it to build a binary classifier that performs sentiment analysis. We'll use the same dataset we used in Chapter 4: the IMDB reviews dataset (*https://oreil.ly/tfjoP*) containing 25,000 positive reviews and 25,000 negative reviews.

Download the dataset if you haven't already and place it in your Jupyter notebooks' *Data* subdirectory. Then create a new notebook and use the following statements to load and shuffle the dataset:

```
import pandas as pd

df = pd.read_csv('Data/reviews.csv', encoding="ISO-8859-1")
df = df.sample(frac=1, random_state=0)
df.head()
```

Remove duplicate rows and check for balance:

```
df = df.drop_duplicates()
df.groupby('Sentiment').describe()
```

Now create the model and include a TextVectorization layer to preprocess input text:

```
from tensorflow.keras.models import import Sequential
from tensorflow.keras.layers import TextVectorization, InputLayer
from tensorflow.keras.layers import Dense, Flatten, Embedding
import tensorflow as tf

max_words = 20000
max_length = 500

model = Sequential()
model.add(InputLayer(input_shape=(1,), dtype=tf.string))
model.add(TextVectorization(max_tokens=max_words,
                            output_sequence_length=max_length))
model.add(Embedding(max_words, 32, input_length=max_length))
model.add(Flatten())
model.add(Dense(128, activation='relu'))
model.add(Dense(1, activation='sigmoid'))
model.compile(loss='binary_crossentropy', optimizer='adam',
              metrics=['accuracy'])
model.summary()
```

Extract the reviews from the DataFrame's Text column and the labels (0 for negative sentiment, 1 for positive) from the Sentiment column, and use the former to fit the TextVectorization layer to the text. Then train the model:

```
x = df['Text']
y = df['Sentiment']
model.layers[0].adapt(x)

hist = model.fit(x, y, validation_split=0.5, epochs=5, batch_size=250)
```

When training is complete, plot the training and validation accuracy:

```
import seaborn as sns
import matplotlib.pyplot as plt
%matplotlib inline
sns.set()

acc = hist.history['accuracy']
val = hist.history['val_accuracy']
epochs = range(1, len(acc) + 1)

plt.plot(epochs, acc, '-', label='Training accuracy')
plt.plot(epochs, val, ':', label='Validation accuracy')
plt.title('Training and Validation Accuracy')
plt.xlabel('Epoch')
plt.ylabel('Accuracy')
plt.legend(loc='lower right')
plt.plot()
```

The model fits to the training text extremely well, but validation accuracy usually peaks between 85% and 90%:

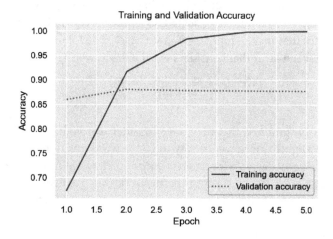

Use the model to score a positive comment for sentiment:

```
text = 'Excellent food and fantastic service!'
model.predict([text])[0][0]
```

Now do the same for a negative comment:

```
text = 'The long lines and poor customer service really turned me off.'
model.predict([text])[0][0]
```

Observe how much simpler the code is. Operationalizing the model is simpler too, because you no longer need a `Tokenizer` fit to the training data to prepare text submitted to the model for predictions. Be aware, however, that a model with a `Text Vectorization` layer can't be saved in Keras's H5 format. It *can* be saved in Tensor-Flow's SavedModel format. The following statement saves the model in the *saved_model* subdirectory of the current directory:

```
model.save('saved_model')
```

Once the model is reloaded, you can pass text directly to it for making predictions.

Factoring Word Order into Predictions

Both of the models you just built are *bag-of-words* models that ignore word order. Such models are common and are often more accurate than other types of models. But that's not always the case. The relative position of the words in a sentence sometimes has meaning. *Credit* and *card* should probably influence a spam classifier one way if they appear far apart in a sentence and another way if they appear together.

One way to improve—or at least *attempt* to improve—on a simple bag-of-words model is to use *n*-grams as described in Chapter 4. An *n*-gram is a collection of *n* words appearing in consecutive order. Keras's `TextVectorization` class features an `ngrams` parameter that makes applying *n*-grams easy. The following statement creates a `TextVectorization` layer that considers word pairs as well as individual words:

```
model.add(TextVectorization(max_tokens=max_words,
                            output_sequence_length=max_length,
                            ngrams=2))
```

One limitation of *n*-grams is that they only consider words that are directly adjacent to each other. A slightly more robust way to factor word position into a classification task is to replace dense layers with `Conv1D` (*https://oreil.ly/4h0M8*) and `MaxPooling1D` (*https://oreil.ly/2KMId*) layers, turning the network into a convolutional neural network. CNNs are most often used to classify images, but one-dimensional convolution layers play well with text sequences. Here's the network presented in the previous example recast as a CNN:

```
from tensorflow.keras.models import Sequential
from tensorflow.keras.layers import Conv1D, MaxPooling1D, GlobalMaxPooling1D
from tensorflow.keras.layers import TextVectorization, InputLayer
```

```
from tensorflow.keras.layers import Dense, Embedding
import tensorflow as tf

model = Sequential()
model.add(InputLayer(input_shape=(1,), dtype=tf.string))
model.add(TextVectorization(max_tokens=max_words,
                            output_sequence_length=max_length))
model.add(Embedding(max_words, 32, input_length=max_length))
model.add(Conv1D(32, 7, activation='relu'))
model.add(MaxPooling1D(5))
model.add(Conv1D(32, 7, activation='relu'))
model.add(GlobalMaxPooling1D())
model.add(Dense(1, activation='sigmoid'))
model.compile(loss='binary_crossentropy', optimizer='adam',
              metrics=['accuracy'])
```

Rather than process individual words, the Conv1D layers in this example extract features from groups of vectors representing words (seven in the first layer and seven more in the second), just as Conv2D layers extract features from blocks of pixels. The MaxPooling1D layer condenses the output from the first Conv1D layer to reveal higher-level structure in input sequences, similar to the way reducing the resolution of an image tends to draw out macro features such as the shape of a person's body while minimizing or filtering out altogether lesser features such as the shape of a person's eyes. A simple CNN like this one sometimes classifies text more accurately than bag-of-words models, and sometimes does not. As is so often the case in machine learning, the only way to know is to try.

Recurrent Neural Networks (RNNs)

Yet another way to factor word position into a classifier is to include recurrent layers in the network. Recurrent layers were originally invented to process time-series data—for example, to look at weather data for the past five days and predict what tomorrow's high temperature will be. If you simply took all the weather data for those five days and treated each day independently, trends evident in the data would be lost. A recurrent layer, however, might detect those trends and factor them into its output. A sequence of vectors output by an embedding layer qualifies as a time series because words in a phrase are ordered consecutively and words used early in a phrase could inform how words that occur later are interpreted.

Figure 13-2 illustrates how a recurrent layer transforms word embeddings into a vector that's influenced by word order. In this example, sequences are input to an embedding layer, which transforms each word (token) in the sequence into a vector of floating-point numbers. These word embeddings are input to a recurrent layer, yielding another output vector. To compute that vector, cells in the recurrent layer loop over the embeddings comprising the sequence. The input to iteration $n + 1$ of the loop is the current embedding vector *and* the output from iteration n—the so-called *hidden state.* The output from the recurrent layer is the output from the final iteration of the loop. The result is different than it would have been had each embedding vector been processed independently because each iteration uses information from the previous iteration to compute an output. Context from a word early in a sequence can carry over to words that occur later on.

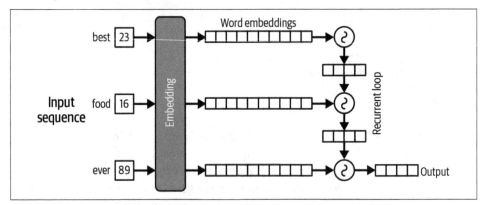

Figure 13-2. Processing word embeddings with a recurrent layer

It's not difficult to build a very simple recurrent layer by hand that works reasonably well with short sequences (say, four or five words), but longer sequences will suffer from a vanishing-gradient effect that means words far apart will exert little influence over one another. One solution is a recurrent layer composed of *Long Short-Term Memory* (LSTM) cells, which were introduced in a 1997 paper titled, appropriately enough, "Long Short-Term Memory" (*https://oreil.ly/W7vrb*).

LSTM cells are miniature neural networks in their own right. As the cells loop over the words in a sequence, they learn (just as the weights connecting neurons in dense layers are learned) which words are important and lend that information precedence in subsequent iterations through the loop. A dense layer doesn't recognize that there's a connection between *blue* and *sky* in the phrase "I like blue, for on a clear and sunny day, it is the color of the sky." An LSTM layer does and can factor that into its output. Its power diminishes, however, as words grow farther apart.

Keras provides a handy implementation of LSTM layers in its LSTM class (*https://oreil.ly/Q6zlb*). It also provides a GRU class (*https://oreil.ly/LtHDE*) implementing gated recurrent unit (GRU) layers, which are simplified LSTM layers that train faster and often yield results that equal or exceed those of LSTM layers. Here's how our spam classifier would look if it were modified to use LSTM:

```
from tensorflow.keras.models import Sequential
from tensorflow.keras.layers import TextVectorization, InputLayer
from tensorflow.keras.layers import Dense, Embedding, LSTM
import tensorflow as tf
```

```
model = Sequential()
model.add(InputLayer(input_shape=(1,), dtype=tf.string))
model.add(TextVectorization(max_tokens=max_words,
                            output_sequence_length=max_length))
model.add(Embedding(max_words, 32, input_length=max_length))
model.add(LSTM(32))
model.add(Dense(1, activation='sigmoid'))
model.compile(loss='binary_crossentropy', optimizer='adam',
              metrics=['accuracy'])
```

LSTMs are compute intensive, especially when dealing with long sequences. If you try this code, you'll find that the model takes *much* longer to train. You'll also find that validation accuracy improves little if at all. This goes back to the fact that bag-of-words models, while simple, also tend to be effective at classifying text.

LSTM layers can be stacked, just like dense layers. The trick is to include a `return_sequences=True` attribute in all LSTM layers except the last so that the previous layers return all the vectors generated (all the hidden state) rather than just the final vector. Google Translate once used two stacks of LSTMs eight layers deep to encode phrases in one language and decode them into another.

Using Pretrained Models to Classify Text

If training your own sentiment analysis model doesn't yield the accuracy you require, you can always turn to a pretrained model. Just as there are pretrained computer-vision models trained with millions of labeled images, there are pretrained sentiment analysis models available that were trained with millions of labeled text samples. Keras doesn't provide a convenient wrapper around these models as it does for pretrained CNNs, but they are relatively easy to consume nonetheless.

Many such models are available from Hugging Face (*https://oreil.ly/ly7sQ*), an AI-focused company whose goal is to advance and democratize AI. Hugging Face originally concentrated on NLP models but has since expanded its library to include other types of models, including image classification models (*https://oreil.ly/2ZQFT*) and object detection models (*https://oreil.ly/A5i4m*). Currently, Hugging Face hosts more than 400 sentiment analysis models (*https://oreil.ly/omrnl*) trained on different types of input ranging from tweets to product reviews in a variety of languages. It even offers models that analyze text for joy, anger, surprise, and other emotions.

All of these models are free for you to use. Care to give it a try? Start by installing Hugging Face's Transformers package (*https://oreil.ly/NoiNp*) in your Python environment (for example, `pip install transformers`). Then fire up a Jupyter notebook and use the following code to load Hugging Face's default sentiment analysis model:

```
from transformers import pipeline

model = pipeline('sentiment-analysis')
```

You'll incur a short delay while the model is downloaded for the first time. Once the download completes, score a sentence for sentiment:

```
model('The long lines and poor customer service really turned me off')
```

Here's the result:

```
[{'label': 'NEGATIVE', 'score': 0.9995430707931519}]
```

Try it with a positive comment:

```
model('Great food and excellent service!')
```

Here's the result:

```
[{'label': 'POSITIVE', 'score': 0.9998843669891357}]
```

In the return value, `label` indicates whether the sentiment is positive or negative, and `score` reveals the model's confidence in the label.

It's just as easy to analyze a text string for emotion by loading a different pretrained model. To demonstrate, try this:

```
model = pipeline('text-classification',
                 model='bhadresh-savani/distilbert-base-uncased-emotion',
                 return_all_scores=True)

model('The long lines and poor customer service really turned me off')
```

Here's what it returned for me:

```
[[{'label': 'sadness', 'score': 0.10837080329656601},
  {'label': 'joy', 'score': 0.002373947761952877},
  {'label': 'love', 'score': 0.0006029471987858415},
  {'label': 'anger', 'score': 0.8861245512962341},
  {'label': 'fear', 'score': 0.0019340706057846546},
  {'label': 'surprise', 'score': 0.00059362960074651182}]]
```

It doesn't get much easier than that! And as you'll see in a moment, pretrained Hugging Face models lend themselves to much more than just text classification.

Neural Machine Translation

In the universe of natural language processing, text classification is a relatively simple task. At the opposite end of the spectrum lies *neural machine translation* (NMT), which uses deep learning to translate text from one language to another. NMT has proven superior to the rules-based machine translation (RBMT) (*https://oreil.ly/V6IlG*) and statistical machine translation (SMT) (*https://oreil.ly/vKhKW*) systems that predated the explosion of deep learning and today is the basis for virtually all state-of-the-art text translation services.

The gist of text classification is that you transform an input sequence into a vector characterizing the sequence, and then you input the vector to a classifier. There are several ways to generate that vector. You can reshape the 2D output from an embedding layer into a 1D vector, or you can feed that output into a recurrent layer or convolution layer in hopes of generating a vector that is more context aware. Whichever route you choose, the goal is simple: convert a string of text into an array of floating-point numbers that uniquely describes it and use a sigmoid or softmax output layer to classify it.

NMT is basically an extension of text classification. You start by converting a text sequence into a vector. But rather than classify the vector, you use it to generate a new sequence. One way to do that is with an LSTM encoder-decoder.

LSTM Encoder-Decoders

Until a few years ago, most NMT models, including the one underlying Google Translate, were LSTM-based sequence-to-sequence models (*https://oreil.ly/5rvCF*) similar to the one in Figure 13-3. In such models, one or more LSTM layers encode a tokenized input sequence representing the phrase to be translated into a vector. A second set of recurrent layers uses that vector as input and decodes it into a tokenized phrase in another language. The model accepts sequences as input and returns sequences as output, hence the term *sequence-to-sequence* model. A softmax output layer at the end outputs a set of probabilities for each token in the output sequence. If the maximum output phrase length that's supported is 20 tokens, for example, and the vocabulary of the output language contains 20,000 words, then the output is 20 sets (one per token) of 20,000 probabilities. The word selected for each output token is the word assigned the highest probability.

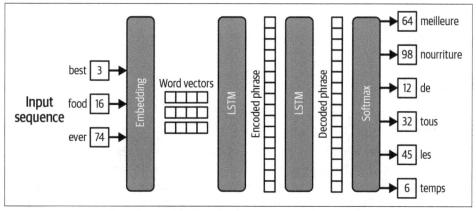

Figure 13-3. LSTM-based encoder-decoder for neural machine translation

LSTM-based sequence-to-sequence models are relatively easy to build with Keras and TensorFlow. This book's GitHub repo contains a notebook (*https://oreil.ly/qieYB*) that uses 50,000 samples to train an LSTM model to translate English to French. The model is defined this way:

```
model = Sequential()
model.add(Embedding(en_vocab_size, 256, input_length=en_max_len, mask_zero=True))
model.add(LSTM(256))
model.add(RepeatVector(fr_max_len))
model.add(LSTM(256, return_sequences=True))
model.add(Dropout(0.4))
model.add(TimeDistributed(Dense(fr_vocab_size, activation='softmax')))
model.compile(loss='sparse_categorical_crossentropy', optimizer='adam',
              metrics=['accuracy'])model.summary(line_length=100)
```

The first layer is an embedding layer that converts word tokens into vectors containing 256 floating-point values. The `mask_zero=True` parameter indicates that zeros in the input sequences denote padding so that the next layer can ignore them. (There is no need to translate those tokens, after all.) Next is an LSTM layer that encodes English phrases input to the model. A second LSTM layer decodes the phrases into dense vectors representing the French equivalents. In between lies a `RepeatVector` layer (*https://oreil.ly/3szHO*) that reshapes the output from the first LSTM layer for input to the second by repeating the output a specified number of times. The final layer is a softmax classification layer that outputs probabilities for each word in the French vocabulary. The `TimeDistributed` wrapper (*https://oreil.ly/Da2jW*) ensures that the model outputs a set of probabilities for each token in the output rather than just one set for the entire sequence.

After 34 epochs of training, the model translates 10 test phrases this way:

```
its fall now => cest maintenant maintenant
im losing => je suis en train
it was quite funny => cetait fut amusant amusant
thats not unusual => ce nest pas inhabituel
i think ill do that => je pense que je le
tom looks different => tom a lair different
its worth a try => ca vaut le coup
fortune smiled on him => la la lui a souri
lets hit the road => taillons la
i love winning => jadore gagner
```

The model isn't perfect, in part due to the limited size of the training set. Real NMT models are trained with hundreds of millions or even billions (*https://oreil.ly/bOZOK*) of phrases. But is this one truly representative of the models used for state-of-the-art text translation? Figure 13-4 is adapted from an image in a 2016 paper (*https://oreil.ly/gXD9A*) written by Google engineers documenting the architecture of Google Translate. The architecture maps closely to that of the model just presented. It's deeper, with eight LSTM layers in the encoder and eight in the decoder. It also employs residual connections between layers to support the network's greater depth.

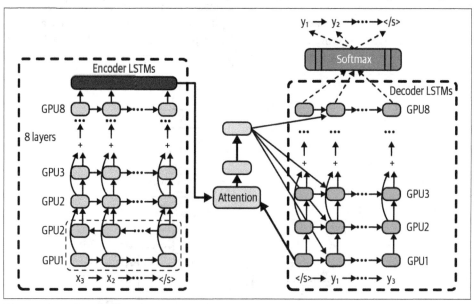

Figure 13-4. Google Translate circa 2016

One difference between our model and the model described in the paper is the block labeled "Attention" between the encoder and decoder. In deep learning, *attention* is a mechanism for focusing a model's attention on the parts of a phrase that are most important, recognizing that one word can have different meanings in different contexts and that the meaning of a word sometimes depends on what's around it.

Attention was introduced to deep learning in a seminal 2014 paper titled "Neural Machine Translation by Jointly Learning to Align and Translate" (*https://oreil.ly/moOX9*), but it wasn't until a few years later that attention took center stage as a way to *replace* LSTM layers rather than supplement them. Enter perhaps the most significant contribution to the field of NMT, and to NLP overall, to date: the transformer model.

Transformer Encoder-Decoders

A landmark 2017 paper titled "Attention Is All You Need" (*https://oreil.ly/OUfbX*) changed the way data scientists approach NMT and other neural text processing tasks. It proposed a better way to perform sequence-to-sequence processing based on *transformer models* that eschew recurrent layers and use attention mechanisms to model the context in which words are used. Today transformer models have almost entirely replaced LSTM-based models.

Figure 13-5 is adapted from an image in the aforementioned paper. On the left is the encoder, which takes text sequences as input and generates dense vector representations of those sequences. On the right is the decoder, which transforms dense vector representations of input sequences into output sequences. At a high level, a transformer model uses the same encoder-decoder architecture as an LSTM-based model. The difference lies in how it does the encoding and decoding.

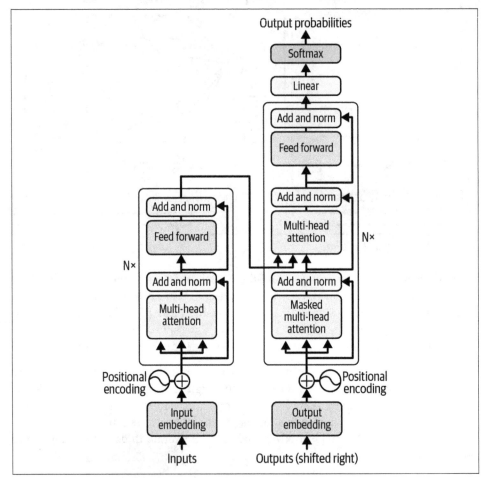

Figure 13-5. The transformer model

The chief innovation introduced by the transformer model is the use of multi-head attention (MHA) layers in place of LSTM layers. MHA layers embody the concept of *self-attention*, which enables a model to analyze an input sequence and focus on the words that are most important as well as the context in which the words are used. In the sentence "We took a walk in the park," for example, the word *park* has a different meaning than it does in "Where did you park the car?" An embedding layer stores one vector representation for *park*, but in a transformer model, the MHA layer modifies the vector output by the embedding layer so that *park* is represented by two different vectors in the two sentences. Not surprisingly, the values used to make embedding vectors context aware are learned during training.

How does self-attention work? The gist is that an MHA layer uses dot products to compute similarity scores for every word pair in a sequence. After normalizing the scores, it uses them to compute weighted versions of each word embedding in the sequence. Then it modifies them again using weights learned during training.

The "multi" in *multi-head attention* denotes the fact that an MHA layer learns several sets of weights rather than just one, not unlike a convolution layer in a CNN. This gives MHA the ability to discern context in long sequences where words might have multiple relationships to one another.

MHA also provides additional context regarding words that refer to other words. An embedding layer represents the pronoun *it* with a single vector, but an MHA layer helps the model understand that in the sentence "I love my car because it is fast," *it* refers to a car. It also adds weight to the word *fast* because it's crucial to the meaning of the sentence. Without it, it's not clear *why* you love your car.

Unlike LSTM layers, MHA layers' ability to model relationships among the words in an input sequence is independent of sequence length. MHA layers also support parallel workloads and therefore train faster on multiple GPUs. They do not, however, encode information regarding the positions of the words in a phrase. To compensate, a transformer uses simple vector addition to add information denoting a word's position in a sequence to each vector output from the embedding layer. This is referred to as *positional encoding* or *positional embedding*. It's denoted by the plus signs labeled "Positional encoding" in Figure 13-5.

Transformers aren't limited to neural machine translation; they're used in virtually all aspects of NLP today. The encoder half of a transformer outputs dense vector representations of the sequences input to it. Text classification can be performed by using a transformer rather than a standalone embedding layer to encode input sequences. Models architected this way frequently outperform bag-of-words models, particularly if the ratio of samples to sample length (the number of training samples divided by the average length of each sample) exceeds 1,500. This so-called "golden constant" was discovered by a team of researchers at Google and documented in a tutorial on text classification (*https://oreil.ly/MSxf5*).

Building a Transformer-Based NMT Model

Keras provides some, but not all, of the building blocks that comprise an end-to-end transformer. It provides a handy implementation of self-attention layers in its `Multi HeadAttention` class (*https://oreil.ly/OgTvn*), for example, but it doesn't implement positional embedding. However, a separate package named KerasNLP (*https://oreil.ly/Wx58Y*) does. Among others, it includes the following classes representing layers in a transformer-based network:

`TransformerEncoder` (*https://oreil.ly/CFsaO*)
> Represents a transformer encoder

`TransformerDecoder` (*https://oreil.ly/vhL4m*)
> Represents a transformer decoder

`TokenAndPositionEmbedding` (*https://oreil.ly/nxCRU*)
> Implements an embedding layer that supports positional embedding

With these classes, transformer-based NMT models are relatively easy to build. You can demonstrate by building an English-to-French translator. Start by installing KerasNLP if it isn't already installed. Then download *en-fr.txt* (*https://oreil.ly/K4mTF*), a data file that contains 50,000 English phrases and their French equivalents, and drop it into the *Data* subdirectory where your Jupyter notebooks are hosted. The file *en-fr.txt* is a subset of a larger file (*https://oreil.ly/WZDMQ*) containing more than 190,000 phrases and their corresponding translations compiled as part of the Tatoeba project (*https://oreil.ly/WHT6F*). The file is tab delimited. Each line contains an English phrase, the equivalent French phrase, and an attribution identifying where the translation came from. We don't need the attributions, so load the dataset into a `DataFrame`, remove the attribution column, and shuffle and reindex the rows:

```
import pandas as pd

df = pd.read_csv('Data/en-fr.txt', names=['en', 'fr', 'attr'],
                 usecols=['en', 'fr'], sep='\t')
df = df.sample(frac=1, random_state=42)
```

```
df = df.reset_index(drop=True)
df.head()
```

Here's the output:

	en	fr
0	You're very clever.	Vous êtes fort ingénieuse.
1	Are there kids?	Y a-t-il des enfants ?
2	Come in.	Entrez !
3	Where's Boston?	Où est Boston ?
4	You see what I mean?	Vous voyez ce que je veux dire ?

The dataset needs to be cleaned before it's used to train a model. Use the following statements to remove numbers and punctuation symbols, convert words with Unicode characters such as *où* into their ASCII equivalents (*ou*), convert characters to lowercase, and insert [start] and [end] tokens at the beginning and end of each French phrase:

```
import re
from unicodedata import normalize

def clean_text(text):
    text = normalize('NFD', text.lower())
    text = re.sub('[^A-Za-z ]+', '', text)
    return text

def clean_and_prepare_text(text):
    text = '[start] ' + clean_text(text) + ' [end]'
    return text

df['en'] = df['en'].apply(lambda row: clean_text(row))
df['fr'] = df['fr'].apply(lambda row: clean_and_prepare_text(row))
df.head()
```

The output looks a little cleaner afterward:

	en	fr
0	youre very clever	[start] vous etes fort ingenieuse [end]
1	are there kids	[start] y atil des enfants [end]
2	come in	[start] entrez [end]
3	wheres boston	[start] ou est boston [end]
4	you see what i mean	[start] vous voyez ce que je veux dire [end]

The next step is to scan the dataset and determine the maximum length of the English phrases and of the French phrases. These lengths will determine the lengths of the sequences input to and output from the model:

```python
en = df['en']
fr = df['fr']

en_max_len = max(len(line.split()) for line in en)
fr_max_len = max(len(line.split()) for line in fr)
sequence_len = max(en_max_len, fr_max_len)

print(f'Max phrase length (English): {en_max_len}')
print(f'Max phrase length (French): {fr_max_len}')
print(f'Sequence length: {sequence_len}')
```

In this example, the longest English phrase contains seven words, while the longest French phrase contains 16 (including the [start] and [end] tokens). The model will be able to translate English phrases up to seven words in length into French phrases up to 14 words in length.

Now fit one `Tokenizer` to the English phrases and another `Tokenizer` to their French equivalents, and generate padded sequences from all the phrases. Note the `filters` parameter passed to the French tokenizer. It configures the tokenizer to remove all the punctuation characters it normally removes *except* for the square brackets used to delimit [start] and [end] tokens:

```python
from tensorflow.keras.preprocessing.text import Tokenizer
from tensorflow.keras.preprocessing.sequence import pad_sequences

en_tokenizer = Tokenizer()
en_tokenizer.fit_on_texts(en)
en_sequences = en_tokenizer.texts_to_sequences(en)
en_x = pad_sequences(en_sequences, maxlen=sequence_len, padding='post')

fr_tokenizer = Tokenizer(filters='!"#$%&()*+,-./:;<=>?@\\^_`{|}~\t\n')
fr_tokenizer.fit_on_texts(fr)
fr_sequences = fr_tokenizer.texts_to_sequences(fr)
fr_y = pad_sequences(fr_sequences, maxlen=sequence_len + 1, padding='post')
```

Next, compute the vocabulary size for each language from the `Tokenizer` instances:

```python
en_vocab_size = len(en_tokenizer.word_index) + 1
fr_vocab_size = len(fr_tokenizer.word_index) + 1

print(f'Vocabulary size (English): {en_vocab_size}')
print(f'Vocabulary size (French): {fr_vocab_size}')
```

The output reveals that the English vocabulary contains 6,033 words, while the French vocabulary contains 12,197. These values will be used to size the model's two embedding layers. The latter will also be used to size the output layer.

Finally, create the features and the labels the model will be trained with. The features are the padded English sequences and the padded French sequences minus the [end] tokens. The labels are the padded French sequences minus the [start] tokens. Package the features in a dictionary so that they can be input to a model that accepts multiple inputs:

```
inputs = { 'encoder_input': en_x, 'decoder_input': fr_y[:, :-1] }
outputs = fr_y[:, 1:]
```

Now let's define a model. This time, we'll use Keras's functional API (*https://oreil.ly/ wmF3Q*) rather than its sequential API. It's necessary because this model has two inputs: one that accepts a tokenized English phrase and another that accepts a tokenized French phrase. We'll also seed the random-number generators used by Keras and TensorFlow to get repeatable results, at least on CPU. This is a departure from all the other examples in this book, but it ensures that when you train the model, you get the same results that I did. Here's the code:

```
import numpy as np
import tensorflow as tf
from tensorflow.keras import Model
from tensorflow.keras.layers import Input, Dense, Dropout
from keras_nlp.layers import TokenAndPositionEmbedding, TransformerEncoder
from keras_nlp.layers import TransformerDecoder

np.random.seed(42)
tf.random.set_seed(42)

num_heads = 8
embed_dim = 256

encoder_input = Input(shape=(None,), dtype='int64', name='encoder_input')
x = TokenAndPositionEmbedding(en_vocab_size, sequence_len,
                              embed_dim)(encoder_input)
encoder_output = TransformerEncoder(embed_dim, num_heads)(x)
encoded_seq_input = Input(shape=(None, embed_dim))

decoder_input = Input(shape=(None,), dtype='int64', name='decoder_input')
x = TokenAndPositionEmbedding(fr_vocab_size, sequence_len, embed_dim,
                              mask_zero=True)(decoder_input)
x = TransformerDecoder(embed_dim, num_heads)(x, encoded_seq_input)
x = Dropout(0.4)(x)

decoder_output = Dense(fr_vocab_size, activation='softmax')(x)
decoder = Model([decoder_input, encoded_seq_input], decoder_output)
decoder_output = decoder([decoder_input, encoder_output])

model = Model([encoder_input, decoder_input], decoder_output)
model.compile(optimizer='adam', loss='sparse_categorical_crossentropy',
              metrics=['accuracy'])
model.summary(line_length=100)
```

The Keras Functional API

The functional API is a richer version of Keras's sequential API. Among other things, it lets you create models with multiple inputs or outputs and models with shared layers like the ones in Faster R-CNN's region proposal network. Here's a simple binary classifier defined with the sequential API:

```
model = Sequential()
model.add(Dense(128, activation='relu', input_dim=3))
model.add(Dense(1, activation='sigmoid'))
```

Here's the same network created with the functional API:

```
input = Input(shape=(3,))
hidden = Dense(128, activation='relu')(input)
output = Dense(1, activation='sigmoid')(hidden)
model = Model(inputs=input, outputs=output)
```

When you create the model, you specify the inputs and outputs, and since the inputs and outputs parameters accept Python lists, it's a simple matter to create a model with multiple inputs and outputs. For a concise introduction to the functional API and examples demonstrating advanced uses, including multiple inputs and outputs and shared layers, see "How to Use the Keras Functional API for Deep Learning" (*https://oreil.ly/fpbgM*) by Jason Brownlee.

The model is designed to operate iteratively. To translate text, you first pass an English phrase to the English input and the word "[start]" to the French input. Then you append the next French word the model predicts to the previous French input and call the model again, and you repeat this process until the entire phrase has been translated—that is, until the next word predicted by the model is "[end]."

Figure 13-6 diagrams the model's architecture. The model includes two embedding layers: one for English sequences and one for French sequences. Both convert word tokens into vectors of 256 floating point values each, and both are instances of KerasNLP's TokenAndPositionEmbedding class, which adds positional information to word embeddings. Output from the English embedding layer passes through an encoder (an instance of TransformerEncoder) before being input along with the output from the French embedding layer to the decoder, which is an instance of TransformerDecoder. The decoder outputs a vector representing the next step in the translation, and a softmax output layer converts that vector into a set of probabilities —one for each word in the French vocabulary—identifying the next token. During training, the mask_zero=True parameter passed to the French embedding layer limits the model to making predictions based on the tokens preceding the one that's being predicted. In other words, given a set of French tokens numbered 0 through n, the model is trained to predict what token $n + 1$ will be without peeking at $n + 1$ in the training text.

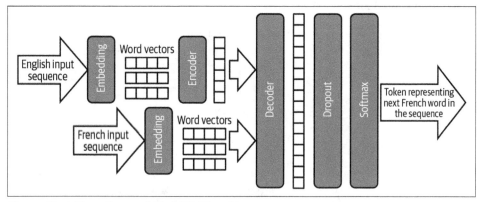

Figure 13-6. Transformer-based NMT architecture

Now call `fit` to train the model, and use an `EarlyStopping` callback to end training if the validation accuracy fails to improve for three consecutive epochs:

```
from tensorflow.keras.callbacks import EarlyStopping

callback = EarlyStopping(monitor='val_accuracy', patience=3,
                         restore_best_weights=True)

hist = model.fit(inputs, outputs, epochs=50, validation_split=0.2,
                 callbacks=[callback])
```

Training typically requires two to three minutes per epoch on CPU. If you don't have a GPU and training is too slow, I recommend running the code in Google Colab (*https://oreil.ly/J1AgM*). (Be sure to go to "Notebook settings" in the Edit menu and select GPU as the hardware accelerator.) When training is complete, plot the per-epoch training and validation accuracy and observe how the latter steadily increases until it levels off:

```
import seaborn as sns
import matplotlib.pyplot as plt
%matplotlib inline
sns.set()

acc = hist.history['accuracy']
val = hist.history['val_accuracy']
epochs = range(1, len(acc) + 1)

plt.plot(epochs, acc, '-', label='Training accuracy')
plt.plot(epochs, val, ':', label='Validation accuracy')
plt.title('Training and Validation Accuracy')
plt.xlabel('Epoch')
plt.ylabel('Accuracy')
```

```
plt.legend(loc='lower right')
plt.plot()
```

The output should look like this:

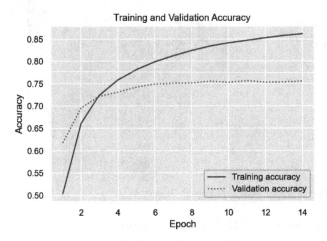

The plot reveals that at the end of 14 epochs, the model was about 75% accurate in translating English samples in the validation data to French. This isn't a very robust measure of accuracy because it literally compares each word in the predicted text to each word in the target text and ignores the fact that a missing or misplaced article such as *le* (French for *the*) doesn't necessarily imply a poor translation. The accuracy of NMT models is typically measured with *bilingual evaluation understudy* (BLEU) scores (*https://oreil.ly/hdxar*). BLEU scores are rather easily computed after the training is complete using packages such as NLTK, but during training, validation accuracy is a reasonable metric for judging when to halt training.

Can the model really translate English to French? Use the following code to define a function that accepts an English phrase and returns a French phrase. Then call it on 10 of the phrases used to validate the model during training and see for yourself. Note that one call to `translate_text` precipitates multiple calls to the model. To translate "hello world," for example, `translate_text` calls the model with the inputs "hello world" and "[start]." Assuming the model predicts that *salut* is the next word, `translate_text` invokes it again with the inputs "hello world" and "[start] salut." It repeats this cycle until the next word predicted by the model is "[end]" denoting the end of the translation.

```
def translate_text(text, model, en_tokenizer, fr_tokenizer, fr_index_lookup,
                   sequence_len):
    input_sequence = en_tokenizer.texts_to_sequences([text])
    padded_input_sequence = pad_sequences(input_sequence, maxlen=sequence_len,
                                          padding='post')
    decoded_text = '[start]'
```

```
    for i in range(sequence_len):
        target_sequence = fr_tokenizer.texts_to_sequences([decoded_text])
        padded_target_sequence = pad_sequences(
            target_sequence, maxlen=sequence_len, padding='post')[:, :-1]

        prediction = model([padded_input_sequence, padded_target_sequence])

        idx = np.argmax(prediction[0, i, :]) - 1
        token = fr_index_lookup[idx]
        decoded_text += ' ' + token

        if token == '[end]':
            break

    return decoded_text[8:-6] # Remove [start] and [end] tokens

fr_vocab = fr_tokenizer.word_index
fr_index_lookup = dict(zip(range(len(fr_vocab)), fr_vocab))
texts = en[40000:40010].values

for text in texts:
    translated = translate_text(text, model, en_tokenizer, fr_tokenizer,
                                fr_index_lookup, sequence_len)
    print(f'{text} => {translated}')
```

Here's the output:

```
its fall now => cest desormais tombe
im losing => je suis en train de perdre
it was quite funny => ce fut assez amusant
thats not unusual => ce nest pas inhabituel
i think ill do that => je pense que je ferai ca
tom looks different => tom a lair different
its worth a try => ca vaut le coup dessayer
fortune smiled on him => la chance lui souri
lets hit the road => cassonsnous
i love winning => jadore gagner
```

If you don't speak French, use Google Translate to translate some of the French phrases to English. According to Google, for example, "la chance lui souri" translates to "Luck smiled on him," while "ce nest pas inhabituel" translates to "it's not unusual." The model isn't perfect, but it's not bad, either. The vocabulary you used is small, so you can't input just any old phrase and expect the model to translate it. But simple phrases that use words in the training text translate reasonably well.

Finish up by using the `translate_text` function to see how the model translates "Hello world" into French:

```
translate_text('Hello world', model, en_tokenizer, fr_tokenizer,
                fr_index_lookup, sequence_len)
```

I haven't had French lessons since high school, but even I know that "Salut le monde" is a reasonable translation of "Hello world."

Using Pretrained Models to Translate Text

Engineers at Microsoft, Google, and Facebook have the resources to collect millions of text translation samples and the hardware to train sophisticated transformer models on them, but you and I do not. The good news is that this needn't stop us from writing software that translates text from one language to another. Hugging Face has published several pretrained transformer models that do a fine job of text translation. Leveraging those models in Python is simplicity itself.

Here's an example that translates English to French:

```
from transformers import pipeline

translator = pipeline('translation_en_to_fr')
translation = translator('Programming is fun!') [0]['translation_text']
print(translation)
```

The same syntax can be used to translate English to German and English to Romanian too. Simply replace `translation_en_to_fr` with `translation_en_to_de` or `translation_en_to_ro` when creating the pipeline.

For other languages, you use a slightly more verbose syntax to load a transformer and a corresponding tokenizer. The following example translates Dutch to English:

```
from transformers import AutoTokenizer, AutoModelForSeq2SeqLM

# Initialize the tokenizer
tokenizer = AutoTokenizer.from_pretrained('Helsinki-NLP/opus-mt-nl-en')

# Initialize the model
model = AutoModelForSeq2SeqLM.from_pretrained('Helsinki-NLP/opus-mt-nl-en')

# Tokenize the input text
text = 'Hallo vrienden, hoe gaat het vandaag?'
tokenized_text = tokenizer.prepare_seq2seq_batch([text], return_tensors='pt')

# Perform translation and decode the output
translation = model.generate(**tokenized_text)
translated_text = tokenizer.batch_decode(
    translation, skip_special_tokens=True)[0]
print(translated_text)
```

You'll find an exhaustive list of Hugging Face translators and tokenizers on the organization's website (*https://oreil.ly/MSuEz*). There are hundreds of them covering dozens of languages.

Bidirectional Encoder Representations from Transformers (BERT)

The introduction of transformers in 2017 laid the groundwork for another landmark innovation in the NLP space: Bidirectional Encoder Representations from Transformers, or BERT for short. Introduced by Google researchers in a 2018 paper titled "BERT: Pre-training of Deep Bidirectional Transformers for Language Understanding" (*https://oreil.ly/2fGXv*), BERT advanced the state of the art by providing pretrained transformers that can be fine-tuned for a variety of NLP tasks.

BERT was instilled with language understanding by training it with more than 2.5 billion words from Wikipedia articles and 800 million words from Google Books. Training required four days on 64 tensor processing units (TPUs). Fine-tuning is accomplished by further training the pretrained model with task-specific samples and a reduced learning rate (Figure 13-7). It's a relatively simple matter, for example, to fine-tune BERT to perform sentiment analysis and outscore bag-of-words models for accuracy. BERT's value lies in the fact that it possesses an innate understanding of the languages it was trained with and can be refined to perform domain-specific tasks.

Aside from the fact that it was trained with a huge volume of samples, the key to BERT's ability to understand human languages is an innovation known as *masked language modeling* (*https://oreil.ly/5OrrT*), or MLM for short. The big idea behind MLM is that a model has a better chance of predicting what word should fill in the blank in the phrase "Every good ____ does fine" than it has at predicting the next word in the phrase "Every good ____." The answer could be *boy*, as in "Every good boy does fine," or it could be *turn*, as in "Every good turn deserves another." Unidirectional models look at the text to the left or the text to the right and attempt to predict what the next word should be. MLM, on the other hand, uses text on the left *and* right to inform its decisions. That's why BERT is a "bidirectional" transformer.

When BERT models are pretrained, a specified percentage of the word tokens in each sequence—usually 15%—are randomly removed or "masked" so that the model can learn to predict them from the words around them. In addition, BERT models are usually pretrained to do next-sentence prediction, which makes them more adept at certain NLP tasks such as answering questions.

BERT has been called the "Swiss Army knife" of NLP. Google uses it to improve search results (*https://oreil.ly/nodc1*) and predict text as you type into a Gmail or Google Doc. Dozens of variations have been published, including DistilBERT (*https://oreil.ly/3u60q*), which retains 97% of the accuracy of the original model while weighing in 40% smaller and running 60% faster. Also available are variations of BERT already fine-tuned for specific tasks such as question answering. Such models can be further refined using domain-specific datasets, or they can be used as is.

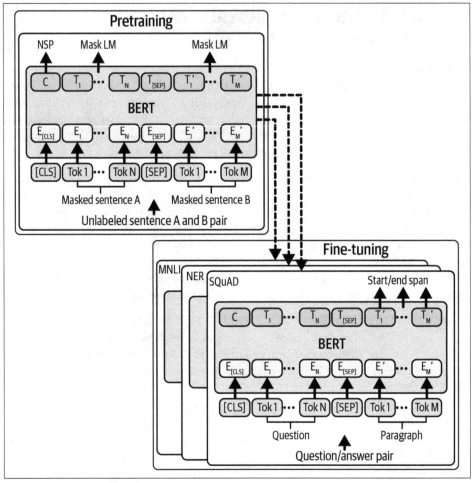

Figure 13-7. Bidirectional Encoder Representations from Transformers (BERT)

Building a BERT-Based Question Answering System

Hugging Face's transformers package contains several pretrained BERT models already fine-tuned for specific tasks. One example is the minilm-uncased-squad2 model (*https://oreil.ly/eBU3a*), which was trained with Stanford's SQuAD 2.0 dataset (*https://oreil.ly/2Cw0C*) to answer questions by extracting text from documents. To get a feel for what models like this one can accomplish, let's use it to build a simple question-answering system.

First, some context. When you ask Google a question like the one in Figure 13-8, it queries a database containing billions of web pages to identify ones that might contain an answer. Then it uses a BERT-based NLP model to extract answers from the pages it identified and rank them based on confidence levels.

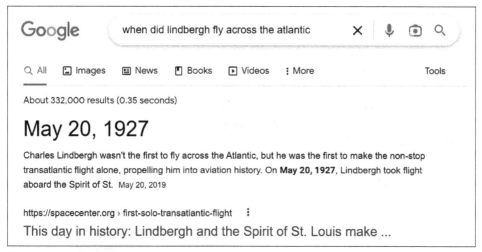

Figure 13-8. Google question answering

Let's load a pretrained BERT model already fine-tuned for question answering and use it to extract answers from passages of text in this book. The model we'll use is a version of the MiniLM model introduced in a 2020 paper titled "MiniLM: Deep Self-Attention Distillation for Task-Agnostic Compression of Pre-trained Transformers" (*https://oreil.ly/37m84*). This version was fine-tuned on SQuAD 2.0, which contains more than 100,000 questions generated by humans paired with answers culled from Wikipedia articles, plus 50,000 questions which lack answers. The MiniLM architecture enables reading comprehension gained from one dataset to be applied to other datasets with little or no retraining.

My eighth-grade history teacher, Mr. Aird, roomed with Charles Lindbergh one summer in the early 1920s. Apparently the world's most famous aviator was something of a daredevil in college, and I'll never forget something Mr. Aird said about him. "In 1927, when I learned that Charles had flown solo across the Atlantic, I wasn't surprised. I have never met a person with less regard for his own life than Charles Lindbergh." ☺

Begin by creating a new Jupyter notebook and using the following statements to load a pretrained MiniLM model from the Hugging Face hub (*https://oreil.ly/sZaah*) and a tokenizer to tokenize text input to the model. (BERT uses a special tokenization format called WordPiece (*https://oreil.ly/h8rZ4*) that is slightly different from the one Keras's `Tokenizer` class and `TextVectorization` layer use. Fortunately, Hugging Face has a solution for that too.) Then compose a pipeline from them. The first time you run this code, you'll experience a momentary delay while the pretrained weights are downloaded. After that, the weights will be cached and loading will be fast:

```
from transformers import AutoTokenizer, TFAutoModelForQuestionAnswering, pipeline

id = 'deepset/minilm-uncased-squad2'
tokenizer = AutoTokenizer.from_pretrained(id)
model = TFAutoModelForQuestionAnswering.from_pretrained(id, from_pt=True)
pipe = pipeline('question-answering', model=model, tokenizer=tokenizer)
```

Hugging Face stores weights for this particular model in PyTorch format. The
from_pt=True parameter converts the weights to TensorFlow format. It's not trivial to
convert neural network weights from one format to another, but the Hugging Face
library reduces it to a simple function parameter.

Now use the pipeline to answer a question by extracting text from a paragraph:

```
question = 'What does NLP stand for?'

context = 'Natural Language Processing, or NLP, encompasses a variety of ' \
          'activities, including text classification, keyword and topic ' \
          'extraction, text summarization, and language translation. The ' \
          'accuracy of NLP models has improved in recent years for a variety ' \
          'of reasons, not the least of which are newer and better ways of ' \
          'converting words and sentences into dense vector representations ' \
          'that incorporate context, and a relatively new neural network ' \
          'architecture called the transformer that can zero in on the most ' \
          'meaningful words and even differentiate between multiple meanings ' \
          'of the same word.'

pipe(question=question, context=context)
```

Is the answer accurate? A human could easily read the paragraph and come up with
the same answer, but the fact that a deep-learning model can do it indicates that the
model displays some level of reading comprehension. Observe that the output contains the answer to the question as well as a confidence score and the starting and
ending indices of the answer in the paragraph:

```
{'score': 0.9793193340301514,
 'start': 0,
 'end': 27,
 'answer': 'Natural Language Processing'}
```

Now try it again with a different question and context:

```
question = 'When was TensorFlow released?'

context = 'Machine learning isn\'t hard when you have a properly ' \
          'engineered dataset to work with. The reason it\'s not ' \
          'hard is libraries such as Scikit-Learn and ML.NET, which ' \
          'reduce complex learning algorithms to a few lines of code. ' \
          'Deep learning isn\'t difficult, either, thanks to libraries ' \
          'such as the Microsoft Cognitive Toolkit (CNTK), Theano, and ' \
          'PyTorch. But the library that most of the world has settled ' \
          'on for building neural networks is TensorFlow, an open source ' \
```

```
                'framework created by Google that was released under the ' \
                'Apache License 2.0 in 2015.'

        pipe(question=question, context=context)['answer']
```

This time, the output is the answer provided by the model rather than the dictionary containing the answer. Once again, is the answer reasonable?

Repeat this process with another question and context from which to extract an answer:

```
        question = 'Is Keras part of TensorFlow?'

        context = 'The learning curve for TensorFlow is rather steep. ' \
                  'Another library, named Keras, provides a simplified ' \
                  'Python interface to TensorFlow and has emerged as the ' \
                  'Scikit of deep learning. Keras is all about neural networks. ' \
                  'It began life as a standalone project in 2015 but was ' \
                  'integrated into TensorFlow in 2019. Any code that you write ' \
                  'using TensorFlow\'s built-in Keras module ultimately executes ' \
                  'in (and is optimized for) TensorFlow. Even Google recommends ' \
                  'using the Keras API.'

        pipe(question=question, context=context)['answer']
```

Perform one final test using the same context as before but a different question:

```
        question = 'Is it better to use Keras or TensorFlow to build neural networks?'
        pipe(question=question, context=context)['answer']
```

The questions posed here were hand-selected to highlight the model's capabilities. It's not difficult to come up with questions that the model can't answer. Nevertheless, you have proven the principle that a pretrained BERT model fine-tuned on SQuAD 2.0 can answer straightforward questions from passages of text presented to it.

Sophisticated question-answering systems employ a retriever-reader architecture in which the retriever searches a data store for relevant documents—ones that might contain an answer to a question—and the reader extracts answers from the documents. The reader is often a BERT instance similar to the one shown earlier. The retriever may be one from the open source Haystack library (*https://oreil.ly/fopHE*) published by Deepset (*https://oreil.ly/OIzma*), a German company focused on NLP solutions. Haystack retrievers interface with a wide range of document stores including Elasticsearch stores (*https://oreil.ly/CHDKf*), which are highly scalable. If you'd like to learn more or build a retriever-reader system of your own, I recommend reading Chapter 7 of *Natural Language Processing with Transformers* by Lewis Tunstall, Leandro von Werra, and Thomas Wolf (O'Reilly).

Fine-Tuning BERT to Perform Sentiment Analysis

State-of-the-art sentiment analysis can be accomplished by fine-tuning pretrained BERT models on sentiment analysis datasets. Let's fine-tune BERT and see if we can create a sentiment analysis model that's more accurate than the bag-of-words model presented earlier in this chapter. If your computer isn't equipped with a GPU, I *highly* recommend running this example in Google Colab. Even on a GPU, it can take an hour or so to run.

If you run this code locally, make sure Hugging Face's Datasets package (*https:// oreil.ly/sYdlg*) is installed. Then create a new Jupyter notebook. If you use Colab instead, create a new notebook and run the following commands in the first cell to install the necessary packages in the Colab environment:

```
!pip install transformers
!pip install datasets
```

Next, use the following statements to load the IMDB dataset from the Datasets package. This is an alternative to loading it from a CSV file. Since we're using Hugging Face models, we may as well load the data from Hugging Face too. Plus, if you're using Colab, this prevents you from having to upload a CSV to the Colab environment. Note that the dataset might take a few minutes to load the first time:

```
from datasets import load_dataset

imdb = load_dataset('imdb')
imdb
```

The value returned by `load_dataset` is a dictionary containing three Hugging Face datasets (*https://oreil.ly/EJ7pp*). Here's the output from the final statement:

```
DatasetDict({
    train: Dataset({
        features: ['text', 'label'],
        num_rows: 25000
    })
    test: Dataset({
        features: ['text', 'label'],
        num_rows: 25000
    })
    unsupervised: Dataset({
        features: ['text', 'label'],
        num_rows: 50000
    })
})
```

`imdb['train']` contains 25,000 samples for training, while `imdb['test']` contains 25,000 samples for testing. Movie reviews are stored in the `text` column of each dataset. Labels are stored in the `label` column.

Next up is tokenizing the input using a BERT WordPiece tokenizer:

```
from transformers import AutoTokenizer

tokenizer = AutoTokenizer.from_pretrained('distilbert-base-uncased')

def tokenize(samples):
    return tokenizer(samples['text'], truncation=True)

tokenized_imdb = imdb.map(tokenize, batched=True)
```

Now that the reviews are tokenized, they need to be converted into TensorFlow datasets (*https://oreil.ly/R35hn*) with Hugging Face's `Dataset.to_tf_dataset` method
(*https://oreil.ly/fcK3H*). The collating function passed to the method dynamically pads
the sequences so that they're all the same length. You can also ask the tokenizer to do
the padding, but padding performed that way is static and requires more memory:

```
from transformers import DataCollatorWithPadding

data_collator = DataCollatorWithPadding(tokenizer=tokenizer, return_tensors='tf')

train_data = tokenized_imdb['train'].to_tf_dataset(
    columns=['attention_mask', 'input_ids', 'label'],
    shuffle=True, batch_size=16, collate_fn=data_collator
)

validation_data = tokenized_imdb['test'].to_tf_dataset(
    columns=['attention_mask', 'input_ids', 'label'],
    shuffle=False, batch_size=16, collate_fn=data_collator
)
```

Now you're ready to fine-tune. Call `fit` on the model as usual, but set the optimizer's
learning rate (the multiplier used to adjust weights and biases during backpropagation) to 0.00002, which is a fraction of `Adam`'s default learning rate of 0.001:

```
from tensorflow.keras.optimizers import Adam
from transformers import TFAutoModelForSequenceClassification

model = TFAutoModelForSequenceClassification.from_pretrained(
    'distilbert-base-uncased', num_labels=2)
model.compile(Adam(learning_rate=2e-5), metrics=['accuracy'])
hist = model.fit(train_data, validation_data=validation_data, epochs=3)
```

Plot the training and validation accuracy to see where the latter topped out:

```
import seaborn as sns
import matplotlib.pyplot as plt
%matplotlib inline
sns.set()

acc = hist.history['accuracy']
val = hist.history['val_accuracy']
epochs = range(1, len(acc) + 1)
```

```
plt.plot(epochs, acc, '-', label='Training accuracy')
plt.plot(epochs, val, ':', label='Validation accuracy')
plt.title('Training and Validation Accuracy')
plt.xlabel('Epoch')
plt.ylabel('Accuracy')
plt.legend(loc='lower right')
plt.plot()
```

Here's how it turned out for me:

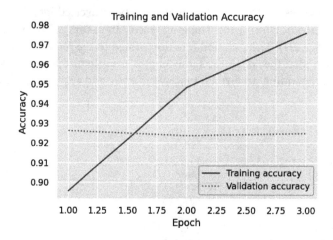

With a little luck, validation accuracy topped out at around 93%—a few points better than the equivalent bag-of-words model. Just imagine what you could do if you trained the model with more than 25,000 reviews. One of Hugging Face's pretrained sentiment analysis models—the twitter-roberta-base model (*https://oreil.ly/uplef*)—was trained with 58 million tweets. Not surprisingly, it does a wonderful job of scoring text for sentiment.

Finish up by defining an `analyze_text` function that returns a sentiment score and using it to score a positive review for sentiment. The model returns an object wrapping a tensor containing unnormalized sentiment scores (one for negative and one for positive), but you can use TensorFlow's `softmax` function (*https://oreil.ly/LY91Q*) to normalize them to values from 0.0 to 1.0:

```
import tensorflow as tf

def analyze_text(text, tokenizer, model):
    tokenized_text = tokenizer(text, padding=True, truncation=True,
                               return_tensors='tf')
    prediction = model(tokenized_text)
    return tf.nn.softmax(prediction[0]).numpy()[0][1]

analyze_text('Great food and excellent service!', tokenizer, model)
```

Try it again with a negative review:

```
analyze_text('The long lines and poor customer service really turned me off.',
            tokenizer, model)
```

Fine-tuning isn't cheap, but it isn't nearly as expensive as training a sophisticated transformer from scratch. The fact that you could train a sentiment analysis model to be this accurate in about an hour of GPU time is a tribute to the power of pretrained BERT models, and to the Google engineers who created them.

Summary

Natural language processing, or NLP, is an area of deep learning that encompasses text classification, question answering, text translation, and other tasks that require computers to process textual data. A key element of every NLP model is an embedding layer, which represents words with arrays of floating-point numbers that model the relationships between words. The vectors for *excellent* and *amazing* in embedding space are close together, for example, while the vectors for *butterfly* and *basketball* are far apart since the words have no semantic relationship. Word embeddings are learned as a model is trained.

Text input to an embedding layer must first be tokenized and turned into sequences of equal length. Keras's `Tokenizer` class does most of the work. Rather than tokenize and sequence text separately, you can include a `TextVectorization` layer in a model to do the tokenization and padding automatically.

One way to classify text is to use a traditional dense layer to classify the vectors output from an embedding layer. An alternative is to use convolution layers or recurrent layers to tease information regarding word position from the embedding vectors and classify the output from those layers.

The use of deep learning to translate text to other languages is known as neural machine translation, or NMT. Until recently, state-of-the-art NMT was performed using LSTM-based encoder-decoder models. Today those models have largely given way to transformer models that use neural attention to focus on the words in a phrase that are most meaningful and model word context. A transformer knows that the word *train* has different meanings in "meet me at the train station" and "it's time to train a model," and it factors word order into its calculations.

BERT is a sophisticated transformer model installed with language understanding when engineers at Google trained it with billions of words. It can be fine-tuned for specific tasks such as question answering and sentiment analysis, and several fine-tuned versions have been published for the public to use. These models can sometimes be used as is. Other times, they can be further refined and adapted to domain-specific tasks. Because fine-tuning requires orders of magnitude less data and compute power than training BERT from scratch, pretrained (and pre–fine-tuned) BERT models have proven a boon to NLP.

Sophisticated NLP models are trained with millions of words or phrases, requiring a substantial investment in data collection and hardware for training. Companies such as Hugging Face publish pretrained models that you can leverage in your code. This is a growing trend in AI: publishing models that are already trained to solve common problems. You'll still build models to solve problems that are domain specific, but many tasks—especially those involving NLP—are not consigned to a particular domain.

Downloading pretrained models isn't the only way to leverage sophisticated AI to solve business problems. Companies such as Microsoft and Amazon train deep-learning models of their own and make them publicly available using REST APIs. Microsoft calls its suite of APIs Azure Cognitive Services. You'll learn all about them in Chapter 14.

Azure Cognitive Services

As a child growing up in the 1960s, I idolized the Apollo astronauts. Swaggering out to the launch pad and riding flame-breathing rockets into space, they were my superheroes. But the group I really wanted to emulate—the people I wanted to *be*—were the engineers in Mission Control. Seated in front of their CRT screens in white shirts and black ties, chatting with the astronauts and poised to spring into action at the first sign of trouble, they were the epitome of cool. They used computers less powerful than today's smartphones to put men on the moon—a scientific achievement that is unsurpassed to this day.

Thanks to deep learning, computers today can perform feats of magic that the engineers in Mission Control could only have dreamed of. They can recognize objects in images, translate text and speech to other languages, identify people in video feeds, turn art into words and words into art, and more. But state-of-the-art deep-learning models are too complex—and too costly—for the average engineer or software developer to build. Microsoft reportedly spent hundreds of thousands of dollars training the ResNet model that won the 2015 ImageNet Large Scale Visual Recognition Challenge. Creating that model required a great deal of expertise, massive amounts of GPU time, and millions of images.

A welcome trend in AI today is *AI as a service*. Microsoft, Amazon, Google, and other tech giants employ professional data scientists who build sophisticated deep-learning models. They train them at their own expense and make them available to anyone who wishes to use them by means of REST APIs (*https://oreil.ly/OGfQf*). If you can write code to send an HTTP request over the internet, you can leverage these APIs to infuse AI into your apps without taking time off from your day job to earn a PhD in deep learning.

Microsoft calls its suite of AI services Azure Cognitive Services (*https://oreil.ly/NhIQs*). Amazon uses the name AWS AI Services (*https://oreil.ly/dmLGH*). Both offer a rich assortment of APIs served by deep-learning models on the backend—models that are continually refined so that they become smarter over time. Need to caption photos uploaded to a website? Microsoft's Computer Vision service (*https://oreil.ly/ K7C1P*) can do that; so can Amazon's Rekognition service (*https://oreil.ly/NGDCv*). How about building a screen reader featuring a lifelike human voice to help the hearing impaired? Amazon Polly (*https://oreil.ly/H4Zfe*) can handle that, as can Azure Cognitive Services' Speech service (*https://oreil.ly/D5M1S*) and Google's text-to-speech API (*https://oreil.ly/8OzPW*). These are just a few examples of the actions cognitive services can perform, often with just a few lines of code.

You saw Azure Cognitive Services at work in Chapter 12 when you used the Custom Vision service (*https://oreil.ly/NAW9S*) to train a custom object detection model. It's one of several services that comprise the Cognitive Services family. This chapter introduces others and demonstrates how to use them and how to build solutions with them. The focus on Azure Cognitive Services isn't meant to imply that they're better than their counterparts from Amazon and Google. I'm simply more familiar with them because I've worked closely with Microsoft for more than two decades. Once you learn how to call Azure Cognitive Services APIs, it's a simple matter to apply that knowledge to cognitive services from other vendors.

Ready to make some magic happen? Let's get started.

Introducing Azure Cognitive Services

The lineup of services that comprise Azure Cognitive Services changes from time to time as new ones are added and old ones are deprecated or matriculated into other Microsoft product lines. Figure 14-1 shows the services that are currently offered and divides them into four categories: vision, language, speech, and decision.

Figure 14-1. Azure Cognitive Services

Vision services bring deep neural networks to bear on computer-vision problems. The Custom Vision service lets you build custom image classification and object detection models. The Face service (*https://oreil.ly/TSV6H*), which Uber uses to verify the identities of its drivers (*https://oreil.ly/SZjaI*), supports state-of-the-art facial recognition systems. The Computer Vision service exposes a rich API featuring a plethora of ways to analyze images and extract information from them. One application for it is captioning photos and generating keywords characterizing their content. Figure 14-2 shows a web app I built called Intellipix. Intellipix captions images uploaded by users and stores keywords describing them in a database so that users can easily pull up all images containing castles, for example, or photos with water in the foreground or background.

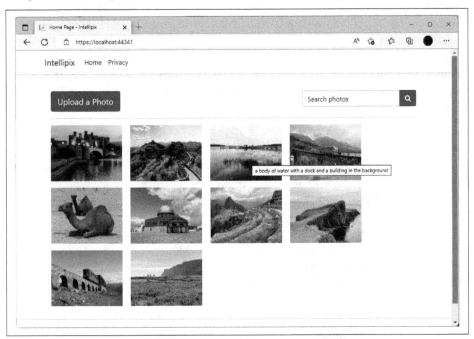

Figure 14-2. Intellipix showing a computer-generated image caption

The language category includes services backed by deep-learning models trained to perform natural language processing (NLP). The Language service (*https://oreil.ly/vOcSJ*) provides APIs for sentiment analysis, named-entity recognition, question answering, key-phrase extraction, language understanding, and more. Among its many uses is building chatbots that respond intelligently to queries regarding a company's product or service offerings. There's also a Translator service (*https://oreil.ly/bHAv2*) that translates text between more than 100 languages and dialects. Volkswagen uses it to translate onscreen instructions in cars (*https://oreil.ly/JWXRF*) to match the language in the owner's manual and to ensure the quality of the documentation that they produce in more than 40 languages.

The Language service is a unified service that subsumes the features of several older services, including the Text Analytics service, the Language Understanding service, and QnA Maker. Microsoft will continue to support the older services for some time, but new development should target the Language service.

The Speech service (*https://oreil.ly/rJLPI*) provides APIs for converting text to speech (*https://oreil.ly/4U9jJ*) and speech to text (*https://oreil.ly/ngZ2v*). KPMG uses the latter to transcribe recorded calls (*https://oreil.ly/GUfGr*) and claims that doing so has saved its customers millions of dollars in compliance costs. Airbus uses it to build voice-enabled apps for pilot training. The Speech service also includes a speaker recognition API (*https://oreil.ly/OLVbn*) that can identify a person's voice, and a speech translation API (*https://oreil.ly/q3CXc*) that can translate speech to dozens of other languages in real time.

Decision services include Anomaly Detector (*https://oreil.ly/OYwFc*), which identifies anomalies in live or recorded data streams and can aggregate inputs from hundreds of disparate sources, such as temperature sensors and pressure gauges, and model relationships between them. (See Chapter 6 for an introduction to anomaly detection.) Airbus uses Anomaly Detector to analyze stresses and strains (*https://oreil.ly/Uci2C*) in aircraft; Siemens uses it to test medical devices (*https://oreil.ly/kdT5N*) for flaws as the final step in production. The Content Moderator service (*https://oreil.ly/Tzmyj*) uses AI optionally supplemented by human intervention to flag offensive content in images and videos, profane text, and other inappropriate (and potentially libelous) content. Last but not least is the Personalizer service (*https://oreil.ly/hn797*), which is perhaps best described as a recommender system on steroids that provides personalized content and experiences to end users.

Keys and Endpoints

Azure Cognitive Services expose REST APIs that are called by transmitting HTTPS requests over the internet. The APIs are language agnostic: they can be called from any programming language that can put a call on the wire. That includes Python, Java, C#, C, C++, JavaScript, Swift, Go, and virtually all other modern programming languages. Calls can also be placed with tools such as Postman (*https://oreil.ly/LoaJw*) and the Linux `curl` command.

Most Azure Cognitive Services are free up to a point, but they're not altogether free. For example, you can submit 5,000 text samples per month to the Language service for sentiment analysis without incurring any costs. More than that, however, and Azure has to know whose Azure subscription to charge. Consequently, before calling an Azure Cognitive Service, you need:

- An endpoint for the service—the URL that's the target of calls
- A subscription key for the service, or some other means of authenticating calls

You can acquire both from the Azure Portal (*https://oreil.ly/r5FA9*). As an example, suppose you wish to perform sentiment analysis on a collection of tweets. You first open the Azure Portal in your browser and log in with your Microsoft account. You then create a Cognitive Services Language resource and specify to whom the Azure subscription costs should be charged, the Azure region in which the service should be located, and a pricing tier, as shown in Figure 14-3. A free tier is generally available if it doesn't already exist for the same service under the same subscription.

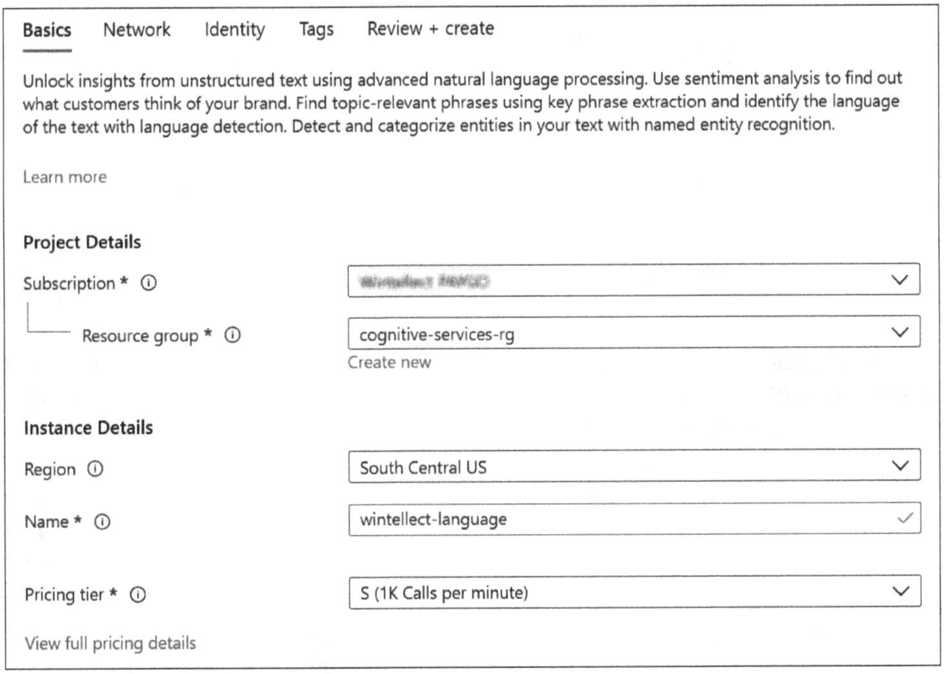

Basics	Network	Identity	Tags	Review + create

Unlock insights from unstructured text using advanced natural language processing. Use sentiment analysis to find out what customers think of your brand. Find topic-relevant phrases using key phrase extraction and identify the language of the text with language detection. Detect and categorize entities in your text with named entity recognition.

Learn more

Project Details

Subscription * ⓘ [⌄]

└──── Resource group * ⓘ [cognitive-services-rg   ⌄]
                          Create new

Instance Details

Region ⓘ [South Central US   ⌄]

Name * ⓘ [wintellect-language   ✓]

Pricing tier * ⓘ [S (1K Calls per minute)   ⌄]

View full pricing details

Figure 14-3. Creating a Language resource in the Azure Portal

Once the Language resource is created, you open it in the Azure Portal and click Keys and Endpoint in the menu on the left side of the page. From there, you can retrieve a subscription key and endpoint for the service (Figure 14-4). The key is a string of letters and numbers that uniquely identifies an Azure subscription. Treat it with care because it can cost you money if it gets out. It's considered a best practice to rotate the key (replace it with a new one) periodically in case it falls into the wrong hands. That's why the portal includes Regenerate Key buttons. Why does the portal provide two keys? So you have a valid key to use after regenerating the other one. With only one key, you'd have to regenerate it and then race to modify any apps that use it, and there would be a dead period in which calls placed by those apps would fail.

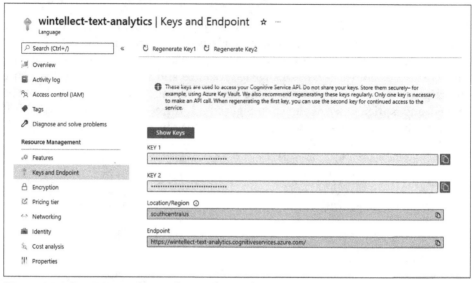

Figure 14-4. Retrieving a key and an endpoint for an Azure Cognitive Services resource

The subscription key in this example is a *single-service* subscription key because it works only with the Language resource that you created. If you create another Cognitive Services resource—for example, one for the Computer Vision service—you get a separate key for it. So that you can avoid managing multiple keys for multiple services, most Azure Cognitive Services support *multiservice keys* that work with a range of services. To get a multiservice key, simply create a Cognitive Services resource rather than a service-specific resource and use the key and endpoint that the portal provides.

Subscription keys aren't the only way to authenticate calls to Azure Cognitive Services. Most Cognitive Services support Azure Active Directory (AAD) authentication, which enhances security by combining AAD identities with role-based access control (RBAC). AAD authentication is particularly compelling for apps that rely on Cognitive Services APIs and are themselves hosted in Azure. For more information, refer to the article "Authenticate Requests to Azure Cognitive Services" (*https://oreil.ly/ofCuI*).

Calling Azure Cognitive Services APIs

Once you have a subscription key and endpoint, you're ready to roll. Here's a snippet of Python code that uses the Language service to evaluate "Programming is fun, but the hours are long" for sentiment. KEY is a placeholder for the Language resource's subscription key, while ENDPOINT is a placeholder for the corresponding endpoint.

The `import` statement imports the Python Requests package (*https://oreil.ly/1pTT8*), which simplifies the HTTP request-response protocol:

```
import requests

input = { 'documents': [
    { 'id': '1000', 'text': 'Programming is fun, but the hours are long' }]
}

headers = {
    'Ocp-Apim-Subscription-Key': KEY,
    'Content-type': 'application/json'
}

uri = ENDPOINT + 'text/analytics/v3.0/sentiment'
response = requests.post(uri, headers=headers, json=input)
results = response.json()

for result in results['documents']:
    print(result['confidenceScores'])
```

The result is as follows:

```
{'positive': 0.94, 'neutral': 0.05, 'negative': 0.01}
```

The service returns three scores: one for positive sentiment, one for neutral sentiment, and one for negative sentiment. Observe that you can submit multiple text samples in one call to the API because the *documents* item in the dictionary passed to the call is a Python list.

It might not be obvious from the code, but input must be JSON encoded and sent in the body of the request. The output comes back as JSON too. The Requests package simplifies JSON encoding and decoding. `requests.post` encodes the input and transmits a POST request (*https://oreil.ly/l1qYK*) to the designated endpoint, while `response.json` converts the JSON returned in the response into a Python dictionary.

To simplify matters, and to insulate you from changes to the underlying APIs as they evolve over time, Microsoft offers free software development kits (SDKs) for most Cognitive Services. The Python package named Azure-ai-textanalytics (*https://oreil.ly/Begdi*) is the Python SDK for sentiment analysis and other text analytics and is formally known as the Azure Cognitive Services Text Analytics client library for Python (*https://oreil.ly/QC8Cr*). It can be installed just like any other Python package—for example, with a `pip install` command. Here's the previous sample rewritten to use the SDK:

```
from azure.core.credentials import AzureKeyCredential
from azure.ai.textanalytics import TextAnalyticsClient

client = TextAnalyticsClient(ENDPOINT, AzureKeyCredential(KEY))
input = [{ 'id': '1000', 'text': 'Programming is fun, but the hours are long' }]
```

```
response = client.analyze_sentiment(input)

for result in response:
    print(result.confidence_scores)
```

The output is exactly the same, but the JSON is hidden away, and you don't have to know what magic string to append to the endpoint because analyze_sentiment (*https://oreil.ly/PW0wN*) does it for you. That method belongs to the SDK's Text AnalyticsClient class (*https://oreil.ly/ICRTV*), and it's one of several methods available for analyzing text.

Another benefit of using the SDKs is more robust error handling. Calls can and sometimes do fail, and a well-written app responds gracefully to such failures. When a call to Azure Cognitive Services fails, code in the SDK throws an exception that you can catch in your code. The following example responds to errors by printing the error message contained in the exception object:

```
from azure.core.credentials import AzureKeyCredential
from azure.ai.textanalytics import TextAnalyticsClient
from azure.core.exceptions import AzureError

try:
    client = TextAnalyticsClient(ENDPOINT, AzureKeyCredential(KEY))
    input = [{ 'id': '1000',
              'text': 'Programming is fun, but the hours are long' }]
    response = client.analyze_sentiment(input)

    for result in response:
        print(result.confidence_scores)

except AzureError as e:
    print(e.message)
```

AzureError is defined in Azure-core (*https://oreil.ly/z0NAS*), which is a library shared by all Azure Cognitive Services Python SDKs. It's the parent class for other exception classes that correspond to specific errors. If, for example, you pass a subscription key that's invalid (perhaps because the corresponding subscription is no longer active or payment is past due), a ClientAuthenticationError exception occurs. You can include as many exception handlers as you'd like to respond to specific types of errors. The preceding example uses AzureError as a catchall.

Cognitive Services SDKs are available for a variety of programming languages, including Python, Java, and C#. Here's the same sample written in C# using the Azure Cognitive Services Text Analytics client library for .NET (*https://oreil.ly/ 3PCW0*), which comes in the form of a NuGet package named Azure.AI.Text-Analytics (*https://oreil.ly/mXRL1*):

```
using Azure;
using Azure.AI.TextAnalytics;
using System;

try
{
    var client = new TextAnalyticsClient(
        new Uri(ENDPOINT),
        new AzureKeyCredential(KEY)
    );

    var response =
        client.AnalyzeSentiment("Programming is fun, but the hours are long");

    var sentiment = response.Value.ConfidenceScores.Positive;
    Console.WriteLine(sentiment);
}

catch (Exception ex)
{
    Console.WriteLine(ex.Message);
}
```

The output from the code is 0.94, which is the same positive-sentiment score output by the Python examples. Different language, different SDK, but same result, and all with very little effort. That's what Azure Cognitive Services are all about.

Azure Cognitive Services Containers

Cognitive Services vastly simplify the process of—and lower the skills barrier for—infusing AI into the apps that you write. But they have drawbacks too:

- Because Azure Cognitive Services run in the cloud, an app that uses them requires an internet connection.

- Calls that travel over the internet incur higher latencies than calls performed locally.

- Azure Cognitive Services APIs evolve over time and sometimes introduce breaking changes, which means code that worked just fine yesterday could behave differently or be inoperative tomorrow.

- The deep-learning models on the backend are continually refined and improved, with the result that the sentiment score for "programming is fun, but the hours are long" could be 0.94 today and 0.85 tomorrow.

Changes to the APIs are rarely abrupt. Microsoft usually warns its customers months in advance before making breaking changes to an API, and in many cases, you can specify the version of the service or API you wish to use and insulate yourself from future changes. These issues might not be a big deal to small firms, but for enterprises, which value stability and reliability above all else, they can be deal breakers.

Microsoft has addressed these issues by making most Azure Cognitive Services available in Docker containers. A containerized service can run on premises or in the cloud, and it locks in API versions and the models that back them. No changes occur unless you update a container image or replace it with a newer version. You can learn more and view the latest list of services that are available in containerized form in "What Are Azure Cognitive Services Containers?" (*https://oreil.ly/pPPFv*). Containerized services are also a solution to security and privacy policies that require data to stay on premises. Microsoft doesn't store data passed to Cognitive Services, but the data does leave your company's domain.

Containerized services are *not* a way to do an end run around the billing department and use Azure Cognitive Services for free. From the documentation:

> The Cognitive Services containers are required to submit metering information for billing purposes. Failure to allow-list various network channels that the Cognitive Services containers rely on will prevent the container from working.

Containers send encrypted usage information back to Microsoft through port 443, and Microsoft uses that information to bill an Azure subscription. Consequently, containers have to be connected to the internet even if they're hosted on premises.

There is one exception. If you want to use Azure Cognitive Services in apps that can't be connected to the internet for technical or compliance reasons, *disconnected containers* can run absent an internet connection, and they don't transmit metering information to Microsoft. Not just anyone can use them, however. You have to submit an application to Microsoft for approval, and one of the requirements is that your organization have an Enterprise Agreement (EA) in place with Microsoft. See "Use Docker Containers in Disconnected Environments" (*https://oreil.ly/5pbPw*) for more information and for an up-to-date list of Azure Cognitive Services that are available for use in disconnected scenarios.

The Computer Vision Service

By now you're probably ready to stop *talking* about Azure Cognitive Services and write some code. So am I. Let's start with perhaps the most feature-rich cognitive service of all. The Computer Vision service is one of three vision services in Azure Cognitive Services. It exposes a set of APIs that support a variety of tasks, including:

- Captioning images
- Generating tags describing the contents of an image
- Identifying objects (and their bounding boxes) in images

- Detecting sensitive or inappropriate content in images

- Extracting text from images

- Detecting faces in images (a subset of the Face service)

- Generating "smart thumbnail" images by using AI to identify the subject of a photo and creating a thumbnail version that's centered on the subject

- Performing spatial analysis on video streams

Spatial analysis is the latest addition to the Computer Vision service. Among other things, it can track people in live video feeds, measure distances between them in real time, and determine who's wearing a mask (and who is not), as shown in Figure 14-5. It's currently in public preview, and you can read all about it in "What Is Spatial Analysis?" (*https://oreil.ly/Y98w6*).

Figure 14-5. Using spatial analysis to verify social distancing and masking (images © Microsoft; used with permission)

The best way to get acquainted with the Computer Vision service is to call a few of its APIs. First install the Azure Cognitive Services Computer Vision SDK for Python (*https://oreil.ly/JYNIA*). (It's in a Python package named Azure-cognitiveservices-vision-computervision (*https://oreil.ly/e4voR*).) Next, download a ZIP file (*https://oreil.ly/lq08y*) containing a few sample images and copy the images into the *Data* sub-directory where your Jupyter notebooks are hosted. Use the Azure Portal to create a Computer Vision resource and obtain a key and an endpoint. Then create a new notebook and run the following code in the first cell after replacing KEY with the key and ENDPOINT with the endpoint:

```
from azure.cognitiveservices.vision.computervision import ComputerVisionClient
from msrest.authentication import CognitiveServicesCredentials

client = ComputerVisionClient(ENDPOINT, CognitiveServicesCredentials(KEY))
```

The ComputerVisionClient class implements several methods (*https://oreil.ly/iPHCn*) that you can call to invoke Computer Vision APIs. Most of those methods come in two versions: one that accepts an image URL and one that accepts an actual image. The `describe_image` method (*https://oreil.ly/eHlOG*), for example, accepts an image URL, while the `describe_image_in_stream` method (*https://oreil.ly/RiYLK*) accepts a stream containing the image. My examples use the `in_stream` methods, but you can easily modify them to pass image URLs rather than images.

Now load one of the sample images you copied from the ZIP file and use `describe_image_in_stream` to caption it:

```
%matplotlib inline
import matplotlib.pyplot as plt

image = plt.imread('Data/dubai.jpg')
fig, ax = plt.subplots(figsize=(12, 8), subplot_kw={'xticks': [], 'yticks': []})
ax.imshow(image)

with open('Data/dubai.jpg', mode='rb') as image:
    result = client.describe_image_in_stream(image)

    for caption in result.captions:
        print(f'{caption.text} ({caption.confidence:.1%})')
```

Confirm that the output is as follows:

```
A man riding a sand dune (53.8%)
```

`describe_image_in_stream` returns zero or more captions, each with a score reflecting the Computer Vision service's confidence that the caption is accurate. In this example, it returned just one caption with a confidence of 53.8%.

In addition to captioning images, the Computer Vision service can generate a list of keywords ("tags") describing an image's content. One use for such tags is to make an image database searchable. Use the `tag_image_in_stream` method (*https://oreil.ly/oJU1X*) to tag the image captioned in the previous example:

```
with open('Data/dubai.jpg', mode='rb') as image:
    result = client.tag_image_in_stream(image)

    for tag in result.tags:
        print(f'{tag.name} ({tag.confidence:.1%})')
```

Here are the tags and the confidence levels assigned to them:

```
dune (99.5%)
sky (99.2%)
outdoor (98.7%)
clothing (98.2%)
desert (98.1%)
sand (97.9%)
aeolian landform (96.9%)
person (96.1%)
singing sand (95.8%)
erg (94.0%)
sahara (93.6%)
nature (93.4%)
footwear (90.9%)
landscape (88.0%)
sand dune (83.5%)
ground (77.5%)
```

The Computer Vision service can also detect objects in images. The following statements load an image and show all the objects that were detected along with bounding boxes and confidence scores:

```
from matplotlib.patches import Rectangle

def annotate_object(name, confidence, bbox, min_confidence=0.5):
    if (confidence > min_confidence):
        x, y, w, h = bbox.x, bbox.y, bbox.w, bbox.h
        rect = Rectangle((x, y), w, h, color='red', fill=False, lw=2)
        ax.add_patch(rect)
        text = f'{name} ({confidence:.1%})'
        ax.text(x + (w / 2), y, text, color='white', backgroundcolor='red',
                ha='center', va='bottom', fontweight='bold',
                bbox=dict(color='red'))

image = plt.imread('Data/xian.jpg')
fig, ax = plt.subplots(figsize=(12, 8),
                       subplot_kw={'xticks': [], 'yticks': []})
ax.imshow(image)

with open('Data/xian.jpg', mode='rb') as image:
```

```
result = client.detect_objects_in_stream(image)

for object in result.objects:
    annotate_object(object.object_property, object.confidence,
                    object.rectangle)
```

detect_objects_in_stream (*https://oreil.ly/bwtVI*) detected the people and the bicycles in the photo, as well as a few other items:

Can the Computer Vision service detect faces in photos too? It certainly can. It can also provide information about age and gender. The following example annotates the faces in a photo with labels denoting age and gender. The ComputerVisionClient class lacks a dedicated method for detecting faces, but you can call the general-purpose analyze_image_in_stream method (*https://oreil.ly/OoC7K*) with a visual _features parameter requesting facial info:

```
from azure.cognitiveservices.vision.computervision.models \
    import VisualFeatureTypes

def annotate_face(face):
    x, y = face.face_rectangle.left, face.face_rectangle.top
    w, h = face.face_rectangle.width, face.face_rectangle.height
    rect = Rectangle((x, y), w, h, color='red', fill=False, lw=2)
    ax.add_patch(rect)
    text = f'{face.gender} ({face.age})'
    ax.text(x + (w / 2), y, text, color='white', backgroundcolor='red',
            ha='center', va='bottom', fontweight='bold',
            bbox=dict(color='red'))

image = plt.imread('Data/amsterdam.jpg')
fig, ax = plt.subplots(figsize=(12, 8), subplot_kw={'xticks': [], 'yticks': []})
ax.imshow(image)

with open('Data/amsterdam.jpg', mode='rb') as image:
```

```
result = client.analyze_image_in_stream(
    image, visual_features=[VisualFeatureTypes.faces])

for face in result.faces:
    annotate_face(face)
```

Here's the output:

The Face service can detect faces and identify ages, genders, emotions, facial landmarks, and more. It also has methods for comparing facial images, identifying faces, and building facial recognition systems. Its API is a superset of the Computer Vision service's face API. As a result of Microsoft's Responsible AI initiative (*https://oreil.ly/3TN8x*), however, the Face service is no longer available to the general public. For more information about the Face service and how to apply for access to it, refer to "What Is the Azure Face Service?" (*https://oreil.ly/kZcSu*).

Suppose you're the proprietor of a public website that accepts photo uploads and you'd like to reject photos containing inappropriate content. Called as follows, analyze_image_in_stream scores a photo for adultness (does the photo contain nudity?), raciness (does it contain bare skin?), and goriness (does it contain blood and gore?) on a scale of 0.0 to 1.0. Here's how it responds to a photo of a young girl cliff-jumping in her bathing suit in Hawaii:

```
image = plt.imread('Data/maui.jpg')
fig, ax = plt.subplots(figsize=(12, 8), subplot_kw={'xticks': [], 'yticks': []})
ax.imshow(image)

with open('Data/maui.jpg', mode='rb') as image:
    result = client.analyze_image_in_stream(image,
            visual_features=[VisualFeatureTypes.adult])
```

```
    print(f'Adultness: {result.adult.adult_score}')
    print(f'Raciness: {result.adult.racy_score}')
    print(f'Goriness: {result.adult.gore_score}')

    print(f'Is adult: {result.adult.is_adult_content}')
    print(f'Is racy: {result.adult.is_racy_content}')
    print(f'Is gory: {result.adult.is_gory_content}')
```

Here's the output:

```
Adultness: 0.02214685082435608
Raciness: 0.4205135107040405
Goriness: 0.00166344463099762797
Is adult: False
Is racy: False
Is gory: False
```

The bikini in the photo yielded a moderate raciness score, but one that's below the threshold of 0.5 required for is_racy_content to be True. Based on the other scores it returned, the Computer Vision service believes that the photo is neither "adult" nor gory.

Yet another capability that the Computer Vision service lends to application developers is using AI to extract text from photos. It's perfect for digitizing printed documents. Here's an example:

```
def draw_box(bbox):
    vals = bbox.split(',')
    x = int(vals[0])
    y = int(vals[1])
    w = int(vals[2])
```

```python
    h = int(vals[3])
    rect = Rectangle((x, y), w, h, color='red', fill=False, lw=2)
    ax.add_patch(rect)

image = plt.imread('Data/1040-es.jpg')
fig, ax = plt.subplots(figsize=(12, 8), subplot_kw={'xticks': [], 'yticks': []})
ax.imshow(image)

with open('Data/1040-es.jpg', mode='rb') as image:
    result = client.recognize_printed_text_in_stream(image)

    for region in result.regions:
        for line in region.lines:
            text = ' '.join([word.text for word in line.words])
            draw_box(line.bounding_box)
            print(text)
```

Here's the output, minus all the lines of text output by the `print` statement:

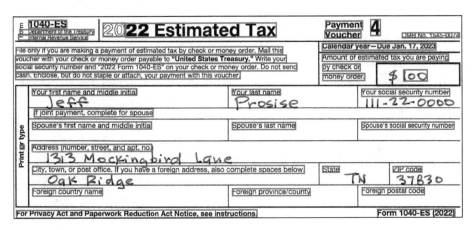

The `recognize_printed_text_in_stream` method (*https://oreil.ly/2KhMK*) only recognizes printed text. A related method named `read_in_stream` (*https://oreil.ly/GNFXO*) recognizes handwritten text too. Let's see how it performs on the same scanned document:

```python
import time
from azure.cognitiveservices.vision.computervision.models \
import OperationStatusCodes

def draw_box(bbox):
    x, y = bbox[0], bbox[1]
    w = bbox[4] - x
    h = bbox[5] - y
    rect = Rectangle((x, y), w, h, color='red', fill=False, lw=2)
    ax.add_patch(rect)
```

```python
image = plt.imread('Data/1040-es.jpg')
fig, ax = plt.subplots(figsize=(12, 8), subplot_kw={'xticks': [], 'yticks': []})
ax.imshow(image)

with open('Data/1040-es.jpg', mode='rb') as image:
    response = client.read_in_stream(image, raw=True)

    location = response.headers["Operation-Location"]
    opid = location[len(location) - 36:]
    results = client.get_read_result(opid)

    while results.status == OperationStatusCodes.running:
        results = client.get_read_result(opid)
        time.sleep(1)

    if results.status == OperationStatusCodes.succeeded:
        for result in results.analyze_result.read_results:
            for line in result.lines:
                draw_box(line.bounding_box)
                print(line.text)
```

This example is a little more involved because `read_in_stream` returns before the call
has completed. We therefore loop until the call completes and then retrieve the
results. The call to `sleep` in each iteration of the `while` loop allows the user interface
to remain responsive while waiting for the call to complete. The results are worth the
wait:

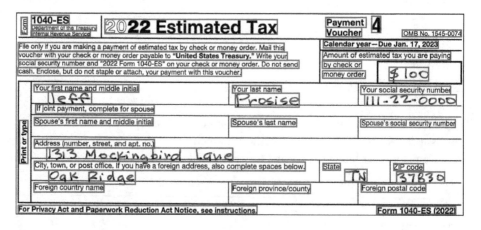

Perhaps you'd prefer to extract *only* handwritten text from a document. You can do
that too, because for each line of text it detects, `read_in_stream` includes a `style`
attribute equal to "handwriting" for handwritten text. To demonstrate, replace the
final two lines in the previous example with these:

```
if (line.appearance.style.name == 'handwriting'):
    draw_box(line.bounding_box)
    print(line.text)
```

This time, only handwritten text is highlighted:

`read_in_stream` does a commendable job of converting the handwritten text into Python strings too:

```
$ 100
Jeff
Prosise
111-22-0000
1313 Mockingbird Lane
Oak Ridge
TN
37830
```

A final note regarding the Computer Vision service is that it doesn't accept images larger than 4 MB. Anything larger produces a `ComputerVisionErrorResponse Exception`. You can catch these exceptions (or `AzureError` exceptions, which are higher up the food chain) and recover gracefully, or you can check an image's size before submitting it and downsize it if necessary.

The Language Service

The Computer Vision service employs deep-learning models similar to the ones you learned about in Chapters 10, 11, and 12. The Language and Translator services use models like the ones in Chapter 13. The latter two services embody NLP: sentiment analysis, neural machine translation, question answering, and more. The `Text AnalyticsClient` class in the Python text analytics SDK (*https://oreil.ly/gkD9c*) provides a convenient interface to many of the Language service's features. You've already seen it used for sentiment analysis. Here's another example that applies sentiment analysis to multiple text samples with one round trip to the Language service:

```python
from azure.core.credentials import AzureKeyCredential
from azure.ai.textanalytics import TextAnalyticsClient

client = TextAnalyticsClient(ENDPOINT, AzureKeyCredential(KEY))

input = [
    { 'id': '1000', 'text': 'Programming is fun, but the hours are long' },
    { 'id': '1001', 'text': 'Great food and excellent service' },
    { 'id': '1002', 'text': 'The product worked as advertised but is ' \
                            'overpriced' },
    { 'id': '1003', 'text': 'Moving to the cloud was the best decision ' \
                            'we ever made' },
    { 'id': '1004', 'text': 'Programming is so fun I\'d do it for free. ' \
                            'Don\'t tell my boss!' }
]
response = client.analyze_sentiment(input)

for result in response:
    text = ''.join([x.text for x in result.sentences])
    print(f'{text} => {result.confidence_scores.positive}')
```

Here's the output:

```
Programming is fun, but the hours are long => 0.94
Great food and excellent service => 1.0
The product worked as advertised but is overpriced => 0.0
Moving to the cloud was the best decision we ever made => 1.0
Programming is so fun I'd do it for free. Don't tell my boss! => 1.0
```

Let's say you're writing an app that collects tweets referencing your company and analyzes them for sentiment. The idea is that if sentiment turns negative, you can give the marketing department a heads-up. It's faster and more efficient to place one call to Azure Cognitive Services and analyze 100 tweets than to make 100 calls analyzing one tweet at a time.

Sentiment analysis is one of several operations supported by the TextAnalytics Client class. Another is named-entity recognition. Suppose you're building a system that sorts and prioritizes support tickets received by your company's help desk. The recognize_entities method (*https://oreil.ly/UtuZx*) extracts entities such as people, places, organizations, dates and times, and quantities from input text. It reveals the entity types as well:

```python
documents = [
    'My printer isn\'t working. Can someone from IT come to my office ' \
    'and have a look?'
]

results = client.recognize_entities(documents)
```

```
for result in results:
    for entity in result.entities:
        print(f'{entity.text} ({entity.category})')
```

The output from this example is as follows:

```
printer (Product)
IT (Skill)
office (Location)
```

A related method named `recognize_pii_entities` (*https://oreil.ly/nOGsr*) extracts personally identifiable information (PII) entities such as bank account info, Social Security numbers, and credit card numbers from text input to it, while the `extract_key_phrases` method (*https://oreil.ly/Au69j*) extracts key phrases:

```
documents = [
    'Natural Language Processing, or NLP, encompasses a variety of ' \
    'activities including text classification, keyword extraction,' \
    'named-entity recognition, question answering, and language '\
    'translation.'
]

results = client.extract_key_phrases(documents)

for result in results:
    for phrase in result.key_phrases:
        print(phrase)
```

Here's the output:

```
Natural Language Processing
language translation
text classification
keyword extraction
question answering
NLP
variety
activities
recognition
```

This example is a simple one given that the text is so brief, but you could pass in hundreds of large documents and use the results to get a snapshot of each document's content, or group documents that contain similar keywords.

`TextAnalyticsClient` provides a wrapper around text analytics APIs. Other Python SDKs unlock additional features of the Language service. For example, the question-answering SDK (*https://oreil.ly/h85Bc*) provides APIs for answering questions from manuals, FAQs, blog posts, and other documents that you provide. For more information, and to see this aspect of the Language service in action, refer to the article titled "Azure Cognitive Language Services Question Answering Client Library for Python" (*https://oreil.ly/J6kqg*).

The Translator Service

The Translator service uses state-of-the-art neural machine translation to translate text between dozens of languages (*https://oreil.ly/hAkuM*). It can also identify written languages. Suppose your objective is to translate into English questions written in other languages and submitted through your company's website. First you need to determine whether the source language is English. If it's not, you want to translate it so that you can respond to the customer's request.

The following code analyzes a text sample and shows the language it's written in. Microsoft doesn't currently offer a Python SDK for the Translator service, but you can use Python's Requests package to simplify calls. To demonstrate, create a Translator resource in the Azure Portal and grab the subscription key (the one labeled "Text Translation"), endpoint, and region. Then run the following code, replacing KEY and ENDPOINT with the key and endpoint and REGION with the Azure region you selected (for example, southcentralus):

```python
import requests

input = [{ 'text': 'Quand votre nouveau livre sera-t-il disponible?' }]

headers = {
    'Ocp-Apim-Subscription-Key': KEY,
    'Ocp-Apim-Subscription-Region': REGION,
    'Content-type': 'application/json'
}

uri = ENDPOINT + 'detect?api-version=3.0&to=en'
response = requests.post(uri, headers=headers, json=input)
results = response.json()

print(results[0]['language'])
```

The output is fr for French. Now that you've determined the source language isn't English, you can translate it this way:

```python
uri = ENDPOINT + 'translate?api-version=3.0&from=fr&to=en'
response = requests.post(uri, headers=headers, json=input)
results = response.json()

print(results[0]['translations'][0]['text'])
```

The translated text is:

```
When will your new book be available?
```

You can omit from=fr from the URL and allow the Translator service to detect the language for you. You can also detect the language and translate the text with one call. If you call the translate endpoint without a from parameter, the return value includes a detectedLanguage item that identifies the language in the source text:

```
[{'detectedLanguage': {'language': 'fr', 'score': 1.0}, 'translations': [{'text':
'When will your new book be available?', 'to': 'en'}]}]
```

If passed a list containing multiple text samples, the Translator service will translate all of them in one call. It also supports *transliteration*: translating text to other alphabets. To see for yourself, use Google Translate to translate "When will your new book be available" to Thai or Hindi. Then paste the Thai or Hindi text over the French text in the previous example and run it again. Be sure to also change from=fr to from=th or from=hi or simply remove the from parameter altogether.

 When you create a Translator resource, you have a choice of creating a global resource or one tied to a specific Azure region. The preceding examples assume that you created it as a regional resource. If you created a global Translator resource instead, you can set the Ocp-Apim-Subscription-Region header to global or omit the header altogether.

When you obtained an endpoint for the Translator service, did you notice that the Azure Portal offered *two* endpoints—one for "Text Translation" and another for "Document Translation?" That's because the Translator service features a second API for translating entire documents, including PDFs. There's even a Python SDK (*https://oreil.ly/uuVmJ*) to help out. The only catch is that documents must first be uploaded to Azure blob storage. For examples showing the document translation API in action, see "Azure Document Translation Client Library for Python" (*https://oreil.ly/RFTbT*). The API is asynchronous and can process batches of documents in one call, so it's ideal not just for translating individual documents, but for translating large volumes of documents at scale.

The Speech Service

One of the more challenging tasks for deep-learning models is processing human speech. Azure Cognitive Services includes a Speech service (*https://oreil.ly/NHBcG*) that converts text to speech, speech to text, and more. It's even capable of captioning recorded videos and live video streams (*https://oreil.ly/1x9gU*) and filtering out profanity as it does. A Python SDK (*https://oreil.ly/Heaxx*) simplifies the code you write and makes it remarkably easy to incorporate speech into your apps.

To demonstrate, install the package named Azure-cognitiveservices-speech (*https://oreil.ly/4cwJo*) containing the Python Speech SDK. Use the Azure Portal to create a Cognitive Services Speech resource and make note of the subscription key and service region. Then create a Jupyter notebook and run the following code in the first cell after replacing KEY with the subscription key and REGION with the region you selected:

```
from azure.cognitiveservices import speech

speech_config = speech.SpeechConfig(KEY, REGION)
speech_config.speech_recognition_language = 'en-US'
```

Now run the following statements, and when prompted, speak into your microphone. This sample creates a `SpeechRecognizer` object (*https://oreil.ly/ZUKOC*) and uses its `recognize_once_async` method (*https://oreil.ly/NY681*) to convert up to 30 seconds of live audio from your PC's default microphone into text. Observe that the text doesn't appear until you've finished speaking:

```
recognizer = speech.SpeechRecognizer(speech_config)

print('Speak into your microphone')
result = recognizer.recognize_once_async().get()

if result.reason == speech.ResultReason.RecognizedSpeech:
    print(result.text)
```

It couldn't be much simpler than that. How about converting text to speech? Here's an example that uses the SDK's `SpeechSynthesizer` class (*https://oreil.ly/3qC4H*) to vocalize a sentence. The synthesized voice belongs to an English speaker named Jenny (`en-US-JennyNeural`), and it's one of more than 300 neural voices (*https://oreil.ly/mXyka*) you can choose from:

```
speech_config.speech_synthesis_voice_name = 'en-US-JennyNeural'
synthesizer = speech.SpeechSynthesizer(speech_config)
synthesizer.speak_text_async('When will your new book be published?').get()
```

All of the "speakers" are multilingual. If you ask a French speaker—for example, `fr-FR-CelesteNeural`—to synthesize an English sentence, the vocalization will feature a French accent.

You can combine a `TranslationRecognizer` object (*https://oreil.ly/R38mY*) with a `SpeechSynthesizer` object to translate speech in real time. The following example takes spoken English as input and plays it back in French using the voice of a native French speaker:

```
speech_config.speech_synthesis_voice_name = 'fr-FR-YvetteNeural'
synthesizer = speech.SpeechSynthesizer(speech_config)

translation_config = speech.translation.SpeechTranslationConfig(KEY, REGION)
translation_config.speech_recognition_language = 'en-US'
translation_config.add_target_language('fr')

recognizer = speech.translation.TranslationRecognizer(translation_config)

print('Speak into your microphone')
result = recognizer.recognize_once_async().get()

if result.reason == speech.ResultReason.TranslatedSpeech:
```

```
text = result.translations['fr']
synthesizer.speak_text_async(text).get()
```

These samples use your PC's default microphone for voice input and default speakers for output. You can specify other sources of input and output by passing an `Audio Config` object (*https://oreil.ly/tj43k*) to the methods that create `SpeechRecognizer`, `SpeechSynthesizer`, and `TranslationRecognizer` objects. Among the options this enables is using a file or stream rather than a microphone as the source of input.

Putting It All Together: Contoso Travel

Imagine you're a web developer and your client is Contoso Travel. To motivate its customers to stay in touch, the travel agency wants its website to include a service that translates signage. The idea is that a customer exploring a faraway land can snap a picture of a sign, upload it to the site, and see a translation in the language of their choice. No typing, no forms to fill out—just "Here's a picture, tell me what it says."

A few years ago, such a website would have been unthinkable for most small businesses. Today it's within the capabilities of anyone who can sling a few lines of code. Let's use Python's Flask framework (*https://oreil.ly/WqqwN*) to build a website that makes your client happy. We'll use Azure Cognitive Services to do the heavy lifting of extracting and translating text, and we'll end up with the product shown in Figure 14-6.

Begin by making sure the required packages are installed, including Flask, Requests, and Azure-cognitiveservices-vision-computervision. Then create a project directory in the location of your choice on your hard disk. Name it Contoso or anything else you'd like. Next, download a ZIP file (*https://oreil.ly/KmFZD*) containing a starter kit for the website and copy its contents into the project directory. Take a moment to examine the files that you copied. These files comprise a website written in Python and Flask. They include the following:

app.py
Holds the Python code that drives the site

templates/index.html
Contains the site's home page

static/main.css
Contains CSS to dress up the home page

static/banner.jpg
Contains the website's banner

static/placeholder.jpg
Contains a placeholder image for photos that have yet to be uploaded

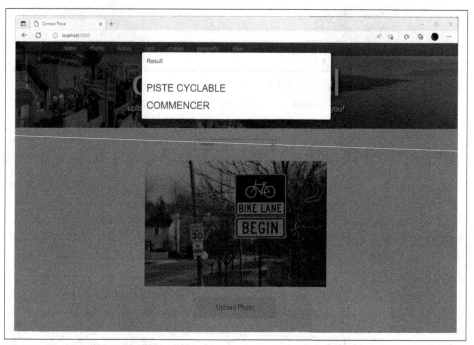

Figure 14-6. The Contoso Travel website translating a road sign

The site contains a single page named *index.html* that's displayed when a user navigates to the site. The code that displays it lives in *app.py*, and it includes logic for uploading photos. You will modify *app.py* to extract and translate text from the photos that users upload.

Does "Contoso" sound familiar? It's the company name used in countless Microsoft samples and tutorials. There's a story behind why it's so popular. When you write content for Microsoft, you're not allowed to make up company names because doing so is likely to get Microsoft sued. Contoso is one of several fictitious company names (*https://oreil.ly/7PszJ*) Microsoft has trademarked over the years. Others that might be familiar include Northwind Traders, Fabrikam, and AdventureWorks.

Open a command prompt or terminal window and cd to the project directory. If you're running Windows, use the following command to create an environment variable named FLASK_ENV that tells Flask to run in development mode:

```
set FLASK_ENV=development
```

If you're running Linux or macOS, use this command instead:

```
export FLASK_ENV=development
```

Running Flask in development mode is helpful when you're developing a website because Flask automatically reloads any files that change while the site is running. If you let Flask default to production mode and change the contents of an HTML file or other asset, you have to restart Flask for the changes to appear in your browser.

Now use the following command to start Flask:

```
flask run
```

Open a browser and go to http://localhost:5000. When Contoso Travel's home page appears, click the Upload Photo button and select a JPG or PNG file on your computer. Confirm that the photo uploads without errors. At this point, no processing is performed on the photo, so the only visual clue that the upload succeeded is that the photo appears in the web page.

Next, use the Azure Portal to create a Computer Vision resource and a Translator resource if you haven't already. Open *app.py* in your favorite code editor and find the following statements near the top of the file:

```
# Define Cognitive Services variables
vision_key = 'VISION_KEY'
vision_endpoint = 'VISION_ENDPOINT'
translator_key = 'TRANSLATOR_KEY'
translator_endpoint = 'TRANSLATOR_ENDPOINT'
translator_region = 'TRANSLATOR_REGION'
```

Replace VISION_KEY and TRANSLATOR_KEY with the services' subscription keys, VISION_ENDPOINT and TRANSLATOR_ENDPOINT with the endpoints, and TRANSLATOR _REGION with the region selected for the Translator service. For Translator, use the endpoint labeled "Text Translation." Be sure to leave the tick marks delimiting the strings intact.

Now add the following function for extracting text from a photo to *app.py*, placing it right after the comment that reads "Function that extracts text from images" near the bottom of the file:

```
def extract_text(endpoint, key, image):
    try:
        client = ComputerVisionClient(endpoint,
                                      CognitiveServicesCredentials(key))
        result = client.recognize_printed_text_in_stream(image)
        lines = []

        for region in result.regions:
            for line in region.lines:
                text = ' '.join([word.text for word in line.words])
                lines.append(text)

        if len(lines) == 0:
            lines.append('Photo contains no text to translate')
```

```
    return lines

except ComputerVisionErrorResponseException as e:
    return ['Error calling the Computer Vision service: ' + e.message]

except Exception as e:
    return ['Error calling the Computer Vision service']
```

Then add the following function for translating text. Place it after the comment that reads "Function that translates text into a specified language" near the bottom of *app.py*:

```
def translate_text(endpoint, region, key, lines, language):
    try:
        headers = {
            'Ocp-Apim-Subscription-Key': key,
            'Ocp-Apim-Subscription-Region': region,
            'Content-type': 'application/json'
        }

        input = []

        for line in lines:
            input.append({ "text": line })

        uri = endpoint + 'translate?api-version=3.0&to=' + language
        response = requests.post(uri, headers=headers, json=input)
        response.raise_for_status() # Raise exception if call failed
        results = response.json()

        translated_lines = []

        for result in results:
            for translated_line in result["translations"]:
                translated_lines.append(translated_line["text"])

        return translated_lines

    except requests.exceptions.HTTPError as e:
        return ['Error calling the Translator service: ' + e.strerror]

    except Exception as e:
        return ['Error calling the Translator service']
```

Find the function named index in *app.py*. This is the function called when the home page is requested or a photo is uploaded. Under the comment that reads "Use the Computer Vision service to extract text from the image," add the following line of code to call extract_text with the photo that the user just uploaded:

```
lines = extract_text(vision_endpoint, vision_key, image)
```

Under the comment that reads "Use the Translator service to translate text extracted from the image," add a line of code to translate the text returned by `extract_text` to the specified language:

```
translated_lines = translate_text(translator_endpoint, translator_region,
                                  translator_key, lines, language)
```

Add the following statements immediately after the comment that reads "Flash the translated text":

```
for translated_line in translated_lines:
    flash(translated_line)
```

These statements use Flask's message-flashing support to display the strings in `translated_lines` in a modal dialog.

Finish up by saving your changes to *app.py*. Return to the browser in which the website is running and refresh the page. (If you closed it, open a new browser instance and navigate to *http://localhost:5000*.) Select a language from the drop-down list at the top of the page. Then upload a photo containing a road sign or any other image with text in it.

Was the text extracted and translated? The app isn't perfect, but it should work as expected most of the time. If you're not satisfied, try replacing calls to `recognize_printed_text_in_stream` with calls to `read_in_stream`. The latter is more aggressive at finding text in photos. While you're at it, you could also use the Translator service to translate error messages into the language the user selected or the Speech service to vocalize translated text. With Azure Cognitive Services lending a hand, the only limit is your imagination.

Summary

Azure Cognitive Services lower the bar for incorporating sophisticated deep-learning models into the apps you write by providing REST APIs that any app with an internet connection can call. Containerized versions of most services allow them to be hosted locally or on premises and also provide an option for running in disconnected scenarios. One of the benefits of containers is locking in a particular version of the models you're using or the APIs that access them so that changes to Azure Cognitive Services won't affect your apps.

The range of services offered by Azure Cognitive Services includes vision services for extracting information from images and video streams, language services for translating text into other languages and performing other NLP tasks, and speech services for incorporating speech into your apps. Free SDKs are available for most services. They simplify the code you write, and they're available for a variety of popular programming languages including Python, Java, and C#.

Machine learning and deep learning are making the world a better place one model at a time. Soon both will be considered part of the essential skill set of every engineer and software developer. Are you up to the task? My hope is that this book gives you the confidence to say yes and the tools to make it happen. It's too late to monitor Apollo missions in Mission Control, but countless other frontiers are waiting to be explored.

Index

E

EarlyStopping class, 214, 345
eigenvectors and eigenvalues, 127
Elasticsearch, 353
elbow method to plot inertias, 12
Embedding class, 323
embedding layer, 319, 323-327
encoder–decoder models, 335-340
ensemble learning, random forests, 37-38, 69, 144
epochs, 189
estimators, 116, 155
Euclidean distance, 23
Excel, adding ML capabilities to, 169-172
expert systems, 7
explained variance, plotting, 130-133
extracting faces from photos, 270-272
extract_key_phrases method, 379

F

F1 score, 57
face detection, 261-272
    with CNNs, 266-270, 280-286
        Computer Vision service, 372
        extracting faces from photos, 270-272
    with Viola-Jones, 262-266
face embedding, 279
Face service, 360, 373
face verification, 280
FaceNet, 272
facial recognition, 272-287
    closed-set versus open-set classification, 286-287
    with CNNs, 273-286
    neural networks, 207-210
    and privacy, 261
    with SVMs, 117-123
    with transfer learning, 273-279
false positive rate (FPR), 58
Fast R-CNN, 292
Faster R-CNN, 292
feature columns, 4
feature maps, 220
filtering noise in images, 133-135
fit method
    in Keras, 189, 216, 246
    in Scikit-Learn, 26, 45
fit_on_texts method, 320
Flask framework, 153, 383

Flatten layer, 250, 324
flow method, 246
flow_from_directory method, 246
folds, 44
FPR (false positive rate), 58
fully connected layers, 179
functional API, Keras, 187, 343-348

G

gamma parameter, SVM kernels, 108-111
GANs (generative adversarial networks), 178
gated recurrent unit (GRU) layer, 332
GBDTs (gradient-boosted decision trees), 38-41
GBMs (gradient-boosting machines), 38-41
generative adversarial networks (GANs), 178
get_weights method, 211
Gini impurity, 35
global minimum, 184
global pooling, 250-251
GlobalAveragePooling2D layer, 250, 255
GlobalMaxPooling2D layer, 250
GlorotUniform initializer, 190
GloVe word vectors, 323
GPUs (graphics processing units), 8, 177, 232
gradient descent, 185
gradient-boosted decision trees (GBDTs), 38-41
gradient-boosting machines (GBMs), 38-41
GradientBoostingClassifier class, 39-41, 70
GradientBoostingRegressor class, 39-41, 50
graphics processing units (GPUs), 8, 177, 232
GridSearchCV, 110, 117, 120-122
GRU (gated recurrent unit) layer, 332
GRU class, 332

H

H2O framework, 165, 167
H5 file format, 212
Haar-like features, 263
handwritten text
    building digit recognition model, 73-76
    extracting from image, 376-377
HashingVectorizer class, 80, 83, 156
Haystack library, 353
hidden layers, 178, 180, 188
hidden state, layer, 331
high recall then precision pattern, 263
high-dimensional data visualization, 137-140
Hugging Face, 333, 348, 350-357
hyperparameter tuning, 108-111

Scaper, soundscape synthesis, 252
scatter function, 10
Scikit-Learn, 10
    hyperparameter optimizers, 110
    versus ML.Net, 165, 167
    multiclass classification feature, 71-73
    Skl2onnx conversion, 162
    sklearn.cluster.KMeans, 11
    sklearn.datasets.fetch_california_housing, 42
    sklearn.datasets.fetch_lfw_people, 118, 207
    sklearn.datasets.load_breast_cancer, 136
    sklearn.datasets.load_digits, 138
    sklearn.datasets.load_iris, 25
    sklearn.decomposition.PCA, 127, 138, 142, 146
    sklearn.ensemble.GradientBoostingClassifier, 39-41, 70
    sklearn.ensemble.GradientBoostingRegressor, 39-41, 50
    sklearn.ensemble.RandomForestClassifier, 37, 69, 70
    sklearn.ensemble.RandomForestRegressor, 37, 50
    sklearn.feature_extraction.text.CountVectorizer, 80-83, 91, 94, 155, 320
    sklearn.feature_extraction.text.HashingVectorizer, 80, 83, 156
    sklearn.feature_extraction.text.TfidfVectorizer, 80, 83
    sklearn.linear_model.Lasso, 31
    sklearn.linear_model.LinearRegression, 31, 43, 49
    sklearn.linear_model.LogisticRegression, 155
    sklearn.linear_model.Perceptron, 180
    sklearn.linear_model.Ridge, 31
    sklearn.manifold.TSNE, 139
    sklearn.metrics.ConfusionMatrixDisplay, 58, 92
    sklearn.metrics.RocCurveDisplay, 92
    sklearn.metrics.roc_auc_score, 58
    sklearn.model_selection.GridSearchCV, 110, 117, 120-122
    sklearn.model_selection.train_test_split, 26, 42-43
    sklearn.naive_bayes.MultinomialNB, 91
    sklearn.neighbors.KNeighborsClassifier, 26
    sklearn.neural_network.MLPClassifier, 180
    sklearn.neural_network.MLPRegressor, 180
    sklearn.pipeline.make_pipeline, 116, 155
    sklearn.preprocessing.LabelEncoder, 16, 60
    sklearn.preprocessing.MinMaxScaler, 114
    sklearn.preprocessing.OneHotEncoder, 61
    sklearn.preprocessing.PolynomialFeatures, 31
    sklearn.preprocessing.StandardScaler, 112, 115, 116, 209
    sklearn.svm.SVC, 42, 101, 110
    sklearn.svm.SVR, 42, 101
    sklearn.tree.DecisionTreeClassifier, 35
    sklearn.tree.DecisionTreeRegressor, 35
    sklearn.utils.shuffle, 42
score method, 26, 42
scree plot, 131
segmentation masks, 293-299
selective search, 291
self-attention, 339
sensitivity metric, 60, 66
sentiment analysis, 5, 79
    Azure Language service, 378
    with BERT, 354
    Hugging Face resources, 333
    with ML.NET, 166
    text classification, 83-87, 328-330
    training and saving pipeline, 155-157
separable convolutions, 233
sequence-to-sequence model, 335
sequences, word embeddings, 319, 320
sequence_to_texts method, 320
sequential API, Keras, 187-191
set_weights method, 211
SGD (stochastic gradient descent), 185
shuffling data, 42, 190, 325
sigmoid activation function, 188, 197-200, 223
sigmoid kernel, 104
similarity matrix, recommender system, 96
simple linear regression, 32
sizing a neural network, 192
SMT (statistical machine translation), 334
soft mask, 297
softmax activation function, 72, 188
    building CNNs, 223
    closed- versus open-set classification, 286-287
    in facial recognition, 209
    LSTM encoder-decoders, 335
    versus sigmoid activation, 205

About the Author

Jeff Prosise is an engineer whose passion is to introduce other engineers and software developers to the wonders of AI and machine learning. Cofounder of Wintellect, he has written nine books and hundreds of magazine articles, trained thousands of developers at Microsoft, and spoken at some of the world's largest software conferences. In another life, Jeff worked on high-powered laser systems and fusion-energy research at Oak Ridge National Laboratory and Lawrence Livermore National Laboratory. In his spare time, he builds and flies large radio-control jets and goes out of his way to get wet in the world's best dive spots. Following the acquisition of his company in 2021, Jeff serves as chief learning officer at Atmosera, where he helps customers infuse AI into their products.

Colophon

The animal on the cover of *Applied Machine Learning and AI for Engineers* is a festive parrot (*Amazona festiva*), also known as a festive amazon. Festive parrots live in the tropical forests, woodlands, and coastal mangroves of several South American countries, including Brazil, Colombia, Ecuador, Peru, and Bolivia. They are rarely found far from water.

Festive parrots are brightly—you might even say *festively*—colored, medium-sized birds. Their plumage is predominantly a striking green, turning slightly yellow toward the edges of their wings. A motley assortment of colors—including red, blue, and sometimes yellow or orange—adorns their faces.

Festive parrots are a highly social species, usually spotted in pairs or small flocks. Large groups of the birds often gather at night for communal roosts or around a localized food source and are known for being incredibly noisy. They enjoy eating fruits such as mangoes and peach palm, with berries, nuts, seeds, flowers, and leaf buds supplementing their diet.

While still relatively common where their forest habitat remains largely intact, festive parrots have been categorized by IUCN as *near threatened* due to continued deforestation and predicted declines in habitat. Many of the animals on O'Reilly covers are endangered; all of them are important to the world.

The cover illustration is by Karen Montgomery, based on an antique line engraving from Wood's *Illustrated Natural History*. The cover fonts are Gilroy Semibold and Guardian Sans. The text font is Adobe Minion Pro; the heading font is Adobe Myriad Condensed; and the code font is Dalton Maag's Ubuntu Mono.